R Graphics Cookbook

Winston Chang

O'REILLY®

Beijing • Cambridge • Farnham • Köln • Sebastopol • Tokyo

R Graphics Cookbook

by Winston Chang

Printed in the United States of America.

Published by O'Reilly Media, Inc., 1005 Gravenstein Highway North, Sebastopol, CA 95472.

O'Reilly books may be purchased for educational, business, or sales promotional use. Online editions are also available for most titles (*http://my.safaribooksonline.com*). For more information, contact our corporate/institutional sales department: 800-998-9938 or *corporate@oreilly.com*.

Editors: Mike Loukides and Courtney Nash
Production Editor: Holly Bauer
Copyeditor: Rachel Head
Proofreader: Jilly Gagnon
Indexer: Lucie Haskins
Cover Designer: Randall Comer
Interior Designer: David Futato
Illustrator: Rebecca Demarest and Robert Romano

December 2012: First Edition

Revision History for the First Edition:

2012-12-04: First release

2013-07-12: Second release

See *http://oreilly.com/catalog/errata.csp?isbn=9781449316952* for release details.

ISBN: 978-1-449-31695-2

[LSI]

Table of Contents

Preface

I started using R several years ago to analyze data I had collected for my research in graduate school. My motivation at first was to escape from the restrictive environments and canned analyses offered by statistical programs like SPSS. And even better, because it's freely available, I didn't need to convince someone to buy me a copy of the software—very important for a poor graduate student! As I delved deeper into R, I discovered that it could also create excellent data graphics.

Each recipe in this book lists a problem and a solution. In most cases, the solutions I offer aren't the only way to do things in R, but they are, in my opinion, the best way. One of the reasons for R's popularity is that there are many available add-on packages, each of which provides some functionality for R. There are many packages for visualizing data in R, but this book primarily uses ggplot2. (Disclaimer: it's now part of my job to do development on ggplot2. However, I wrote much of this book before I had any idea that I would start a job related to ggplot2.)

This book isn't meant to be a comprehensive manual of all the different ways of creating data visualizations in R, but hopefully it will help you figure out how to make the graphics you have in mind. Or, if you're not sure what you want to make, browsing its pages may give you some ideas about what's possible.

Recipes

This book is intended for readers who have at least a basic understanding of R. The recipes in this book will show you how to do specific tasks. I've tried to use examples that are simple, so that you can understand how they work and transfer the solutions over to your own problems.

Software and Platform Notes

Most of the recipes here use the ggplot2 graphing package. Some of the recipes require the most recent version of ggplot2, 0.9.3, and this in turn requires a relatively recent version of R. You can always get the latest version of R from the main R project site (*http://www.r-project.org*).

If you are not familiar with ggplot2, see Appendix A for a brief introduction to the package.

Once you've installed R, you can install the necessary packages. In addition to ggplot2, you'll also want to install the gcookbook package, which contains data sets for many of the examples in this book. To install them both, run:

```
install.packages("ggplot2")
install.packages("gcookbook")
```

You may be asked to choose a mirror site for CRAN, the Comprehensive R Archive Network. Any of the sites should work, but it's a good idea to choose one close to you because it will likely be faster than one far away. Once you've installed the packages, run this in each R session in which you want to use ggplot2:

```
library(ggplot2)
```

The recipes in this book will assume that you've already loaded ggplot2, so they won't show this line.

If you see an error like this, it means that you forgot to load ggplot2:

```
Error: could not find function "ggplot"
```

The major platforms for R are Mac OS X, Linux, and Windows, and all the recipes in this book should work on all of these platforms. There are some platform-specific differences when it comes to creating bitmap output files, and these differences are covered in Chapter 14.

Conventions Used in This Book

The following typographical conventions are used in this book:

Italic
: Indicates new terms, URLs, email addresses, filenames, and file extensions.

`Constant width`
: Used for program listings, as well as within paragraphs to refer to program elements such as variable or function names, databases, data types, environment variables, statements, and keywords.

`Constant width bold`
: Shows commands or other text that should be typed literally by the user.

`Constant width italic`
: Shows text that should be replaced with user-supplied values or by values determined by context.

This icon signifies a tip, suggestion, or general note.

This icon indicates a warning or caution.

Using Code Examples

This book is here to help you get your job done. In general, you may use the code in this book in your programs and documentation. You do not need to contact us for permission unless you're reproducing a significant portion of the code. For example, writing a program that uses several chunks of code from this book does not require permission. Selling or distributing a CD-ROM of examples from O'Reilly books does require permission. Answering a question by citing this book and quoting example code does not require permission. Incorporating a significant amount of example code from this book into your product's documentation does require permission.

We appreciate, but do not require, attribution. An attribution usually includes the title, author, publisher, and ISBN. For example: "*R Graphics Cookbook* by Winston Chang (O'Reilly). Copyright 2013 Winston Chang, 978-1-449-31695-2."

If you feel your use of code examples falls outside fair use or the permission given above, feel free to contact us at *permissions@oreilly.com*.

Safari® Books Online

Safari Books Online (*www.safaribooksonline.com*) is an on-demand digital library that delivers expert content in both book and video form from the world's leading authors in technology and business.

Technology professionals, software developers, web designers, and business and creative professionals use Safari Books Online as their primary resource for research, problem solving, learning, and certification training.

Safari Books Online offers a range of product mixes and pricing programs for organizations, government agencies, and individuals. Subscribers have access to thousands of books, training videos, and prepublication manuscripts in one fully searchable database from publishers like O'Reilly Media, Prentice Hall Professional, Addison-Wesley Professional, Microsoft Press, Sams, Que, Peachpit Press, Focal Press, Cisco Press, John Wiley & Sons, Syngress, Morgan Kaufmann, IBM Redbooks, Packt, Adobe Press, FT Press, Apress, Manning, New Riders, McGraw-Hill, Jones & Bartlett, Course Technology, and dozens more. For more information about Safari Books Online, please visit us online.

How to Contact Us

Please address comments and questions concerning this book to the publisher:

O'Reilly Media, Inc.
1005 Gravenstein Highway North
Sebastopol, CA 95472
800-998-9938 (in the United States or Canada)
707-829-0515 (international or local)
707-829-0104 (fax)

We have a web page for this book, where we list errata, examples, and any additional information. You can access this page at:

http://oreil.ly/R_Graphics_Cookbook

To comment or ask technical questions about this book, send email to:

bookquestions@oreilly.com

For more information about our books, courses, conferences, and news, see our website at *http://www.oreilly.com*.

Find us on Facebook: *http://facebook.com/oreilly*

Follow us on Twitter: *http://twitter.com/oreillymedia*

Watch us on YouTube: *http://www.youtube.com/oreillymedia*

Acknowledgments

No book is the product of a single person. There are many people who helped make this book possible, directly and indirectly. I'd like to thank the R community for creating R and for fostering a dynamic ecosystem around it. Thanks to Hadley Wickham for creating the software that this book revolves around, for pointing O'Reilly in my direction when they were considering a book about R graphics, and for opening up many opportunities for me to deepen my knowledge of R.

Thanks to the technical reviewers for this book: Paul Teetor, Hadley Wickham, Dennis Murphy, and Erik Iverson. Their depth of knowledge and attention to detail has greatly improved this book. I'd like to thank the editors at O'Reilly who have shepherded this book along: Mike Loukides, for guiding me through the early stages, and Courtney Nash, for pulling me through to the end. I also owe a big thanks to Holly Bauer and the rest of the production team at O'Reilly, for putting up with many last-minute edits, and for handling the unusual features of this book.

Finally, I would like to thank my wife, Sylia, for her support and understanding—and not just with regard to the book.

CHAPTER 1

R Basics

This chapter covers the basics: installing and using packages and loading data.

If you want to get started quickly, most of the recipes in this book require the ggplot2 and gcookbook packages to be installed on your computer. To do this, run:

```
install.packages(c("ggplot2", "gcookbook"))
```

Then, in each R session, before running the examples in this book, you can load them with:

```
library(ggplot2)
library(gcookbook)
```

Appendix A provides an introduction to the ggplot2 graphing package, for readers who are not already familiar with its use.

Packages in R are collections of functions and/or data that are bundled up for easy distribution, and installing a package will extend the functionality of R on your computer. If an R user creates a package and thinks that it might be useful for others, that user can distribute it through a package repository. The primary repository for distributing R packages is called CRAN (the Comprehensive R Archive Network), but there are others, such as Bioconductor and Omegahat.

1.1. Installing a Package

Problem

You want to install a package from CRAN.

Solution

Use `install.packages()` and give it the name of the package you want to install. To install ggplot2, run:

```
install.packages("ggplot2")
```

At this point you may be prompted to select a download mirror. You can either choose the one nearest to you, or, if you want to make sure you have the most up-to-date version of your package, choose the Austria site, which is the primary CRAN server.

Discussion

When you tell R to install a package, it will automatically install any other packages that the first package depends on.

CRAN is a repository of packages for R, and it is mirrored on servers around the globe. It's the default repository system used by R. There are other package repositories; Bioconductor, for example, is a repository of packages related to analyzing genomic data.

1.2. Loading a Package

Problem

You want to load an installed package.

Solution

Use `library()` and give it the name of the package you want to install. To load ggplot2, run:

```
library(ggplot2)
```

The package must already be installed on the computer.

Discussion

Most of the recipes in this book require loading a package before running the code, either for the graphing capabilities (as in the ggplot2 package) or for example data sets (as in the MASS and gcookbook packages).

One of R's quirks is the package/library terminology. Although you use the `library()` function to load a package, a package is not a library, and some longtime R users will get irate if you call it that.

A *library* is a directory that contains a set of packages. You might, for example, have a system-wide library as well as a library for each user.

1.3. Loading a Delimited Text Data File

Problem

You want to load data from a delimited text file.

Solution

The most common way to read in a file is to use comma-separated values (CSV) data:

```
data <- read.csv("datafile.csv")
```

Discussion

Since data files have many different formats, there are many options for loading them. For example, if the data file does *not* have headers in the first row:

```
data <- read.csv("datafile.csv", header=FALSE)
```

The resulting data frame will have columns named `V1`, `V2`, and so on, and you will probably want to rename them manually:

```
# Manually assign the header names
names(data) <- c("Column1","Column2","Column3")
```

You can set the delimiter with `sep`. If it is space-delimited, use `sep=" "`. If it is tab-delimited, use `\t`, as in:

```
data <- read.csv("datafile.csv", sep="\t")
```

By default, strings in the data are treated as factors. Suppose this is your data file, and you read it in using `read.csv()`:

```
"First","Last","Sex","Number"
"Currer","Bell","F",2
"Dr.","Seuss","M",49
"","Student",NA,21
```

The resulting data frame will store `First` and `Last` as *factors*, though it makes more sense in this case to treat them as strings (or *characters* in R terminology). To differentiate this, set `stringsAsFactors=FALSE`. If there are any columns that should be treated as factors, you can then convert them individually:

```
data <- read.csv("datafile.csv", stringsAsFactors=FALSE)

# Convert to factor
data$Sex <- factor(data$Sex)

str(data)

'data.frame':   3 obs. of  4 variables:
```

```
 $ First : chr  "Currer" "Dr." ""
 $ Last  : chr  "Bell" "Seuss" "Student"
 $ Sex   : Factor w/ 2 levels "F","M": 1 2 NA
 $ Number: int  2 49 21
```

Alternatively, you could load the file with strings as factors, and then convert individual columns from factors to characters.

See Also

`read.csv()` is a convenience wrapper function around `read.table()`. If you need more control over the input, see `?read.table`.

1.4. Loading Data from an Excel File

Problem

You want to load data from an Excel file.

Solution

The xlsx package has the function `read.xlsx()` for reading Excel files. This will read the first sheet of an Excel spreadsheet:

```
# Only need to install once
install.packages("xlsx")

library(xslx)
data <- read.xlsx("datafile.xlsx", 1)
```

For reading older Excel files in the *.xls* format, the gdata package has the function `read.xls()`:

```
# Only need to install once
install.packages("gdata")

library(gdata)
# Read first sheet
data <- read.xls("datafile.xls")
```

Discussion

With `read.xlsx()`, you can load from other sheets by specifying a number for `shee tIndex` or a name for `sheetName`:

```
data <- read.xlsx("datafile.xls", sheetIndex=2)

data <- read.xlsx("datafile.xls", sheetName="Revenues")
```

With `read.xls()`, you can load from other sheets by specifying a number for `sheet`:

```
data <- read.xls("datafile.xls", sheet=2)
```

Both the xlsx and gdata packages require other software to be installed on your computer. For xlsx, you need to install Java on your machine. For gdata, you need Perl, which comes as standard on Linux and Mac OS X, but not Windows. On Windows, you'll need ActiveState Perl. The Community Edition can be obtained for free (*http://www.activestate.com/activeperl*).

If you don't want to mess with installing this stuff, a simpler alternative is to open the file in Excel and save it as a standard format, such as CSV.

See Also

See `?read.xls` and `?read.xlsx` for more options controlling the reading of these files.

1.5. Loading Data from an SPSS File

Problem

You want to load data from an SPSS file.

Solution

The foreign package has the function `read.spss()` for reading SPSS files. To load data from the first sheet of an SPSS file:

```
# Only need to install the first time
install.packages("foreign")

library(foreign)
data <- read.spss("datafile.sav")
```

Discussion

The foreign package also includes functions to load from other formats, including:

- `read.octave()`: Octave and MATLAB
- `read.systat()`: SYSTAT
- `read.xport()`: SAS XPORT
- `read.dta()`: Stata

See Also

See `ls("package:foreign")` for a full list of functions in the package.

CHAPTER 2

Quickly Exploring Data

Although I've used the ggplot2 package for most of the graphics in this book, it is not the only way to make graphs. For very quick exploration of data, it's sometimes useful to use the plotting functions in base R. These are installed by default with R and do not require any additional packages to be installed. They're quick to type, are straightforward to use in simple cases, and run very quickly.

If you want to do anything beyond very simple graphs, though, it's generally better to switch to ggplot2. This is in part because ggplot2 provides a unified interface and set of options, instead of the grab bag of modifiers and special cases required in base graphics. Once you learn how ggplot2 works, you can use that knowledge for everything from scatter plots and histograms to violin plots and maps.

Each recipe in this section shows how to make a graph with base graphics. Each recipe also shows how to make a similar graph with the `qplot()` function in ggplot2, which has a syntax similar to the base graphics functions. For each `qplot()` graph, there is also an equivalent using the more powerful `ggplot()` function.

If you already know how to use base graphics, having these examples side by side will help you transition to using ggplot2 for when you want to make more sophisticated graphics.

2.1. Creating a Scatter Plot

Problem

You want to create a scatter plot.

Solution

To make a scatter plot (Figure 2-1), use `plot()` and pass it a vector of *x* values followed by a vector of *y* values:

```
plot(mtcars$wt, mtcars$mpg)
```

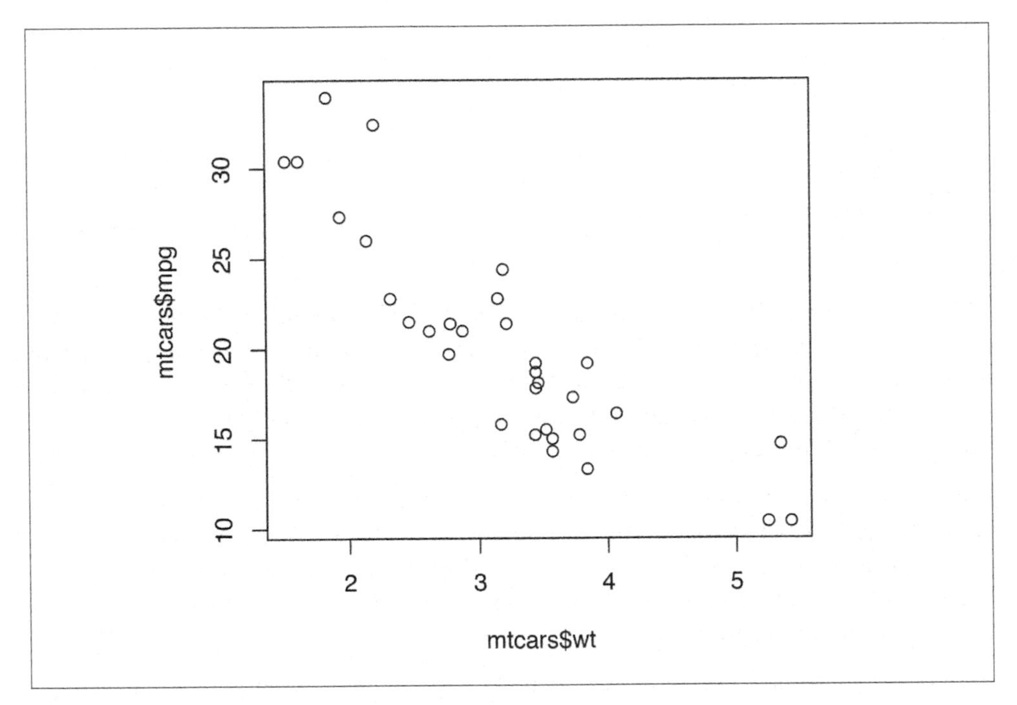

Figure 2-1. Scatter plot with base graphics

With the ggplot2 package, you can get a similar result using `qplot()` (Figure 2-2):

```
library(ggplot2)
qplot(mtcars$wt, mtcars$mpg)
```

If the two vectors are already in the same data frame, you can use the following syntax:

```
qplot(wt, mpg, data=mtcars)
# This is equivalent to:
ggplot(mtcars, aes(x=wt, y=mpg)) + geom_point()
```

See Also

See Chapter 5 for more in-depth information about creating scatter plots.

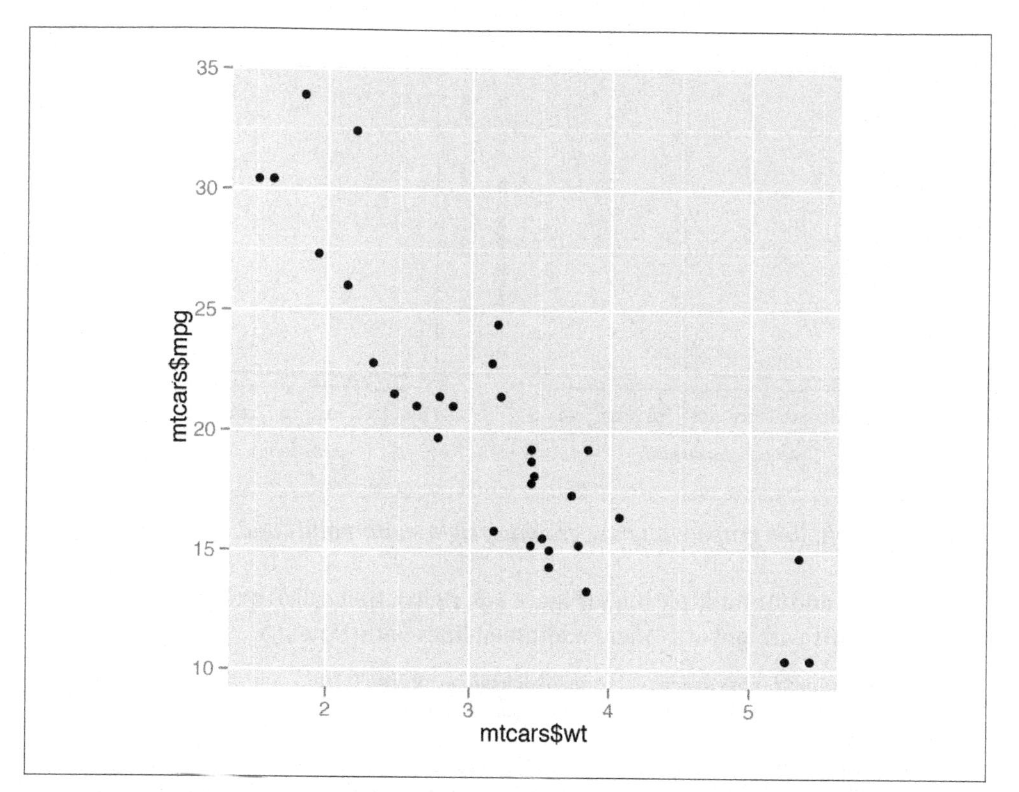

Figure 2-2. Scatter plot with qplot() from ggplot2

2.2. Creating a Line Graph

Problem

You want to create a line graph.

Solution

To make a line graph using `plot()` (Figure 2-3, left), pass it a vector of *x* values and a vector of *y* values, and use `type="l"`:

```
plot(pressure$temperature, pressure$pressure, type="l")
```

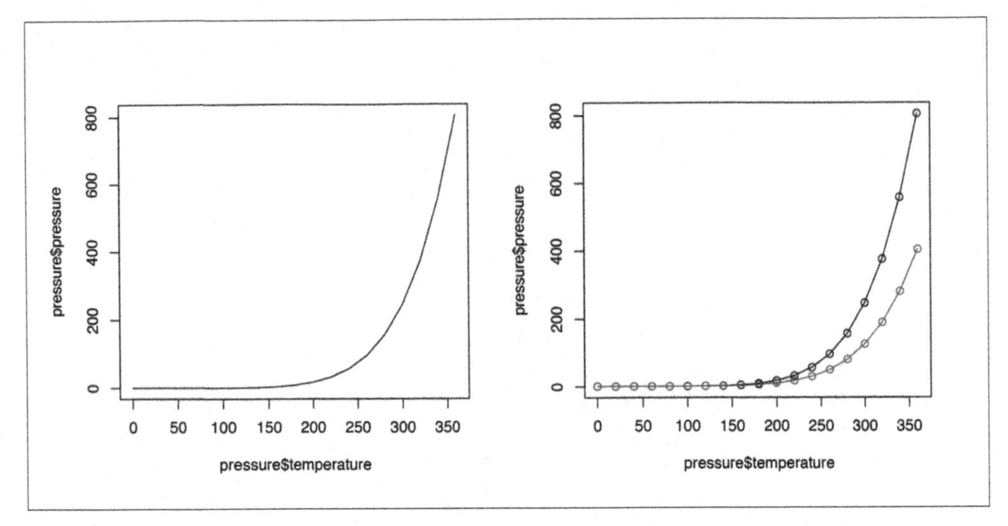

Figure 2-3. Left: line graph with base graphics; right: with points and another line

To add points and/or multiple lines (Figure 2-3, right), first call `plot()` for the first line, then add points with `points()` and additional lines with `lines()`:

```
plot(pressure$temperature, pressure$pressure, type="l")
points(pressure$temperature, pressure$pressure)

lines(pressure$temperature, pressure$pressure/2, col="red")
points(pressure$temperature, pressure$pressure/2, col="red")
```

With ggplot2, you can get a similar result using `qplot()` with `geom="line"` (Figure 2-4):

```
library(ggplot2)
qplot(pressure$temperature, pressure$pressure, geom="line")
```

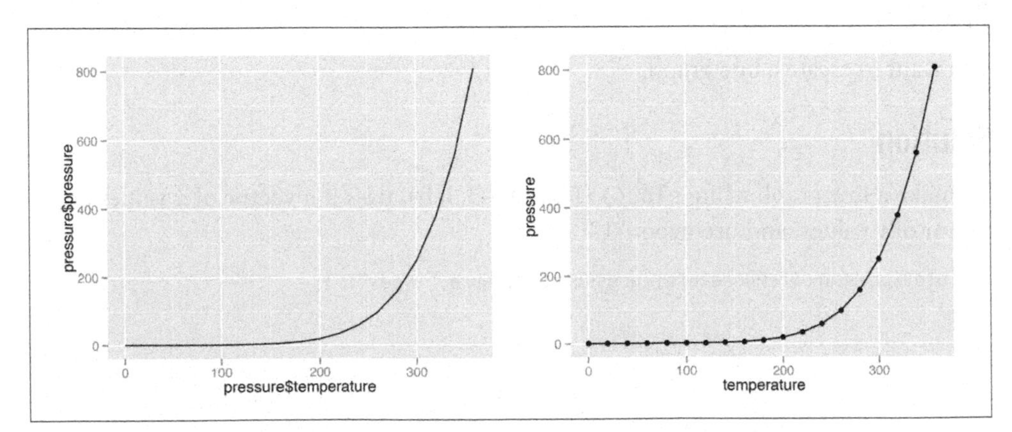

Figure 2-4. Left: line graph with qplot() from ggplot2; right: with points added

If the two vectors are already in the same data frame, you can use the following syntax:

```
qplot(temperature, pressure, data=pressure, geom="line")
# This is equivalent to:
ggplot(pressure, aes(x=temperature, y=pressure)) + geom_line()


# Lines and points together
qplot(temperature, pressure, data=pressure, geom=c("line", "point"))
# Equivalent to:
ggplot(pressure, aes(x=temperature, y=pressure)) + geom_line() + geom_point()
```

See Also

See Chapter 4 for more in-depth information about creating line graphs.

2.3. Creating a Bar Graph

Problem

You want to make a bar graph.

Solution

To make a bar graph of values (Figure 2-5), use `barplot()` and pass it a vector of values for the height of each bar and (optionally) a vector of labels for each bar. If the vector has names for the elements, the names will automatically be used as labels:

```
barplot(BOD$demand, names.arg=BOD$Time)
```

Sometimes "bar graph" refers to a graph where the bars represent the *count* of cases in each category. This is similar to a histogram, but with a discrete instead of continuous x-axis. To generate the count of each unique value in a vector, use the `table()` function:

```
table(mtcars$cyl)

 4  6  8
11  7 14
# There are 11 cases of the value 4, 7 cases of 6, and 14 cases of 8
```

Simply pass the table to `barplot()` to generate the graph of counts:

```
# Generate a table of counts
barplot(table(mtcars$cyl))
```

With the ggplot2 package, you can get a similar result using `qplot()` (Figure 2-6). To plot a bar graph of *values*, use `geom="bar"` and `stat="identity"`. Notice the difference in the output when the *x* variable is continuous and when it is discrete:

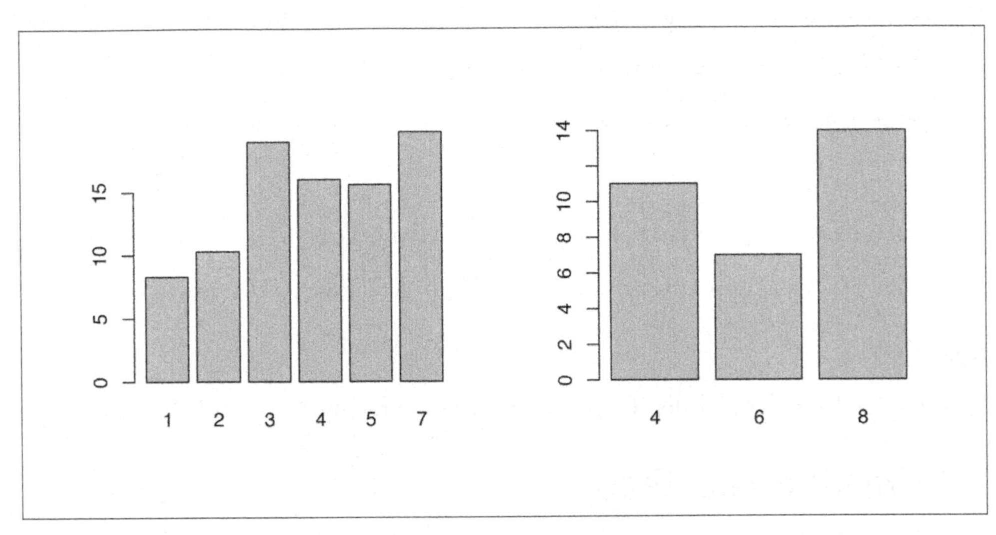

Figure 2-5. Left: bar graph of values with base graphics; right: bar graph of counts

```
library(ggplot2)
qplot(BOD$Time, BOD$demand, geom="bar", stat="identity")

# Convert the x variable to a factor, so that it is treated as discrete
qplot(factor(BOD$Time), BOD$demand, geom="bar", stat="identity")
```

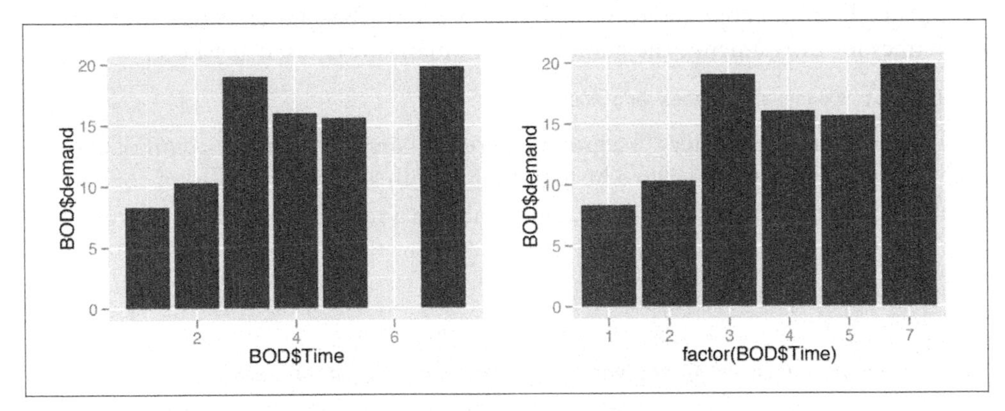

Figure 2-6. Left: bar graph of values with qplot() with continuous x variable; right: with x variable converted to a factor (notice that there is no entry for 6)

`qplot()` can also be used to graph the *counts* in each category (Figure 2-7). This is in fact the default way that ggplot2 creates bar graphs, and requires less typing than a bar graph of values. Once again, notice the difference between a continuous x-axis and a discrete one.

```
# cyl is continuous here
qplot(mtcars$cyl)

# Treat cyl as discrete
qplot(factor(mtcars$cyl))
```

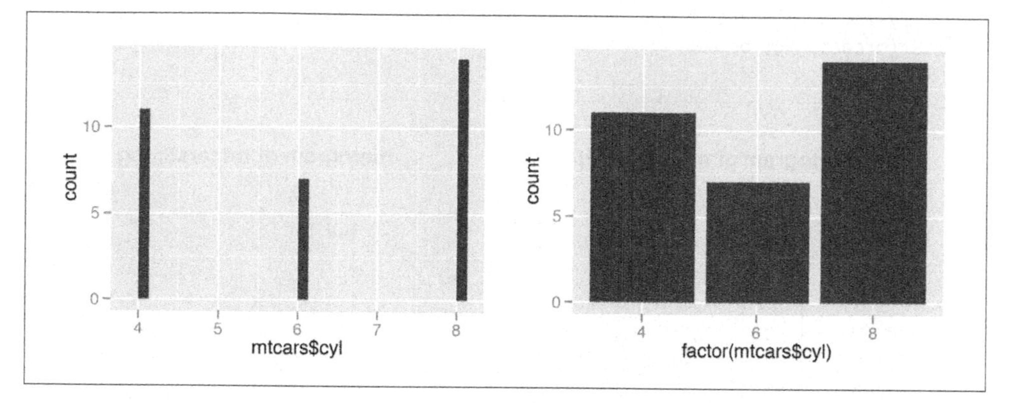

Figure 2-7. Left: bar graph of counts with qplot() with continuous x variable; right: with x variable converted to a factor

If the vector is in a data frame, you can use the following syntax:

```
# Bar graph of values. This uses the BOD data frame, with the
#"Time" column for x values and the "demand" column for y values.
qplot(Time, demand, data=BOD, geom="bar", stat="identity")
# This is equivalent to:
ggplot(BOD, aes(x=Time, y=demand)) + geom_bar(stat="identity")


# Bar graph of counts
qplot(factor(cyl), data=mtcars)
# This is equivalent to:
ggplot(mtcars, aes(x=factor(cyl))) + geom_bar()
```

See Also

See Chapter 3 for more in-depth information about creating bar graphs.

2.4. Creating a Histogram

Problem

You want to view the distribution of one-dimensional data with a histogram.

Solution

To make a histogram (Figure 2-8), use `hist()` and pass it a vector of values:

```
hist(mtcars$mpg)

# Specify approximate number of bins with breaks
hist(mtcars$mpg, breaks=10)
```

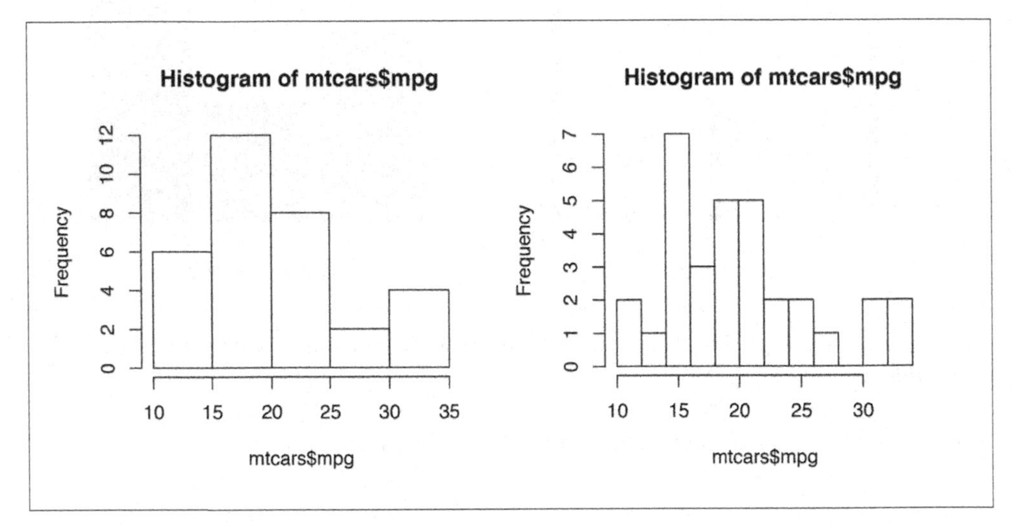

Figure 2-8. Left: histogram with base graphics; right: with more bins. Notice that because the bins are narrower, there are fewer items in each bin.

With the ggplot2 package, you can get a similar result using `qplot()` (Figure 2-9):

```
qplot(mtcars$mpg)
```

If the vector is in a data frame, you can use the following syntax:

```
library(ggplot2)
qplot(mpg, data=mtcars, binwidth=4)
# This is equivalent to:
ggplot(mtcars, aes(x=mpg)) + geom_histogram(binwidth=4)
```

See Also

For more in-depth information about creating histograms, see Recipes 6.1 and 6.2.

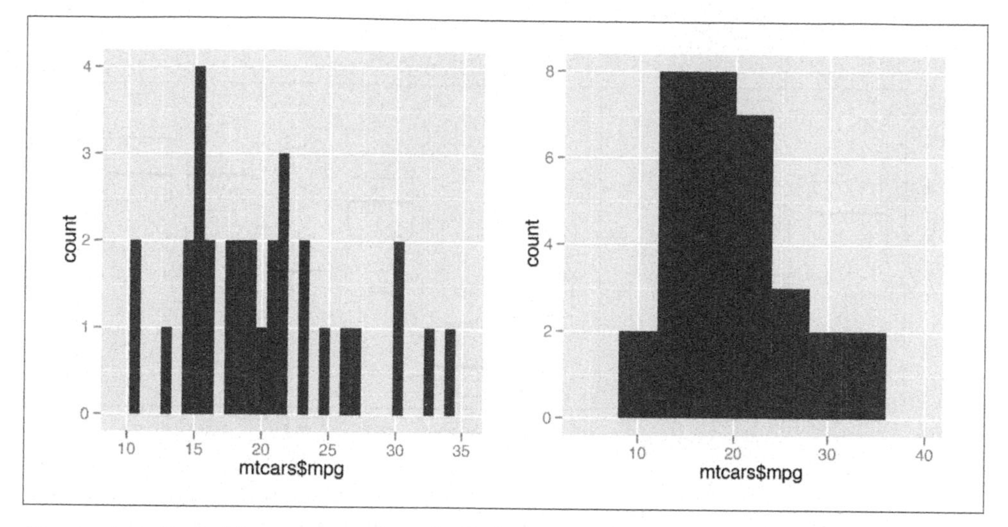

Figure 2-9. Left: histogram with qplot() from ggplot2, with default bin width; right: with wider bins

2.5. Creating a Box Plot

Problem

You want to create a box plot for comparing distributions.

Solution

To make a box plot (Figure 2-10), use `plot()` and pass it a factor of *x* values and a vector of *y* values. When *x* is a factor (as opposed to a numeric vector), it will automatically create a box plot:

```
plot(ToothGrowth$supp, ToothGrowth$len)
```

If the two vectors are in the same data frame, you can also use formula syntax. With this syntax, you can combine two variables on the x-axis, as in Figure 2-10:

```
# Formula syntax
boxplot(len ~ supp, data = ToothGrowth)

# Put interaction of two variables on x-axis
boxplot(len ~ supp + dose, data = ToothGrowth)
```

With the ggplot2 package, you can get a similar result using `qplot()` (Figure 2-11), with `geom="boxplot"`:

```
library(ggplot2)
qplot(ToothGrowth$supp, ToothGrowth$len, geom="boxplot")
```

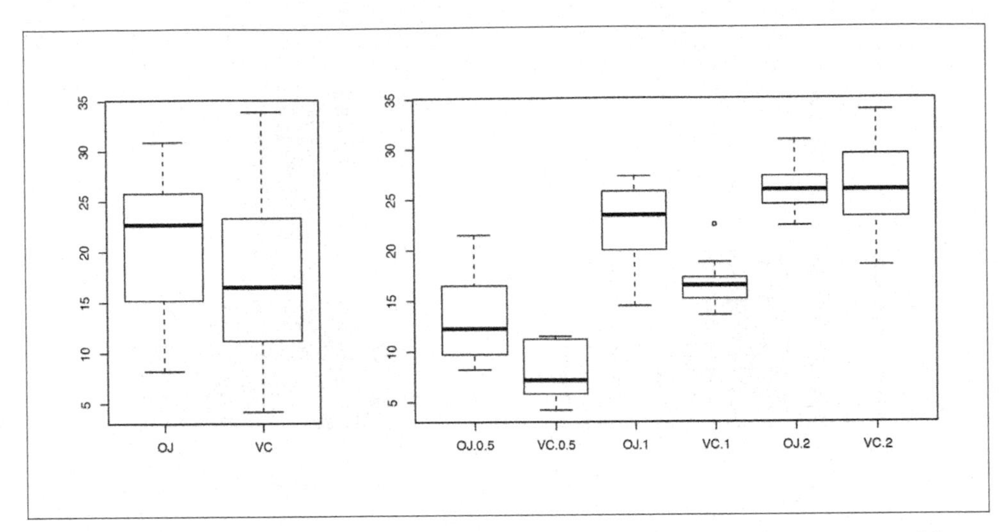

Figure 2-10. Left: box plot with base graphics; right: with multiple grouping variables

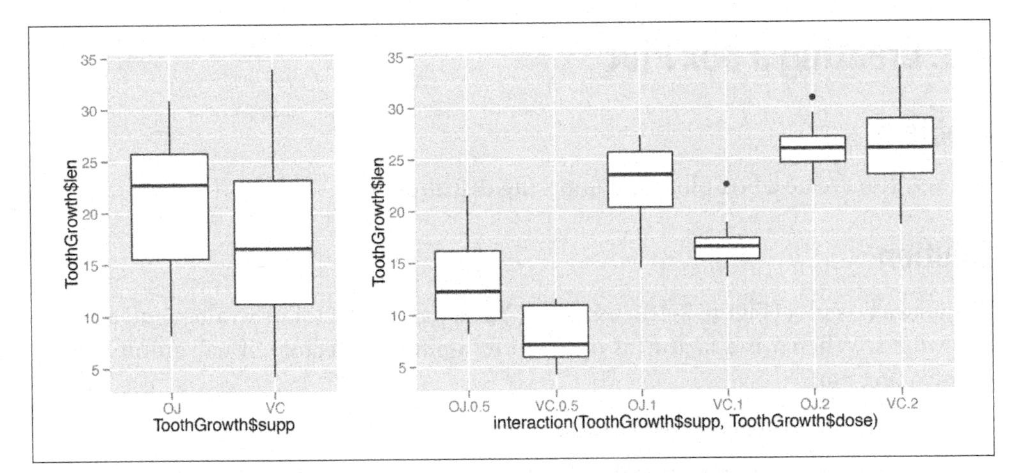

Figure 2-11. Left: box plot with qplot(); right: with multiple grouping variables

If the two vectors are already in the same data frame, you can use the following syntax:

```
qplot(supp, len, data=ToothGrowth, geom="boxplot")
# This is equivalent to:
ggplot(ToothGrowth, aes(x=supp, y=len)) + geom_boxplot()
```

It's also possible to make box plots for multiple variables, by combining the variables with `interaction()`, as in Figure 2-11. In this case, the `dose` variable is numeric, so we must convert it to a factor to use it as a grouping variable:

```
# Using three separate vectors
qplot(interaction(ToothGrowth$supp, ToothGrowth$dose), ToothGrowth$len,
      geom="boxplot")

# Alternatively, get the columns from the data frame
qplot(interaction(supp, dose), len, data=ToothGrowth, geom="boxplot")
# This is equivalent to:
ggplot(ToothGrowth, aes(x=interaction(supp, dose), y=len)) + geom_boxplot()
```

You may have noticed that the box plots from base graphics are ever-so-slightly different from those from ggplot2. This is because they use slightly different methods for calculating quantiles. See `?geom_boxplot` and `?boxplot.stats` for more information on how they differ.

See Also

For more on making basic box plots, see Recipe 6.6.

2.6. Plotting a Function Curve

Problem

You want to plot a function curve.

Solution

To plot a function curve, as in Figure 2-12, use `curve()` and pass it an expression with the variable `x`:

```
curve(x^3 - 5*x, from=-4, to=4)
```

You can plot any function that takes a numeric vector as input and returns a numeric vector, including functions that you define yourself. Using `add=TRUE` will add a curve to the previously created plot:

```
# Plot a user-defined function
myfun <- function(xvar) {
    1/(1 + exp(-xvar + 10))
}
curve(myfun(x), from=0, to=20)
# Add a line:
curve(1-myfun(x), add = TRUE, col = "red")
```

With the ggplot2 package, you can get a similar result using `qplot()` (Figure 2-13), by using `stat="function"` and `geom="line"` and passing it a function that takes a numeric vector as input and returns a numeric vector:

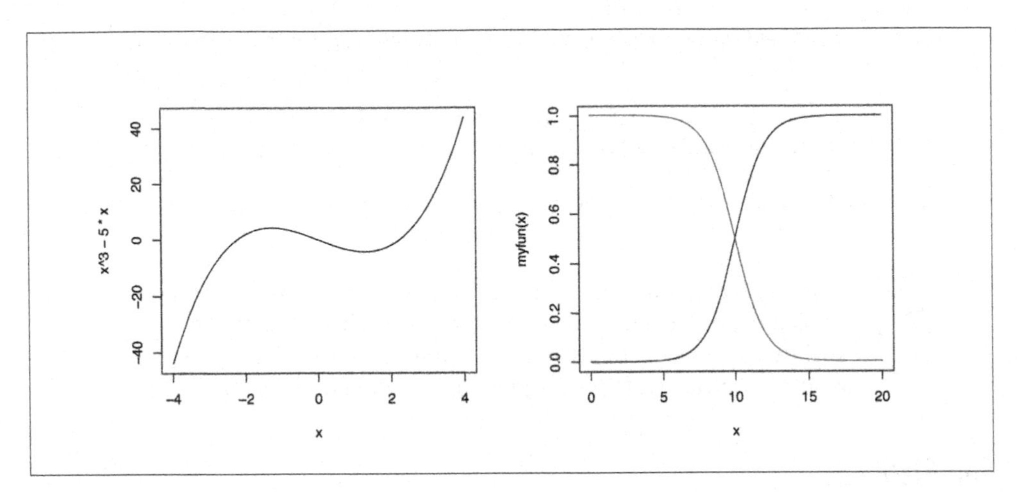

Figure 2-12. Left: function curve with base graphics; right: with user-defined function

```
library(ggplot2)
# This sets the x range from 0 to 20
qplot(c(0,20), fun=myfun, stat="function", geom="line")
# This is equivalent to:
ggplot(data.frame(x=c(0, 20)), aes(x=x)) + stat_function(fun=myfun, geom="line")
```

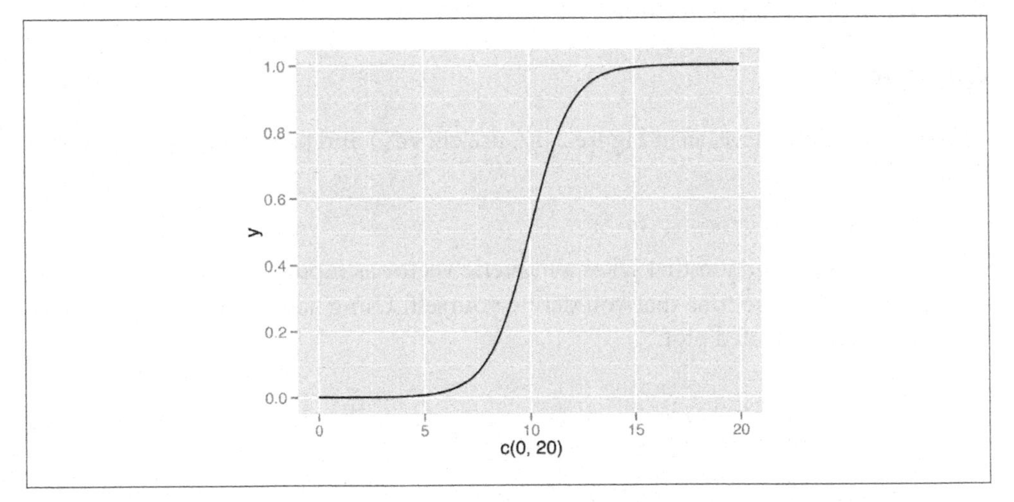

Figure 2-13. A function curve with qplot()

See Also

See Recipe 13.2 for more in-depth information about plotting function curves.

CHAPTER 3
Bar Graphs

Bar graphs are perhaps the most commonly used kind of data visualization. They're typically used to display numeric values (on the y-axis), for different categories (on the x-axis). For example, a bar graph would be good for showing the prices of four different kinds of items. A bar graph generally wouldn't be as good for showing prices over time, where time is a continuous variable—though it can be done, as we'll see in this chapter.

There's an important distinction you should be aware of when making bar graphs: sometimes the bar heights represent *counts* of cases in the data set, and sometimes they represent *values* in the data set. Keep this distinction in mind—it can be a source of confusion since they have very different relationships to the data, but the same term is used for both of them. In this chapter I'll discuss this more, and present recipes for both types of bar graphs.

3.1. Making a Basic Bar Graph

Problem

You have a data frame where one column represents the *x* position of each bar, and another column represents the vertical (y) height of each bar.

Solution

Use `ggplot()` with `geom_bar(stat="identity")` and specify what variables you want on the x- and y-axes (Figure 3-1):

```
library(gcookbook) # For the data set
ggplot(pg_mean, aes(x=group, y=weight)) + geom_bar(stat="identity")
```

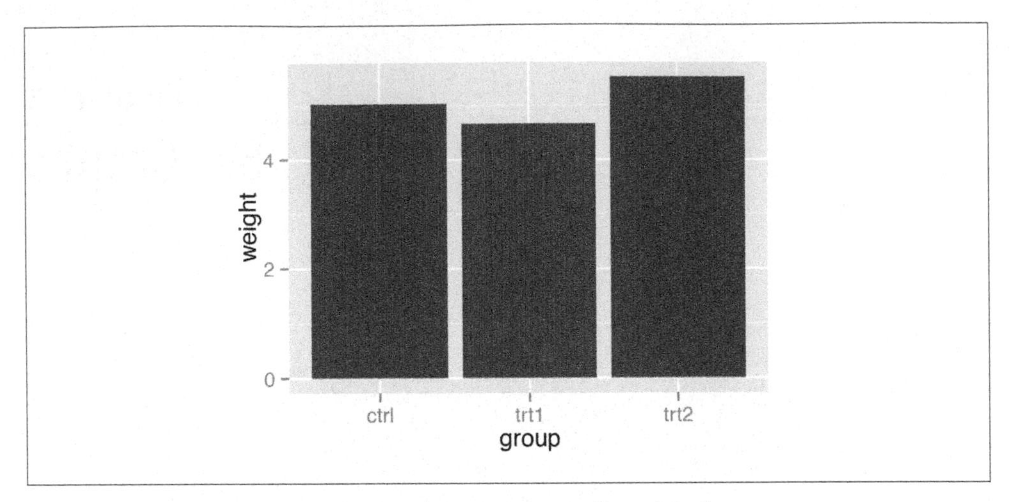

Figure 3-1. Bar graph of values (with stat="identity") with a discrete x-axis

Discussion

When x is a continuous (or numeric) variable, the bars behave a little differently. Instead of having one bar at each actual x value, there is one bar at each possible x value between the minimum and the maximum, as in Figure 3-2. You can convert the continuous variable to a discrete variable by using `factor()`:

```
# There's no entry for Time == 6
BOD

 Time demand
    1    8.3
    2   10.3
    3   19.0
    4   16.0
    5   15.6
    7   19.8

# Time is numeric (continuous)
str(BOD)

'data.frame':   6 obs. of  2 variables:
 $ Time  : num  1 2 3 4 5 7
 $ demand: num  8.3 10.3 19 16 15.6 19.8
 - attr(*, "reference")= chr "A1.4, p. 270"

ggplot(BOD, aes(x=Time, y=demand)) + geom_bar(stat="identity")

# Convert Time to a discrete (categorical) variable with factor()
ggplot(BOD, aes(x=factor(Time), y=demand)) + geom_bar(stat="identity")
```

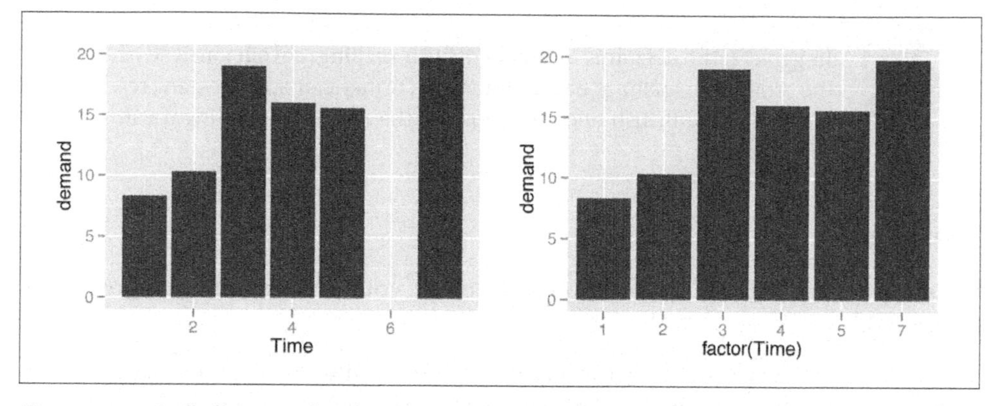

Figure 3-2. Left: bar graph of values (with stat="identity") with a continuous x-axis; right: with x variable converted to a factor (notice that the space for 6 is gone)

In these examples, the data has a column for `x` values and another for `y` values. If you instead want the height of the bars to represent the *count* of cases in each group, see Recipe 3.3.

By default, bar graphs use a very dark grey for the bars. To use a color fill, use `fill`. Also, by default, there is no outline around the fill. To add an outline, use `colour`. For Figure 3-3, we use a light blue fill and a black outline:

```
ggplot(pg_mean, aes(x=group, y=weight)) +
    geom_bar(stat="identity", fill="lightblue", colour="black")
```

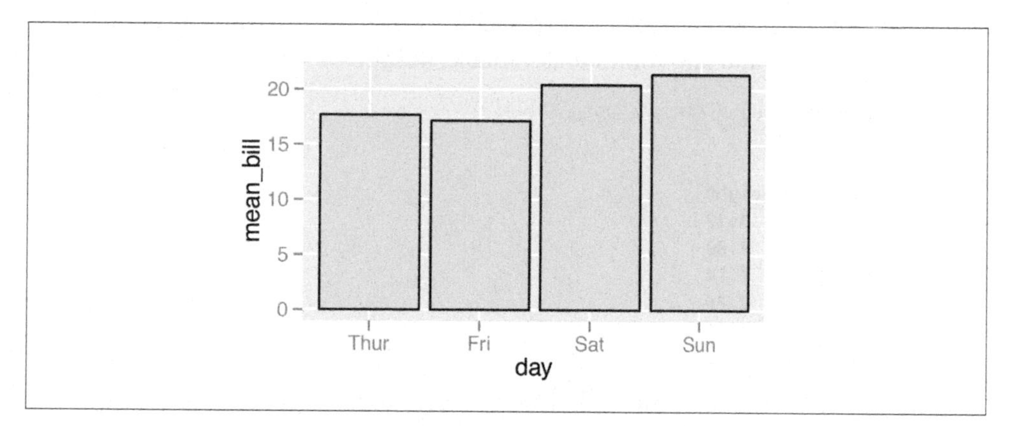

Figure 3-3. A single fill and outline color for all bars

In ggplot2, the default is to use the British spelling, `colour`, instead of the American spelling, `color`. Internally, American spellings are remapped to the British ones, so if you use the American spelling it will still work.

See Also

If you want the height of the bars to represent the count of cases in each group, see Recipe 3.3.

To reorder the levels of a factor based on the values of another variable, see Recipe 15.9. To manually change the order of factor levels, see Recipe 15.8.

For more information about using colors, see Chapter 12.

3.2. Grouping Bars Together

Problem

You want to group bars together by a second variable.

Solution

Map a variable to `fill`, and use `geom_bar(position="dodge")`.

In this example we'll use the `cabbage_exp` data set, which has two categorical variables, `Cultivar` and `Date`, and one continuous variable, `Weight`:

```
library(gcookbook) # For the data set
cabbage_exp

 Cultivar Date Weight
      c39  d16   3.18
      c39  d20   2.80
      c39  d21   2.74
      c52  d16   2.26
      c52  d20   3.11
      c52  d21   1.47
```

We'll map `Date` to the *x* position and map `Cultivar` to the fill color (Figure 3-4):

```
ggplot(cabbage_exp, aes(x=Date, y=Weight, fill=Cultivar)) +
    geom_bar(position="dodge", stat="identity")
```

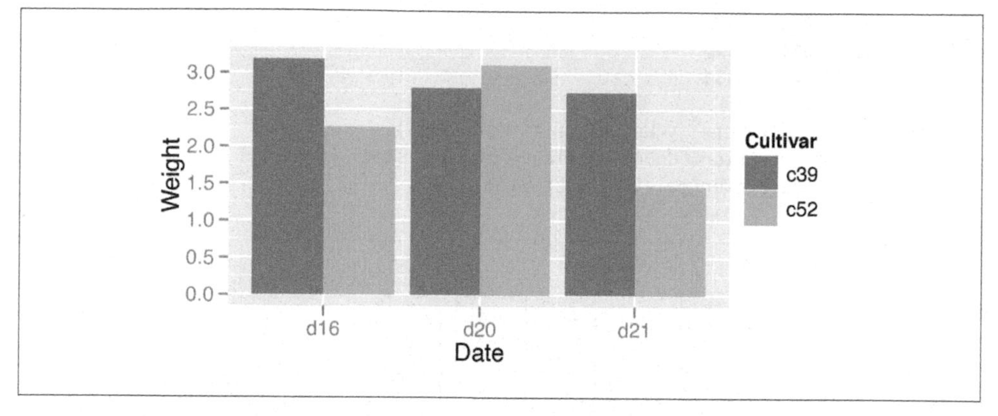

Figure 3-4. Graph with grouped bars

Discussion

The most basic bar graphs have one categorical variable on the x-axis and one continuous variable on the y-axis. Sometimes you'll want to use another categorical variable to divide up the data, in addition to the variable on the x-axis. You can produce a grouped bar plot by mapping that variable to `fill`, which represents the fill color of the bars. You must also use `position="dodge"`, which tells the bars to "dodge" each other horizontally; if you don't, you'll end up with a stacked bar plot (Recipe 3.7).

As with variables mapped to the x-axis of a bar graph, variables that are mapped to the fill color of bars must be categorical rather than continuous variables.

To add a black outline, use `colour="black"` inside `geom_bar()`. To set the colors, you can use `scale_fill_brewer()` or `scale_fill_manual()`. In Figure 3-5 we'll use the `Pastel1` palette from `RColorBrewer`:

```
ggplot(cabbage_exp, aes(x=Date, y=Weight, fill=Cultivar)) +
    geom_bar(position="dodge", colour="black", stat="identity") +
    scale_fill_brewer(palette="Pastel1")
```

Other aesthetics, such as `colour` (the color of the outlines of the bars) or `linestyle`, can also be used for grouping variables, but `fill` is probably what you'll want to use.

Note that if there are any missing combinations of the categorical variables, that bar will be missing, and the neighboring bars will expand to fill that space. If we remove the last row from our example data frame, we get Figure 3-6:

```
ce <- cabbage_exp[1:5, ]    # Copy the data without last row
ce
```

```
 Cultivar Date Weight
      c39  d16   3.18
      c39  d20   2.80
```

```
c39  d21   2.74
c52  d16   2.26
c52  d20   3.11
```

```
ggplot(ce, aes(x=Date, y=Weight, fill=Cultivar)) +
    geom_bar(position="dodge", colour="black", stat="identity") +
    scale_fill_brewer(palette="Pastel1")
```

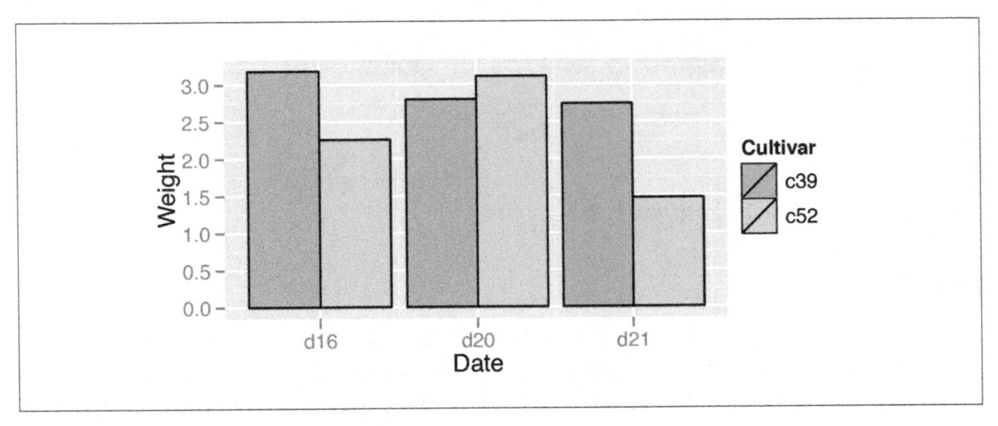

Figure 3-5. Grouped bars with black outline and a different color palette

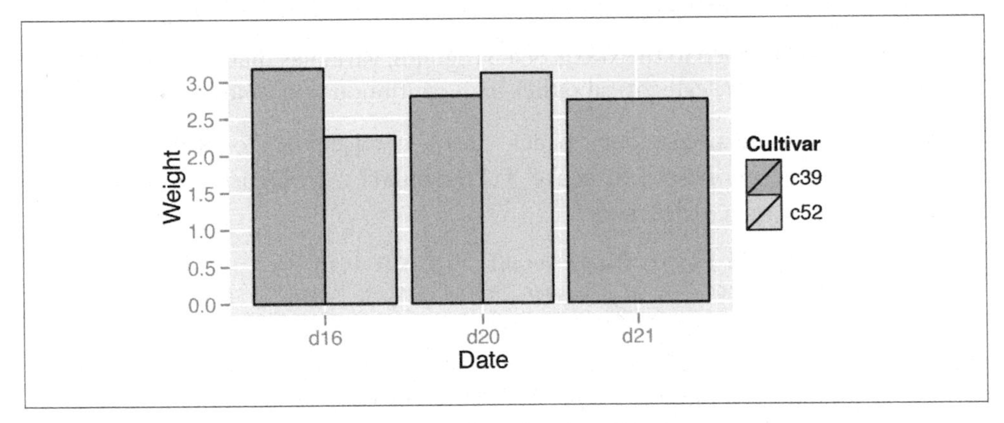

Figure 3-6. Graph with a missing bar—the other bar fills the space

If your data has this issue, you can manually make an entry for the missing factor level combination with an `NA` for the *y* variable.

See Also

For more on using colors in bar graphs, see Recipe 3.4.

To reorder the levels of a factor based on the values of another variable, see Recipe 15.9.

3.3. Making a Bar Graph of Counts

Problem

Your data has one row representing each case, and you want plot counts of the cases.

Solution

Use `geom_bar()` without mapping anything to `y` (Figure 3-7):

```
ggplot(diamonds, aes(x=cut)) + geom_bar()
# Equivalent to using geom_bar(stat="bin")
```

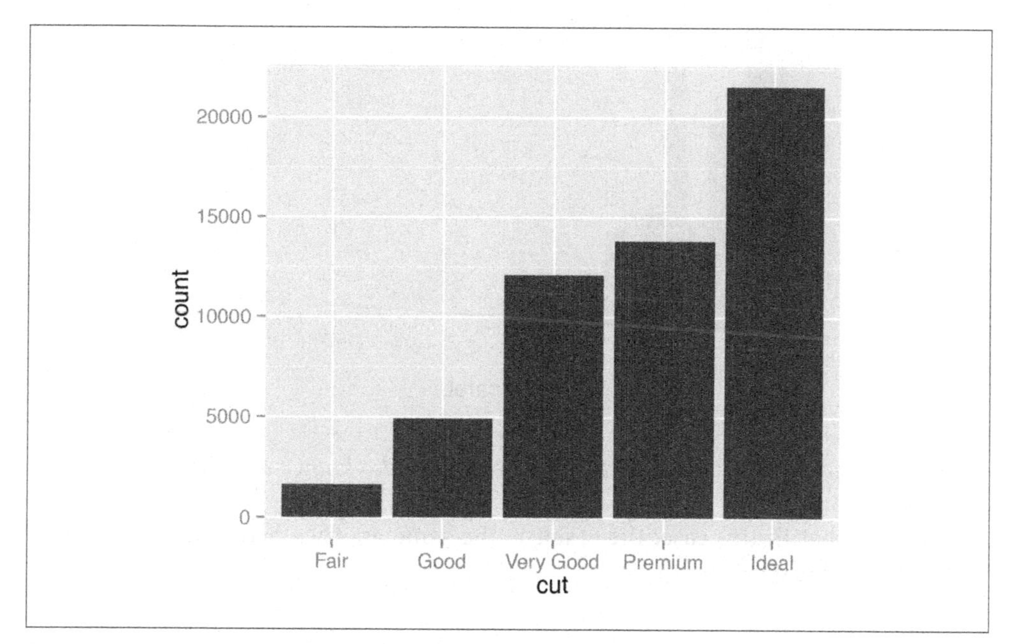

Figure 3-7. Bar graph of counts

Discussion

The `diamonds` data set has 53,940 rows, each of which represents information about one diamond:

```
diamonds

      carat       cut color clarity depth table price    x    y    z
1      0.23     Ideal     E     SI2  61.5    55   326 3.95 3.98 2.43
```

```
2       0.21   Premium     E     SI1  59.8     61   326 3.89 3.84 2.31
3       0.23      Good     E     VS1  56.9     65   327 4.05 4.07 2.31
...
53939  0.86   Premium     H     SI2  61.0     58  2757 6.15 6.12 3.74
53940  0.75     Ideal     D     SI2  62.2     55  2757 5.83 5.87 3.64
```

With `geom_bar()`, the default behavior is to use `stat="bin"`, which counts up the number of cases for each group (each *x* position, in this example). In the graph we can see that there are about 23,000 cases with an ideal cut.

In this example, the variable on the x-axis is discrete. If we use a continuous variable on the x-axis, we'll get a histogram, as shown in Figure 3-8:

```
ggplot(diamonds, aes(x=carat)) + geom_bar()
```

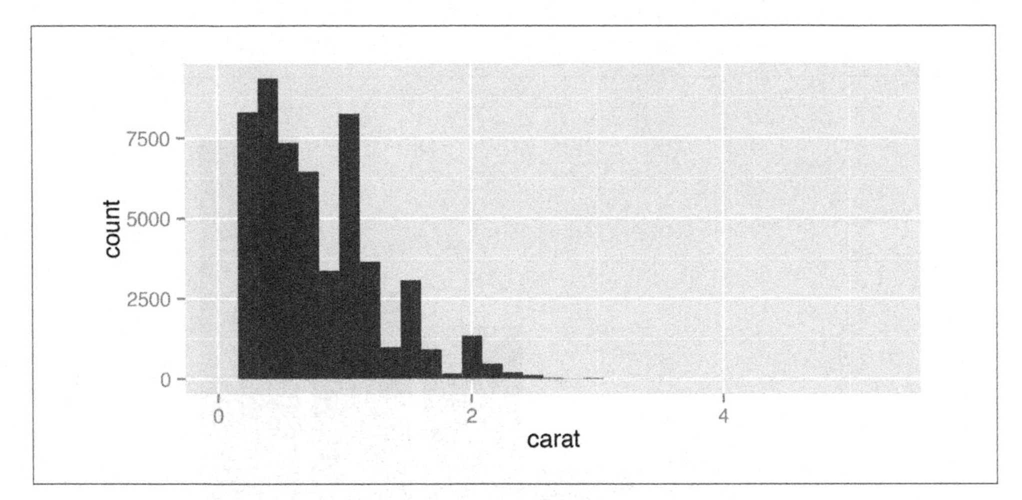

Figure 3-8. Bar graph of counts on a continuous axis, also known as a histogram

It turns out that in this case, the result is the same as if we had used `geom_histogram()` instead of `geom_bar()`.

See Also

If, instead of having `ggplot()` count up the number of rows in each group, you have a column in your data frame representing the `y` values, see Recipe 3.1.

You could also get the same graphical output by calculating the counts before sending the data to `ggplot()`. See Recipe 15.17 for more on summarizing data.

For more about histograms, see Recipe 6.1.

3.4. Using Colors in a Bar Graph

Problem

You want to use different colors for the bars in your graph.

Solution

Map the appropriate variable to the `fill` aesthetic.

We'll use the `uspopchange` data set for this example. It contains the percentage change in population for the US states from 2000 to 2010. We'll take the top 10 fastest-growing states and graph their percentage change. We'll also color the bars by region (Northeast, South, North Central, or West).

First, we'll take the top 10 states:

```
library(gcookbook) # For the data set
upc <- subset(uspopchange, rank(Change)>40)
upc
```

```
          State Abb Region Change
        Arizona  AZ   West   24.6
       Colorado  CO   West   16.9
        Florida  FL  South   17.6
        Georgia  GA  South   18.3
          Idaho  ID   West   21.1
         Nevada  NV   West   35.1
 North Carolina  NC  South   18.5
 South Carolina  SC  South   15.3
          Texas  TX  South   20.6
           Utah  UT   West   23.8
```

Now we can make the graph, mapping `Region` to `fill` (Figure 3-9):

```
ggplot(upc, aes(x=Abb, y=Change, fill=Region)) + geom_bar(stat="identity")
```

Discussion

The default colors aren't very appealing, so you may want to set them, using `scale_fill_brewer()` or `scale_fill_manual()`. With this example, we'll use the latter, and we'll set the outline color of the bars to black, with `colour="black"` (Figure 3-10). Note that *setting* occurs outside of `aes()`, while *mapping* occurs within `aes()`:

```
ggplot(upc, aes(x=reorder(Abb, Change), y=Change, fill=Region)) +
    geom_bar(stat="identity", colour="black") +
    scale_fill_manual(values=c("#669933", "#FFCC66")) +
    xlab("State")
```

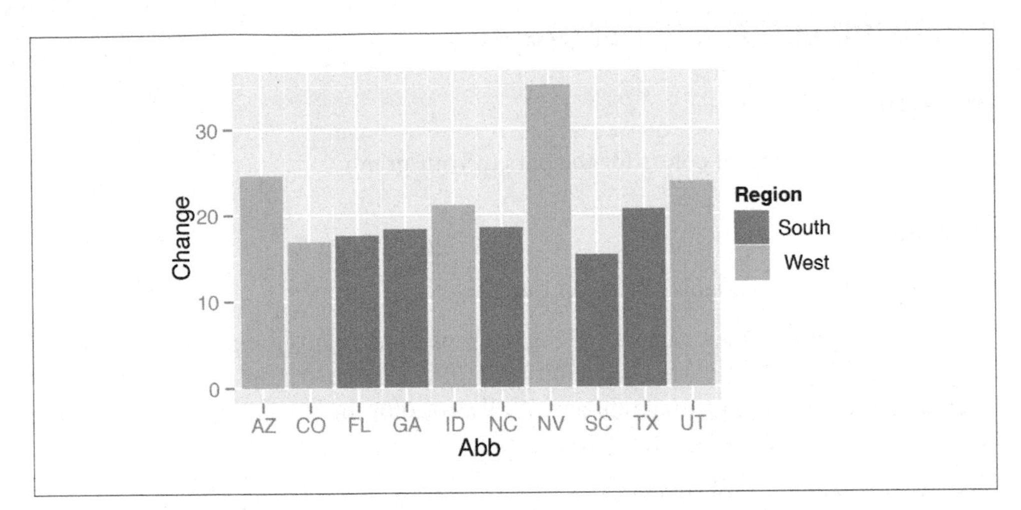

Figure 3-9. A variable mapped to fill

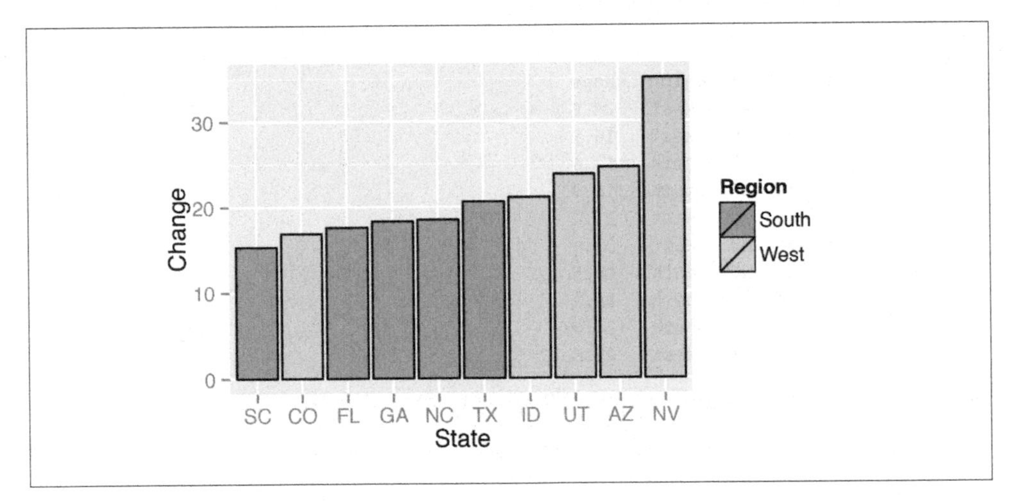

Figure 3-10. Graph with different colors, black outlines, and sorted by percentage change

This example also uses the `reorder()` function, as in this particular case it makes sense to sort the bars by their height, instead of in alphabetical order.

See Also

For more about using the `reorder()` function to reorder the levels of a factor based on the values of another variable, see Recipe 15.9.

For more information about using colors, see Chapter 12.

3.5. Coloring Negative and Positive Bars Differently

Problem

You want to use different colors for negative and positive-valued bars.

Solution

We'll use a subset of the `climate` data and create a new column called `pos`, which indicates whether the value is positive or negative:

```
library(gcookbook) # For the data set
csub <- subset(climate, Source=="Berkeley" & Year >= 1900)
csub$pos <- csub$Anomaly10y >= 0

csub

   Source Year Anomaly1y Anomaly5y Anomaly10y Unc10y
 Berkeley 1900        NA        NA     -0.171  0.108 FALSE
 Berkeley 1901        NA        NA     -0.162  0.109 FALSE
 Berkeley 1902        NA        NA     -0.177  0.108 FALSE
 ...
 Berkeley 2002        NA        NA      0.856  0.028  TRUE
 Berkeley 2003        NA        NA      0.869  0.028  TRUE
 Berkeley 2004        NA        NA      0.884  0.029  TRUE
```

Once we have the data, we can make the graph and map `pos` to the fill color, as in Figure 3-11. Notice that we use `position="identity"` with the bars. This will prevent a warning message about stacking not being well defined for negative numbers:

```
ggplot(csub, aes(x=Year, y=Anomaly10y, fill=pos)) +
  geom_bar(stat="identity", position="identity")
```

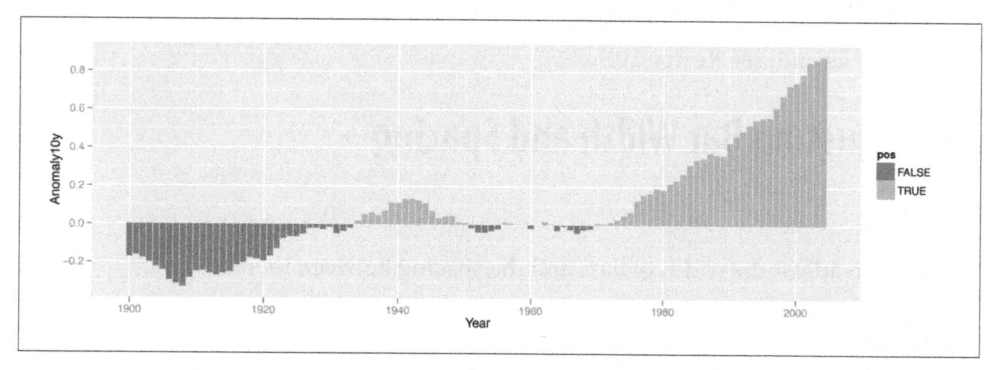

Figure 3-11. Different colors for positive and negative values

Discussion

There are a few problems with the first attempt. First, the colors are probably the reverse of what we want: usually, blue means cold and red means hot. Second, the legend is redundant and distracting.

We can change the colors with `scale_fill_manual()` and remove the legend with `guide=FALSE`, as shown in Figure 3-12. We'll also add a thin black outline around each of the bars by setting `colour` and specifying `size`, which is the thickness of the outline, in millimeters:

```
ggplot(csub, aes(x=Year, y=Anomaly10y, fill=pos)) +
    geom_bar(stat="identity", position="identity", colour="black", size=0.25) +
    scale_fill_manual(values=c("#CCEEFF", "#FFDDDD"), guide=FALSE)
```

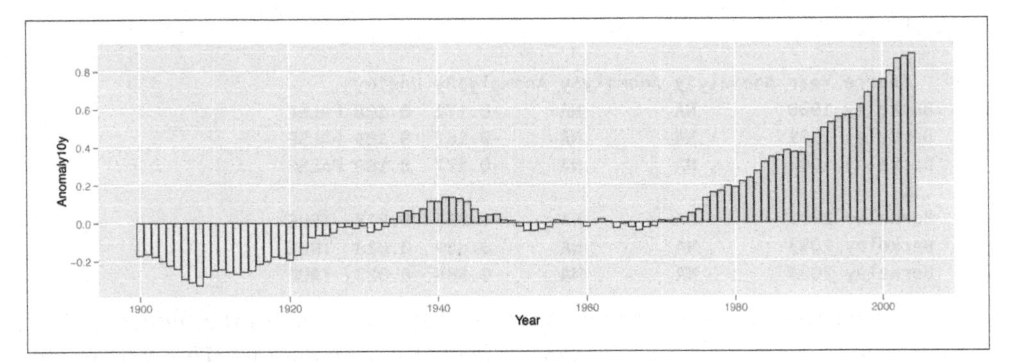

Figure 3-12. Graph with customized colors and no legend

See Also

To change the colors used, see Recipes 12.3 and 12.4.

To hide the legend, see Recipe 10.1.

3.6. Adjusting Bar Width and Spacing

Problem

You want to adjust the width of bars and the spacing between them.

Solution

To make the bars narrower or wider, set `width` in `geom_bar()`. The default value is 0.9; larger values make the bars wider, and smaller values make the bars narrower (Figure 3-13).

For example, for standard-width bars:

```
library(gcookbook) # For the data set

ggplot(pg_mean, aes(x=group, y=weight)) + geom_bar(stat="identity")
```

For narrower bars:

```
ggplot(pg_mean, aes(x=group, y=weight)) + geom_bar(stat="identity", width=0.5)
```

And for wider bars (these have the maximum width of 1):

```
ggplot(pg_mean, aes(x=group, y=weight)) + geom_bar(stat="identity", width=1)
```

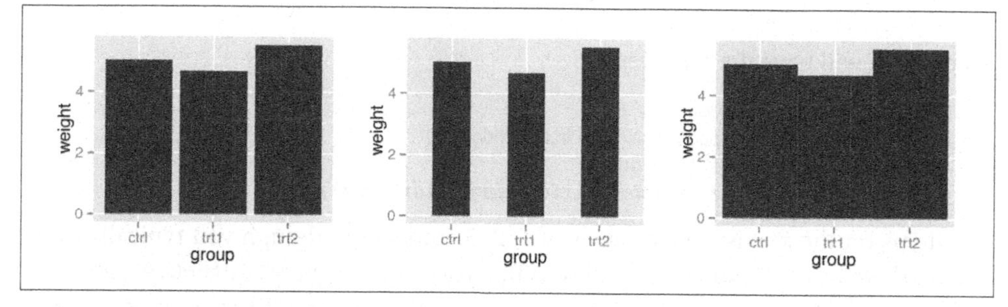

Figure 3-13. Different bar widths

For grouped bars, the default is to have no space between bars within each group. To add space between bars within a group, make `width` smaller and set the value for `position_dodge` to be larger than `width` (Figure 3-14).

For a grouped bar graph with narrow bars:

```
ggplot(cabbage_exp, aes(x=Date, y=Weight, fill=Cultivar)) +
    geom_bar(stat="identity", width=0.5, position="dodge")
```

And with some space between the bars:

```
ggplot(cabbage_exp, aes(x=Date, y=Weight, fill=Cultivar)) +
    geom_bar(stat="identity", width=0.5, position=position_dodge(0.7))
```

The first graph used `position="dodge"`, and the second graph used `position=position_dodge()`. This is because `position="dodge"` is simply shorthand for `position=position_dodge()` with the default value of 0.9, but when we want to set a specific value, we need to use the more verbose command.

Discussion

The default value of `width` is 0.9, and the default value used for `position_dodge()` is the same. To be more precise, the value of `width` in `position_dodge()` is `NULL`, which tells ggplot2 to use the same value as the `width` from `geom_bar()`.

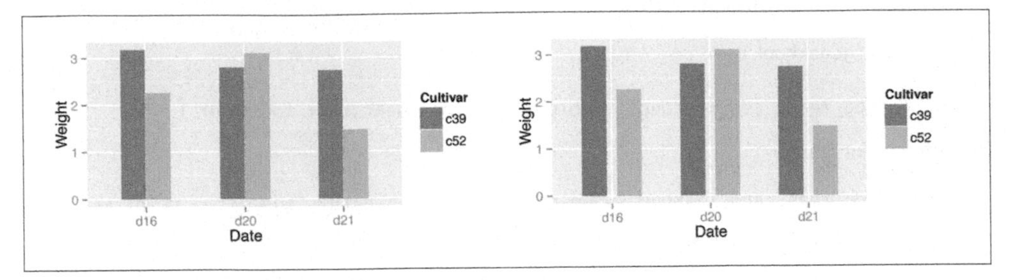

Figure 3-14. Left: bar graph with narrow grouped bars; right: with space between the bars

All of these will have the same result:

```
geom_bar(position="dodge")
geom_bar(width=0.9, position=position_dodge())
geom_bar(position=position_dodge(0.9))
geom_bar(width=0.9, position=position_dodge(width=0.9))
```

The items on the x-axis have `x` values of 1, 2, 3, and so on, though you typically don't refer to them by these numerical values. When you use `geom_bar(width=0.9)`, it makes each group take up a total width of 0.9 on the x-axis. When you use `position_dodge(width=0.9)`, it spaces the bars so that the *middle* of each bar is right where it would be if the bar width were 0.9 and the bars were touching. This is illustrated in Figure 3-15. The two graphs both have the same dodge width of 0.9, but while the top has a bar width of 0.9, the bottom has a bar width of 0.2. Despite the different bar widths, the middles of the bars stay aligned.

If you make the entire graph wider or narrower, the bar dimensions will scale proportionally. To see how this works, you can just resize the window in which the graphs appear. For information about controlling this when writing to a file, see Chapter 14.

3.7. Making a Stacked Bar Graph

Problem

You want to make a stacked bar graph.

Solution

Use `geom_bar()` and map a variable `fill`. This will put `Date` on the x-axis and use `Cultivar` for the fill color, as shown in Figure 3-16:

```
library(gcookbook) # For the data set
ggplot(cabbage_exp, aes(x=Date, y=Weight, fill=Cultivar)) +
    geom_bar(stat="identity")
```

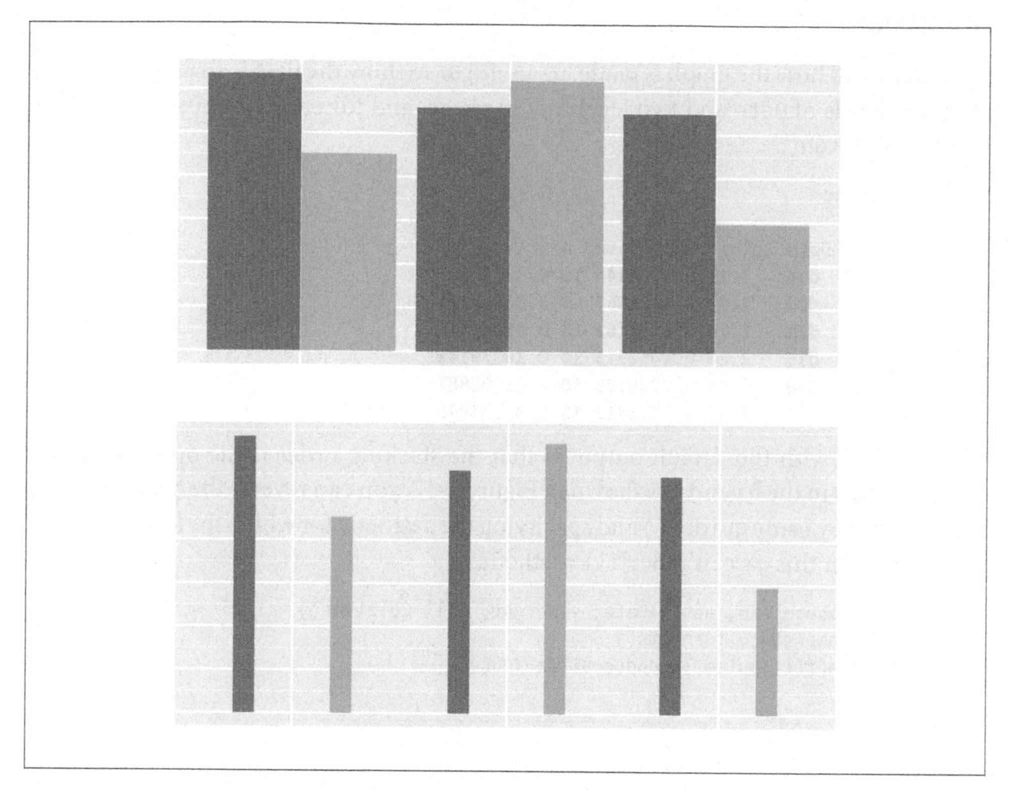

Figure 3-15. Same dodge width of 0.9, but different bar widths of 0.9 (top) and 0.2 (bottom)

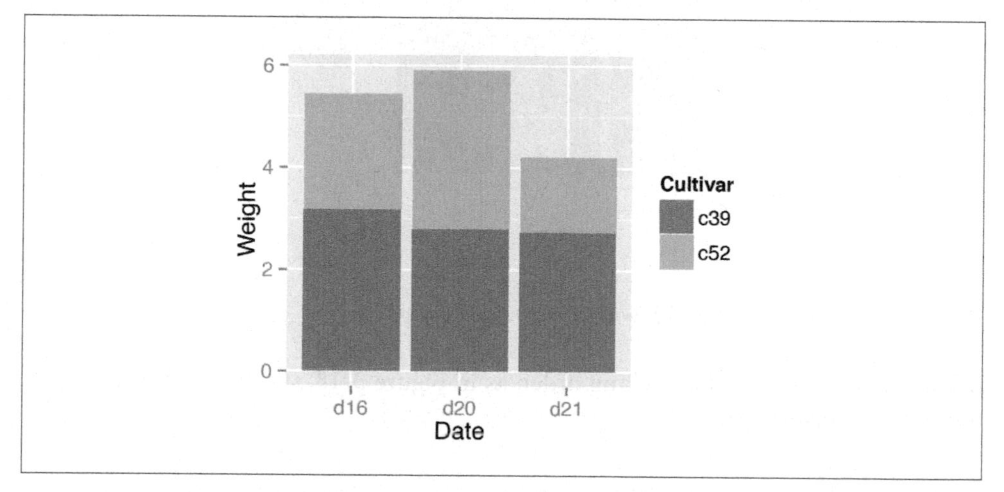

Figure 3-16. Stacked bar graph

Discussion

To understand how the graph is made, it's useful to see how the data is structured. There are three levels of `Date` and two levels of `Cultivar`, and for each combination there is a value for `Weight`:

```
cabbage_exp
```

```
 Cultivar Date Weight        sd  n         se
      c39  d16   3.18 0.9566144 10 0.30250803
      c39  d20   2.80 0.2788867 10 0.08819171
      c39  d21   2.74 0.9834181 10 0.31098410
      c52  d16   2.26 0.4452215 10 0.14079141
      c52  d20   3.11 0.7908505 10 0.25008887
      c52  d21   1.47 0.2110819 10 0.06674995
```

One problem with the default output is that the stacking order is the opposite of the order of items in the legend. As shown in Figure 3-17, you can reverse the order of items in the legend by using `guides()` and specifying the aesthetic for which the legend should be reversed. In this case, it's the `fill` aesthetic:

```
ggplot(cabbage_exp, aes(x=Date, y=Weight, fill=Cultivar)) +
    geom_bar(stat="identity") +
    guides(fill=guide_legend(reverse=TRUE))
```

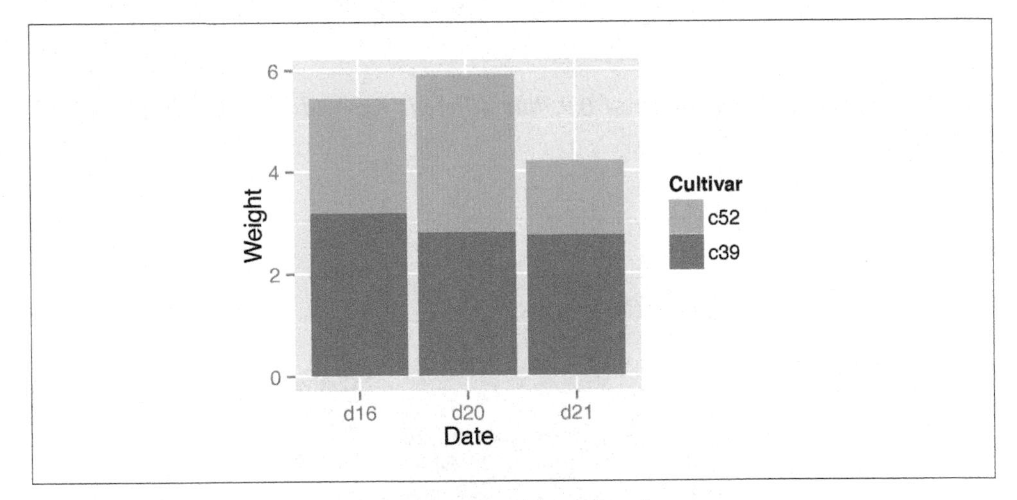

Figure 3-17. Stacked bar graph with reversed legend order

If you'd like to reverse the stacking order, as in Figure 3-18, specify `order=desc()` in the aesthetic mapping:

```
library(plyr) # Needed for desc()
ggplot(cabbage_exp, aes(x=Date, y=Weight, fill=Cultivar, order=desc(Cultivar))) +
    geom_bar(stat="identity")
```

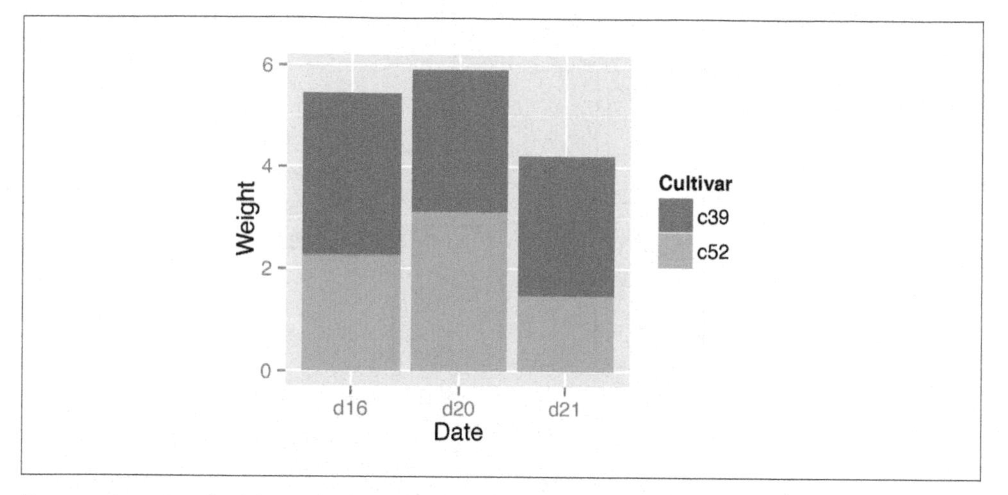

Figure 3-18. Stacked bar graph with reversed stacking order

It's also possible to modify the column of the data frame so that the factor levels are in a different order (see Recipe 15.8). Do this with care, since the modified data could change the results of other analyses.

For a more polished graph, we'll keep the reversed legend order, use `scale_fill_brewer()` to get a different color palette, and use `colour="black"` to get a black outline (Figure 3-19):

```
ggplot(cabbage_exp, aes(x=Date, y=Weight, fill=Cultivar)) +
    geom_bar(stat="identity", colour="black") +
    guides(fill=guide_legend(reverse=TRUE)) +
    scale_fill_brewer(palette="Pastel1")
```

See Also

For more on using colors in bar graphs, see Recipe 3.4.

To reorder the levels of a factor based on the values of another variable, see Recipe 15.9. To manually change the order of factor levels, see Recipe 15.8.

3.8. Making a Proportional Stacked Bar Graph

Problem

You want to make a stacked bar graph that shows proportions (also called a 100% stacked bar graph).

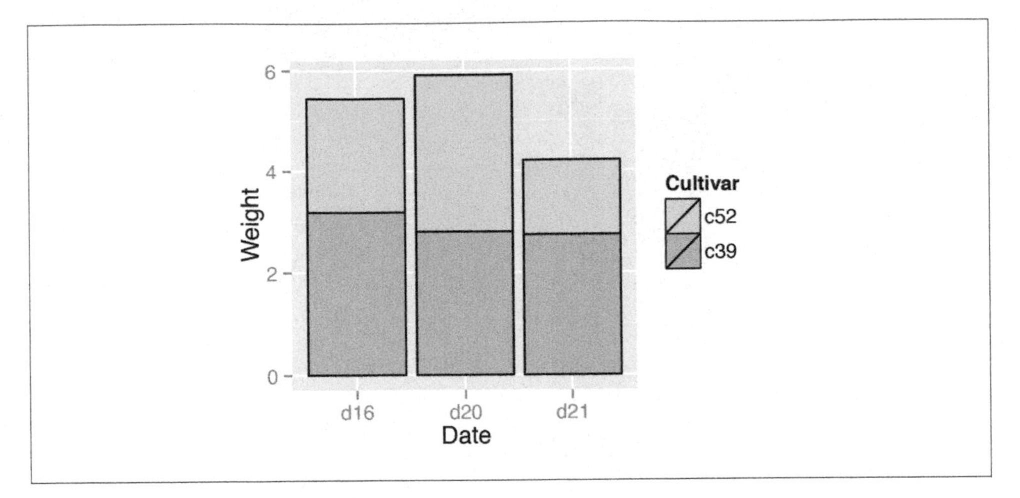

Figure 3-19. Stacked bar graph with reversed legend, new palette, and black outline

Solution

First, scale the data to 100% within each stack. This can be done by using `ddply()` from the plyr package, with `transform()`. Then plot the resulting data, as shown in Figure 3-20:

```
library(gcookbook) # For the data set
library(plyr)
# Do a group-wise transform(), splitting on "Date"
ce <- ddply(cabbage_exp, "Date", transform,
            percent_weight = Weight / sum(Weight) * 100)

ggplot(ce, aes(x=Date, y=percent_weight, fill=Cultivar)) +
    geom_bar(stat="identity")
```

Discussion

To calculate the percentages within each `Weight` group, we used the `ddply()` function. In the example here, the `ddply()` function splits the input data frame, `cabbage_exp`, by the specified variable, `Weight`, and applies a function, `transform()`, to each piece. (Any remaining arguments in the `ddply()` call are passed along to the function.)

This is what `cabbage_exp` looks like, and what the `ddply()` call does to it:

```
cabbage_exp

 Cultivar Date Weight        sd  n         se
      c39  d16   3.18 0.9566144 10 0.30250803
      c39  d20   2.80 0.2788867 10 0.08819171
      c39  d21   2.74 0.9834181 10 0.31098410
```

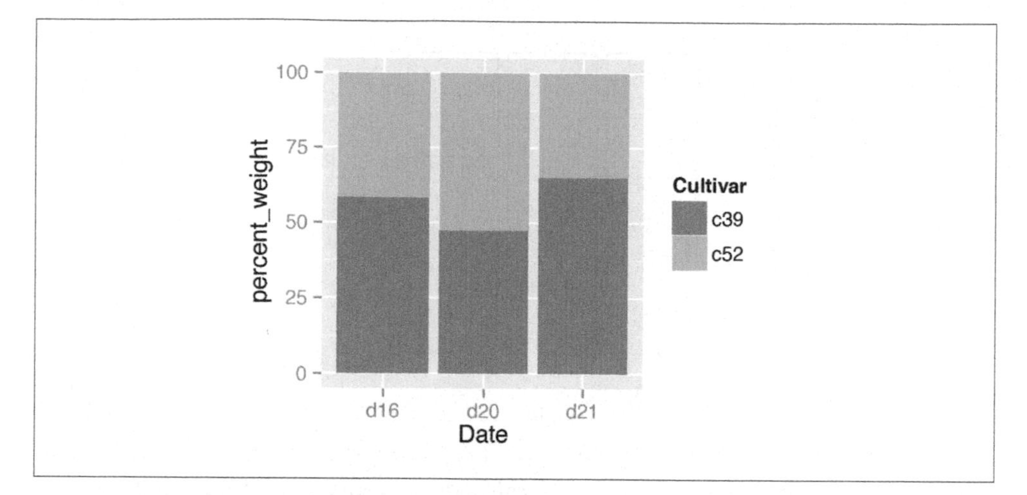

Figure 3-20. Proportional stacked bar graph

```
      c52  d16    2.26 0.4452215 10 0.14079141
      c52  d20    3.11 0.7908505 10 0.25008887
      c52  d21    1.47 0.2110819 10 0.06674995
```

```
ce <- ddply(cabbage_exp, "Date", transform,
      percent_weight = Weight / sum(Weight) * 100)
```

```
 Cultivar Date Weight        sd  n         se percent_weight
      c39  d16   3.18 0.9566144 10 0.30250803       58.45588
      c52  d16   2.26 0.4452215 10 0.14079141       41.54412
      c39  d20   2.80 0.2788867 10 0.08819171       47.37733
      c52  d20   3.11 0.7908505 10 0.25008887       52.62267
      c39  d21   2.74 0.9834181 10 0.31098410       65.08314
      c52  d21   1.47 0.2110819 10 0.06674995       34.91686
```

Once the percentages are computed, making the graph is the same as with a regular stacked bar graph.

As with regular stacked bar graphs, it makes sense to reverse the legend order, change the color palette, and add an outline. This is shown in (Figure 3-21):

```
ggplot(ce, aes(x=Date, y=percent_weight, fill=Cultivar)) +
    geom_bar(stat="identity", colour="black") +
    guides(fill=guide_legend(reverse=TRUE)) +
    scale_fill_brewer(palette="Pastel1")
```

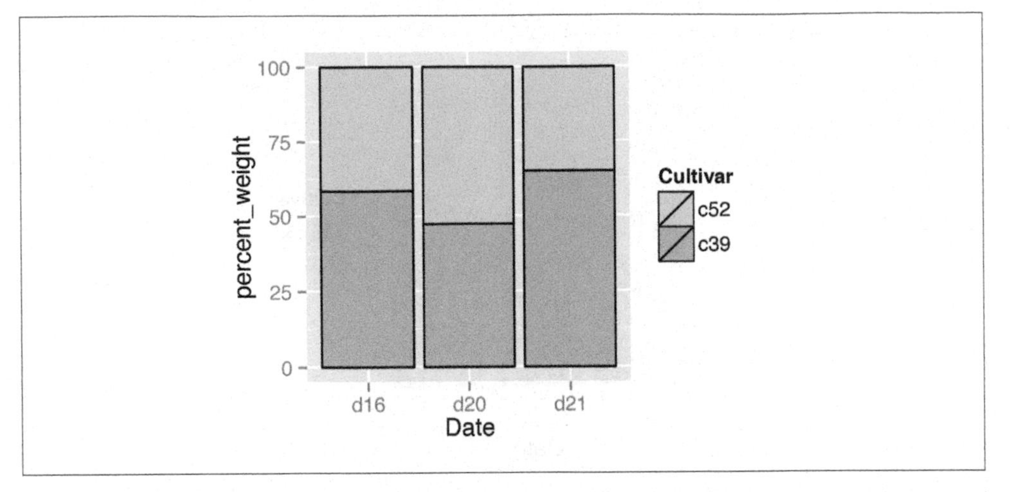

Figure 3-21. Proportional stacked bar graph with reversed legend, new palette, and black outline

See Also

For more on transforming data by groups, see Recipe 15.16.

3.9. Adding Labels to a Bar Graph

Problem

You want to add labels to the bars in a bar graph.

Solution

Add `geom_text()` to your graph. It requires a mapping for `x`, `y`, and the text itself. By setting `vjust` (the vertical justification), it is possible to move the text above or below the tops of the bars, as shown in Figure 3-22:

```
library(gcookbook) # For the data set

# Below the top
ggplot(cabbage_exp, aes(x=interaction(Date, Cultivar), y=Weight)) +
    geom_bar(stat="identity") +
    geom_text(aes(label=Weight), vjust=1.5, colour="white")

# Above the top
ggplot(cabbage_exp, aes(x=interaction(Date, Cultivar), y=Weight)) +
    geom_bar(stat="identity") +
    geom_text(aes(label=Weight), vjust=-0.2)
```

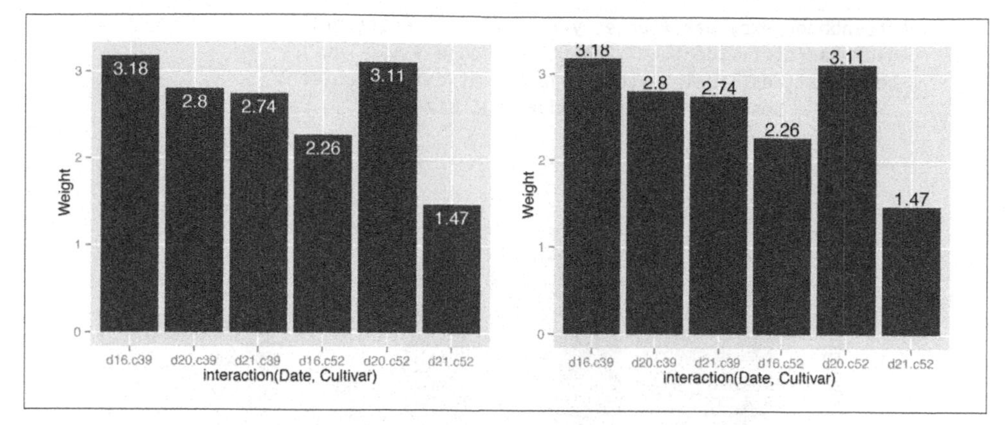

Figure 3-22. Left: labels under the tops of bars; right: labels above bars

Notice that when the labels are placed atop the bars, they may be clipped. To remedy this, see Recipe 8.2.

Discussion

In Figure 3-22, the *y* coordinates of the labels are centered at the top of each bar; by setting the vertical justification (`vjust`), they appear below or above the bar tops. One drawback of this is that when the label is above the top of the bar, it can go off the top of the plotting area. To fix this, you can manually set the *y* limits, or you can set the *y* positions of the text *above* the bars and not change the vertical justification. One drawback to changing the text's *y* position is that if you want to place the text fully above or below the bar top, the value to add will depend on the *y* range of the data; in contrast, changing `vjust` to a different value will always move the text the same distance relative to the height of the bar:

```
# Adjust y limits to be a little higher
ggplot(cabbage_exp, aes(x=interaction(Date, Cultivar), y=Weight)) +
    geom_bar(stat="identity") +
    geom_text(aes(label=Weight), vjust=-0.2) +
    ylim(0, max(cabbage_exp$Weight) * 1.05)

# Map y positions slightly above bar top - y range of plot will auto-adjust
ggplot(cabbage_exp, aes(x=interaction(Date, Cultivar), y=Weight)) +
    geom_bar(stat="identity") +
    geom_text(aes(y=Weight+0.1, label=Weight))
```

For grouped bar graphs, you also need to specify `position=position_dodge()` and give it a value for the dodging width. The default dodge width is 0.9. Because the bars are narrower, you might need to use `size` to specify a smaller font to make the labels fit. The default value of `size` is 5, so we'll make it smaller by using 3 (Figure 3-23):

```
ggplot(cabbage_exp, aes(x=Date, y=Weight, fill=Cultivar)) +
    geom_bar(stat="identity", position="dodge") +
    geom_text(aes(label=Weight), vjust=1.5, colour="white",
              position=position_dodge(.9), size=3)
```

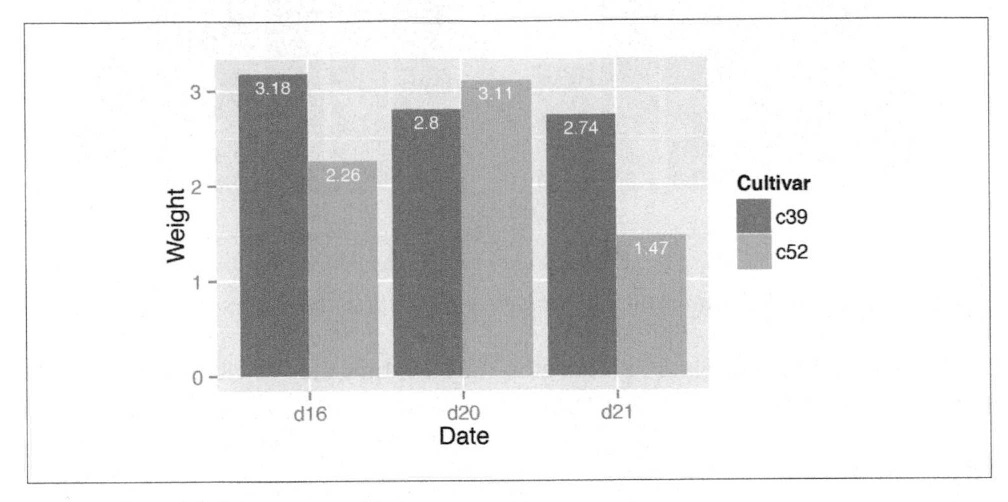

Figure 3-23. Labels on grouped bars

Putting labels on stacked bar graphs requires finding the cumulative sum for each stack. To do this, first make sure the data is sorted properly—if it isn't, the cumulative sum might be calculated in the wrong order. We'll use the `arrange()` function from the plyr package, which automatically gets loaded with ggplot2:

```
library(plyr)
# Sort by the day and sex columns
ce <- arrange(cabbage_exp, Date, Cultivar)
```

Once we make sure the data is sorted properly, we'll use `ddply()` to chunk it into groups by `Date`, then calculate a cumulative sum of `Weight` within each chunk:

```
# Get the cumulative sum
ce <- ddply(ce, "Date", transform, label_y=cumsum(Weight))
ce

 Cultivar Date Weight        sd  n         se label_y
      c39  d16   3.18 0.9566144 10 0.30250803    3.18
      c52  d16   2.26 0.4452215 10 0.14079141    5.44
      c39  d20   2.80 0.2788867 10 0.08819171    2.80
      c52  d20   3.11 0.7908505 10 0.25008887    5.91
      c39  d21   2.74 0.9834181 10 0.31098410    2.74
      c52  d21   1.47 0.2110819 10 0.06674995    4.21

ggplot(ce, aes(x=Date, y=Weight, fill=Cultivar)) +
```

```
    geom_bar(stat="identity") +
    geom_text(aes(y=label_y, label=Weight), vjust=1.5, colour="white")
```

The result is shown in Figure 3-24.

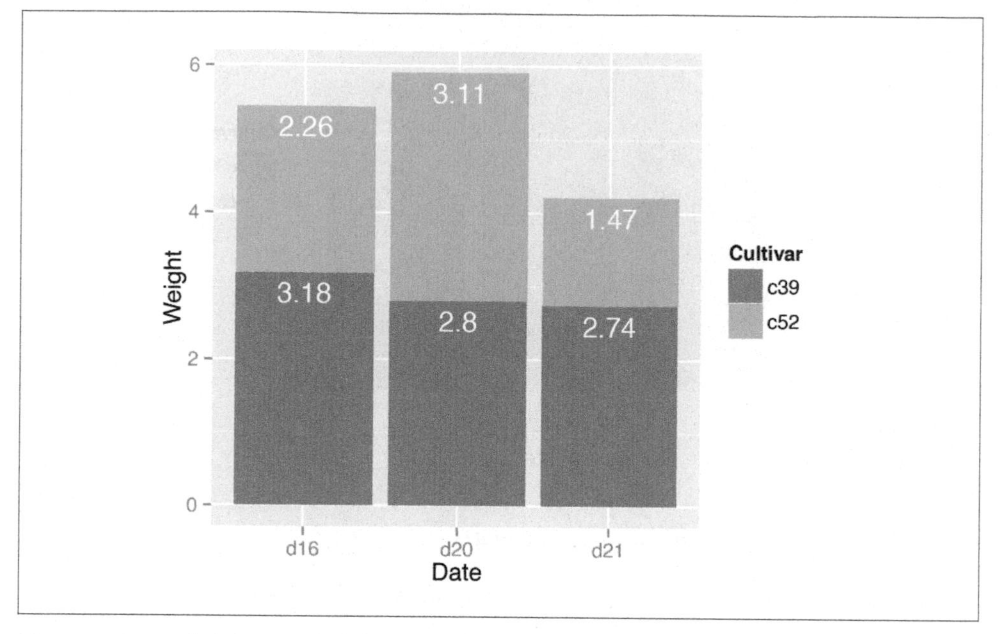

Figure 3-24. Labels on stacked bars

When using labels, changes to the stacking order are best done by modifying the order of levels in the factor (see Recipe 15.8) before taking the cumulative sum. The other method of changing stacking order, by specifying `breaks` in a scale, won't work properly, because the order of the cumulative sum won't be the same as the stacking order.

To put the labels in the middle of each bar (Figure 3-25), there must be an adjustment to the cumulative sum, and the *y* offset in `geom_bar()` can be removed:

```
ce <- arrange(cabbage_exp, Date, Cultivar)

# Calculate y position, placing it in the middle
ce <- ddply(ce, "Date", transform, label_y=cumsum(Weight)-0.5*Weight)

ggplot(ce, aes(x=Date, y=Weight, fill=Cultivar)) +
    geom_bar(stat="identity") +
    geom_text(aes(y=label_y, label=Weight), colour="white")
```

For a more polished graph (Figure 3-26), we'll change the legend order and colors, add labels in the middle with a smaller font using `size`, add a "kg" using `paste`, and make sure there are always two digits after the decimal point by using `format`:

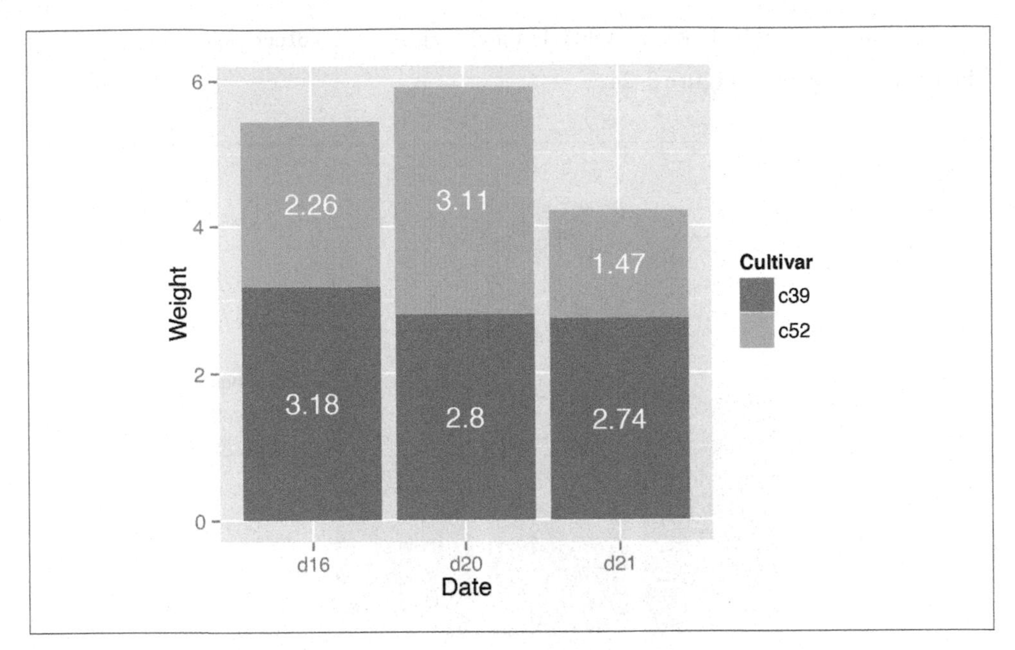

Figure 3-25. Labels in the middle of stacked bars

```
ggplot(ce, aes(x=Date, y=Weight, fill=Cultivar)) +
    geom_bar(stat="identity", colour="black") +
    geom_text(aes(y=label_y, label=paste(format(Weight, nsmall=2), "kg")),
              size=4) +
    guides(fill=guide_legend(reverse=TRUE)) +
    scale_fill_brewer(palette="Pastel1")
```

See Also

To control the appearance of the text, see Recipe 9.2.

For more on transforming data by groups, see Recipe 15.16.

3.10. Making a Cleveland Dot Plot

Problem

You want to make a Cleveland dot plot.

Solution

Cleveland dot plots are sometimes used instead of bar graphs because they reduce visual clutter and are easier to read.

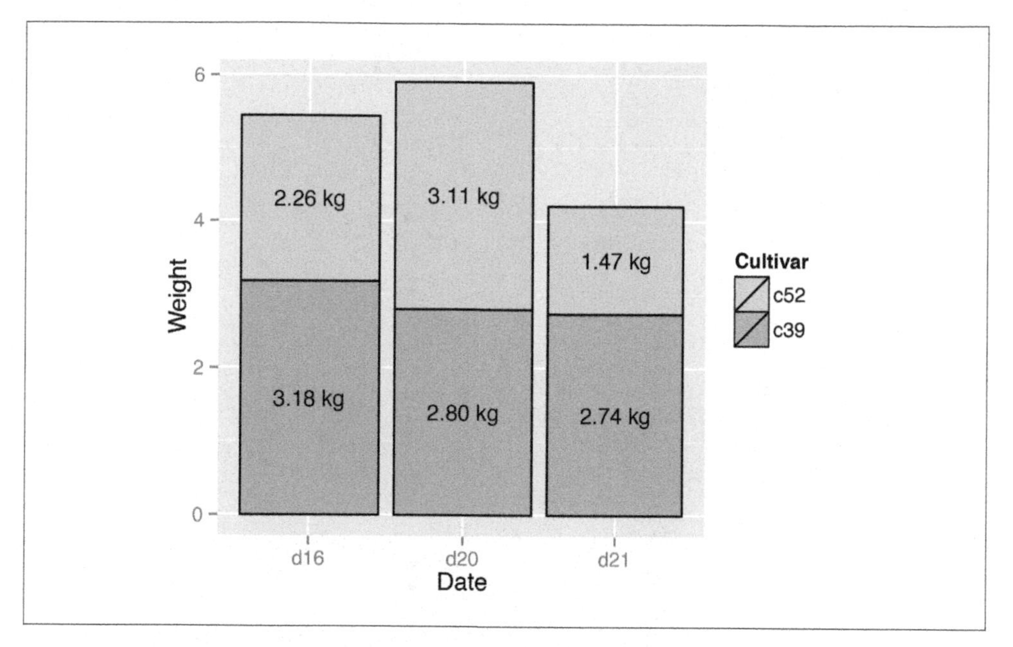

Figure 3-26. Customized stacked bar graph with labels

The simplest way to create a dot plot (as shown in Figure 3-27) is to use `geom_point()`:

```
library(gcookbook) # For the data set
tophit <- tophitters2001[1:25, ] # Take the top 25 from the tophitters data set

ggplot(tophit, aes(x=avg, y=name)) + geom_point()
```

Discussion

The `tophitters2001` data set contains many columns, but we'll focus on just three of them for this example:

```
tophit[, c("name", "lg", "avg")]

          name lg    avg
  Larry Walker NL 0.3501
 Ichiro Suzuki AL 0.3497
  Jason Giambi AL 0.3423
...
   Jeff Conine AL 0.3111
   Derek Jeter AL 0.3111
```

In Figure 3-27 the names are sorted alphabetically, which isn't very useful in this graph. Dot plots are often sorted by the value of the continuous variable on the horizontal axis.

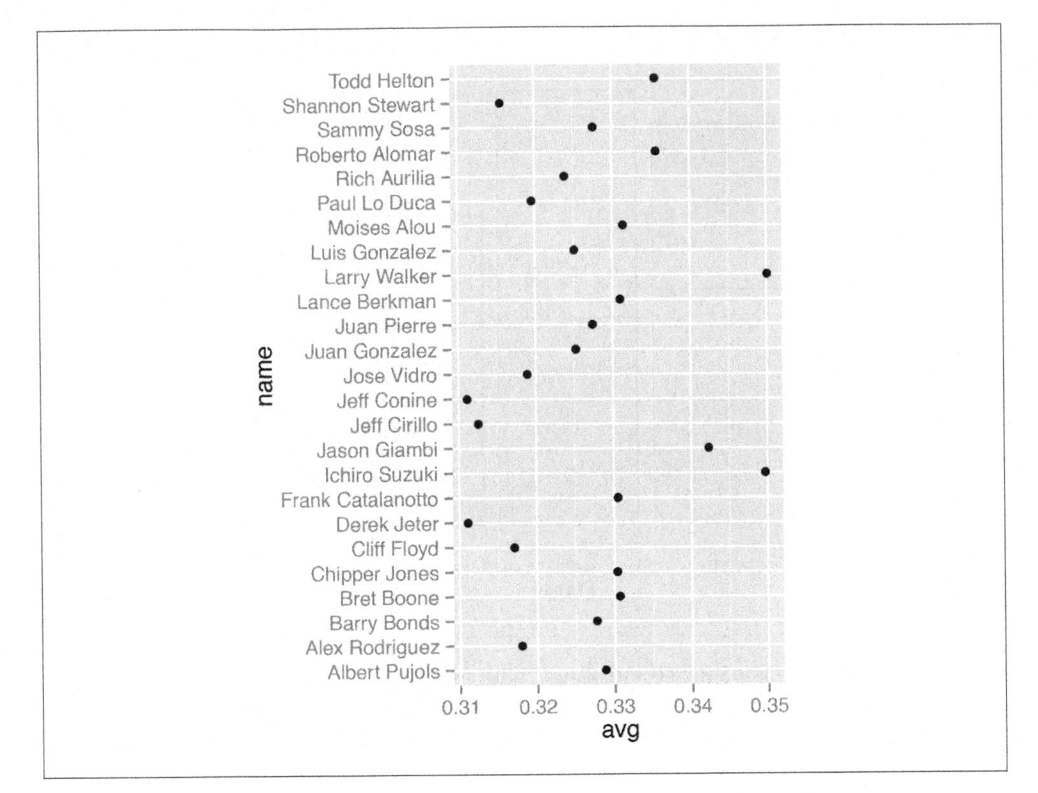

Figure 3-27. Basic dot plot

Although the rows of `tophit` happen to be sorted by `avg`, that doesn't mean that the items will be ordered that way in the graph. By default, the items on the given axis will be ordered however is appropriate for the data type. `name` is a character vector, so it's ordered alphabetically. If it were a factor, it would use the order defined in the factor levels. In this case, we want `name` to be sorted by a different variable, `avg`.

To do this, we can use `reorder(name, avg)`, which takes the `name` column, turns it into a factor, and sorts the factor levels by `avg`. To further improve the appearance, we'll make the vertical grid lines go away by using the theming system, and turn the horizontal grid lines into dashed lines (Figure 3-28):

```
ggplot(tophit, aes(x=avg, y=reorder(name, avg))) +
    geom_point(size=3) +                          # Use a larger dot
    theme_bw() +
    theme(panel.grid.major.x = element_blank(),
          panel.grid.minor.x = element_blank(),
          panel.grid.major.y = element_line(colour="grey60", linetype="dashed"))
```

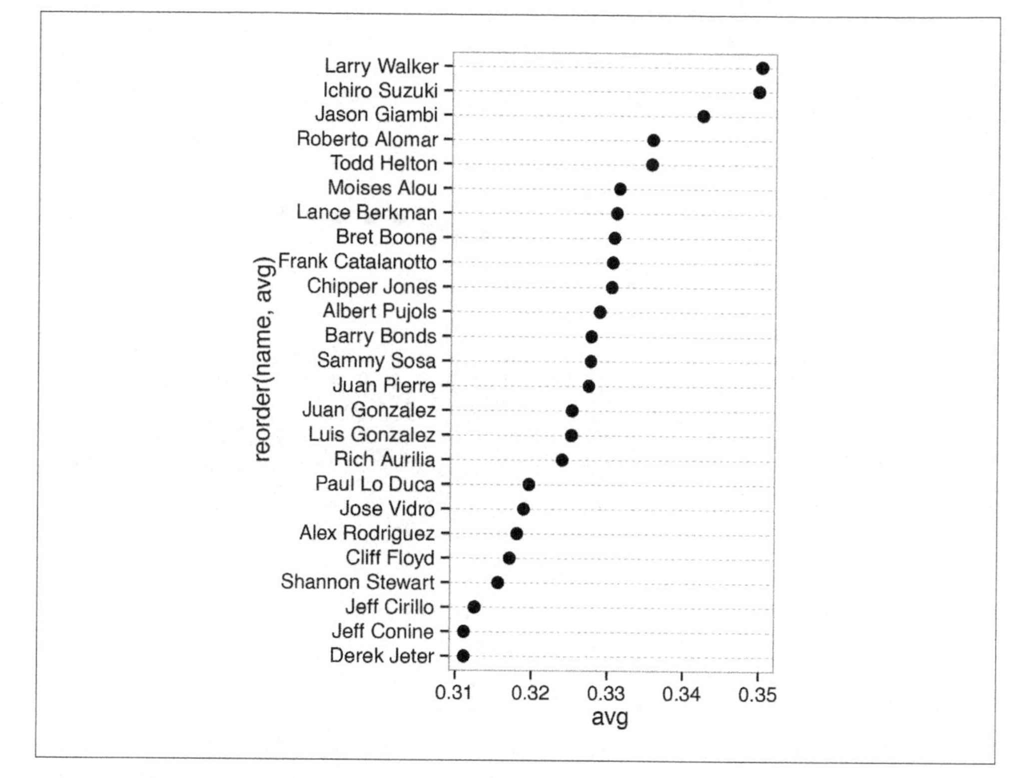

Figure 3-28. Dot plot, ordered by batting average

It's also possible to swap the axes so that the names go along the x-axis and the values go along the y-axis, as shown in Figure 3-29. We'll also rotate the text labels by 60 degrees:

```
ggplot(tophit, aes(x=reorder(name, avg), y=avg)) +
    geom_point(size=3) +                        # Use a larger dot
    theme_bw() +
    theme(axis.text.x = element_text(angle=60, hjust=1),
          panel.grid.major.y = element_blank(),
          panel.grid.minor.y = element_blank(),
          panel.grid.major.x = element_line(colour="grey60", linetype="dashed"))
```

It's also sometimes desirable to group the items by another variable. In this case we'll use the factor `lg`, which has the levels `NL` and `AL`, representing the National League and the American League. This time we want to sort first by `lg` and then by `avg`. Unfortunately, the `reorder()` function will only order factor levels by one other variable; to order the factor levels by two variables, we must do it manually:

```
# Get the names, sorted first by lg, then by avg
nameorder <- tophit$name[order(tophit$lg, tophit$avg)]
```

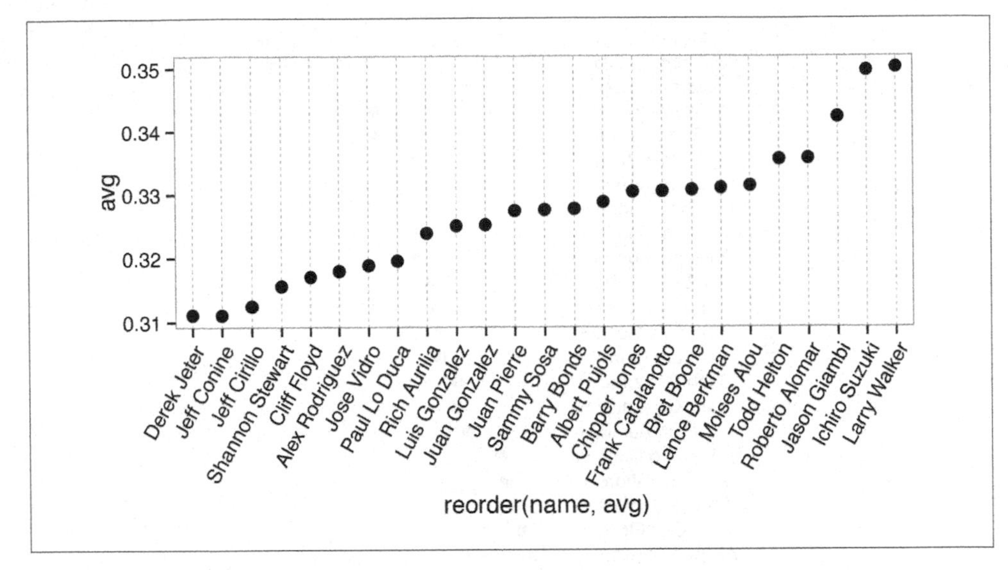

Figure 3-29. Dot plot with names on x-axis and values on y-axis

```
# Turn name into a factor, with levels in the order of nameorder
tophit$name <- factor(tophit$name, levels=nameorder)
```

To make the graph (Figure 3-30), we'll also add a mapping of `lg` to the color of the points. Instead of using grid lines that run all the way across, this time we'll make the lines go only up to the points, by using `geom_segment()`. Note that `geom_segment()` needs values for `x`, `y`, `xend`, and `yend`:

```
ggplot(tophit, aes(x=avg, y=name)) +
    geom_segment(aes(yend=name), xend=0, colour="grey50") +
    geom_point(size=3, aes(colour=lg)) +
    scale_colour_brewer(palette="Set1", limits=c("NL","AL")) +
    theme_bw() +
    theme(panel.grid.major.y = element_blank(),   # No horizontal grid lines
          legend.position=c(1, 0.55),             # Put legend inside plot area
          legend.justification=c(1, 0.5))
```

Another way to separate the two groups is to use facets, as shown in Figure 3-31. The order in which the facets are displayed is different from the sorting order in Figure 3-30; to change the display order, you must change the order of factor levels in the `lg` variable:

```
ggplot(tophit, aes(x=avg, y=name)) +
    geom_segment(aes(yend=name), xend=0, colour="grey50") +
    geom_point(size=3, aes(colour=lg)) +
    scale_colour_brewer(palette="Set1", limits=c("NL","AL"), guide=FALSE) +
    theme_bw() +
```

```
theme(panel.grid.major.y = element_blank()) +
facet_grid(lg ~ ., scales="free_y", space="free_y")
```

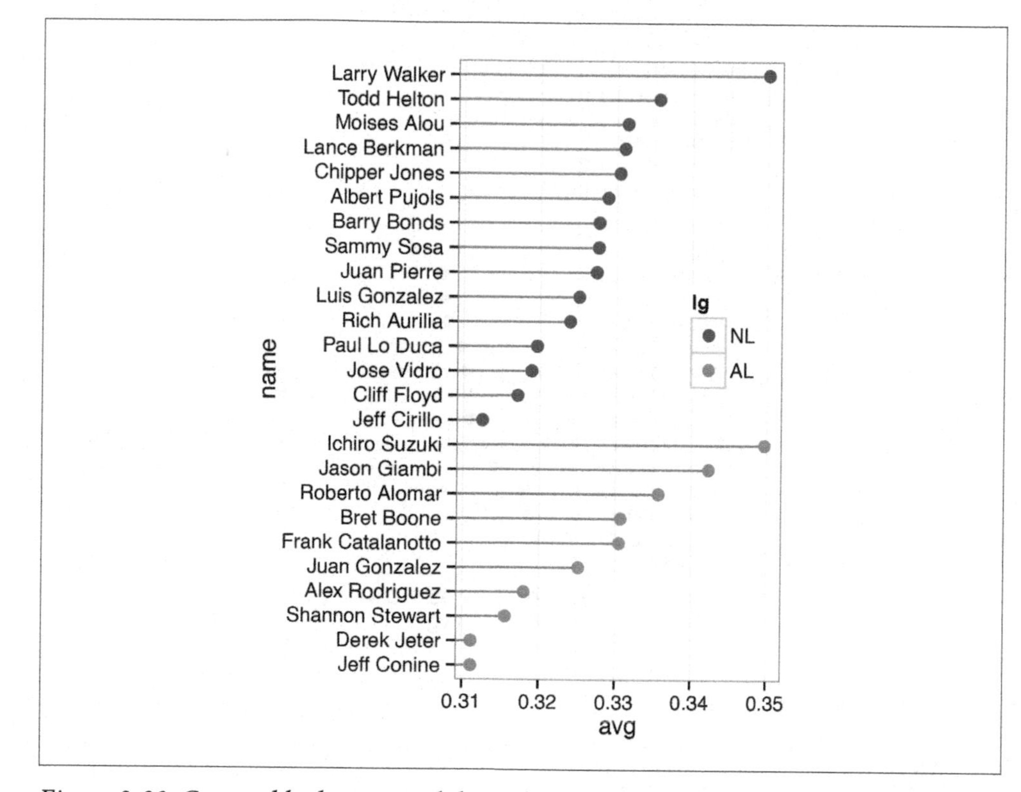

Figure 3-30. Grouped by league, with lines that stop at the point

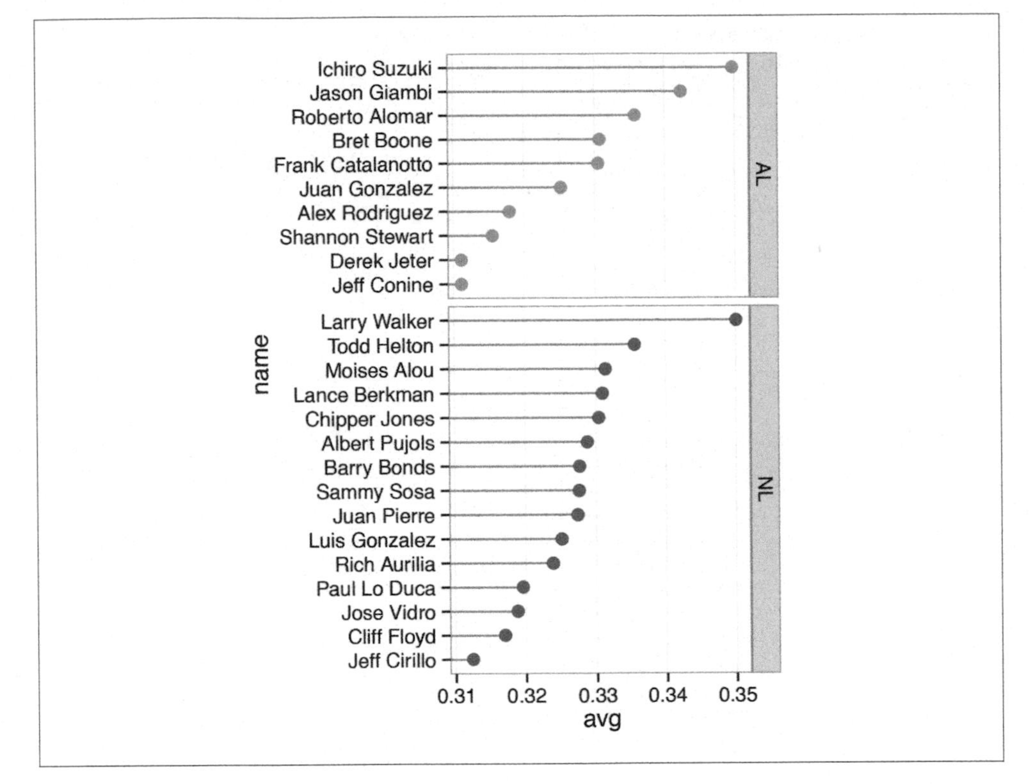

Figure 3-31. Faceted by league

See Also

For more on changing the order of factor levels, see Recipe 15.8. Also see Recipe 15.9 for details on changing the order of factor levels based on some other values.

For more on moving the legend, see Recipe 10.2. To hide grid lines, see Recipe 9.6.

CHAPTER 4
Line Graphs

Line graphs are typically used for visualizing how one continuous variable, on the y-axis, changes in relation to another continuous variable, on the x-axis. Often the *x* variable represents time, but it may also represent some other continuous quantity, like the amount of a drug administered to experimental subjects.

As with bar graphs, there are exceptions. Line graphs can also be used with a discrete variable on the x-axis. This is appropriate when the variable is ordered (e.g., "small", "medium", "large"), but not when the variable is unordered (e.g., "cow", "goose", "pig"). Most of the examples in this chapter use a continuous *x* variable, but we'll see one example where the variable is converted to a factor and thus treated as a discrete variable.

4.1. Making a Basic Line Graph

Problem

You want to make a basic line graph.

Solution

Use `ggplot()` with `geom_line()`, and specify what variables you mapped to `x` and `y` (Figure 4-1):

```
ggplot(BOD, aes(x=Time, y=demand)) + geom_line()
```

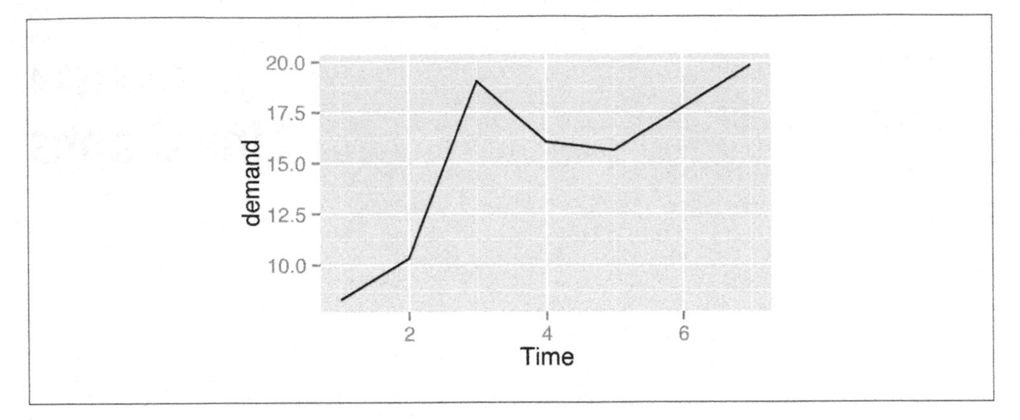

Figure 4-1. Basic line graph

Discussion

In this sample data set, the *x* variable, `Time`, is in one column and the *y* variable, `demand`, is in another:

```
BOD

 Time demand
    1    8.3
    2   10.3
    3   19.0
    4   16.0
    5   15.6
    7   19.8
```

Line graphs can be made with discrete (categorical) or continuous (numeric) variables on the x-axis. In the example here, the variable `demand` is numeric, but it could be treated as a categorical variable by converting it to a factor with `factor()` (Figure 4-2). When the *x* variable is a factor, you must also use `aes(group=1)` to ensure that `ggplot()` knows that the data points belong together and should be connected with a line (see Recipe 4.3 for an explanation of why `group` is needed with factors):

```
BOD1 <- BOD # Make a copy of the data
BOD1$Time <- factor(BOD1$Time)
ggplot(BOD1, aes(x=Time, y=demand, group=1)) + geom_line()
```

In the `BOD` data set there is no entry for Time=6, so there is no level 6 when `Time` is converted to a factor. Factors hold categorical values, and in that context, 6 is just another value. It happens to not be in the data set, so there's no space for it on the x-axis.

With ggplot2, the default *y* range of a line graph is just enough to include the *y* values in the data. For some kinds of data, it's better to have the *y* range start from zero. You can use `ylim()` to set the range, or you can use `expand_limits()` to expand the range

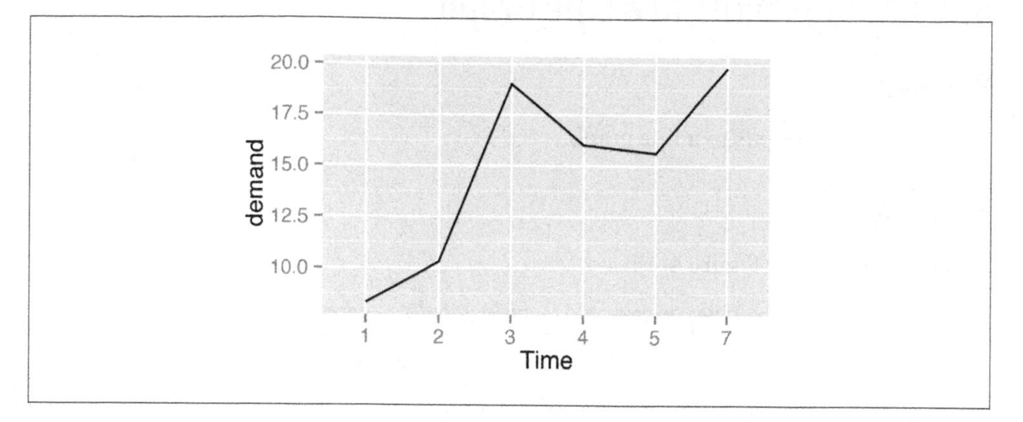

Figure 4-2. Basic line graph with a factor on the x-axis (notice that no space is allocated on the x-axis for 6)

to include a value. This will set the range from zero to the maximum value of the `demand` column in `BOD` (Figure 4-3):

```
# These have the same result
ggplot(BOD, aes(x=Time, y=demand)) + geom_line() + ylim(0, max(BOD$demand))
ggplot(BOD, aes(x=Time, y=demand)) + geom_line() + expand_limits(y=0)
```

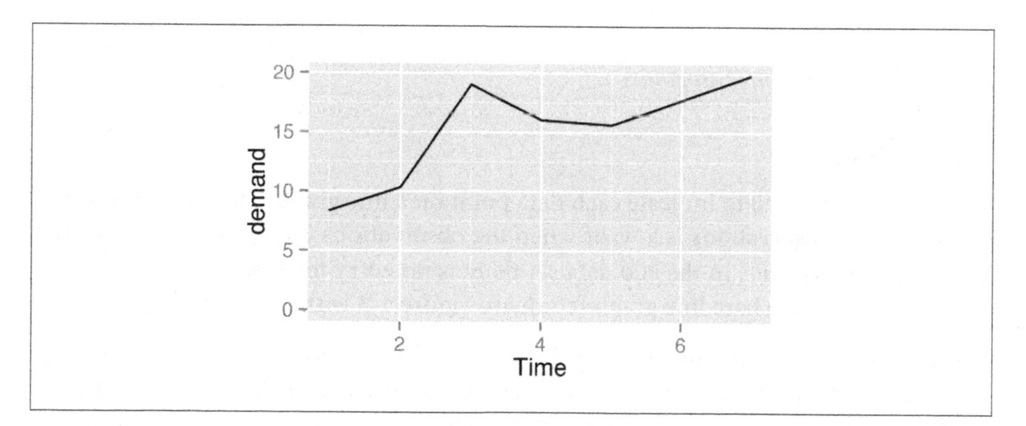

Figure 4-3. Line graph with manually set y range

See Also

See Recipe 8.2 for more on controlling the range of the axes.

4.2. Adding Points to a Line Graph

Problem

You want to add points to a line graph.

Solution

Add `geom_point()` (Figure 4-4):

```
ggplot(BOD, aes(x=Time, y=demand)) + geom_line() + geom_point()
```

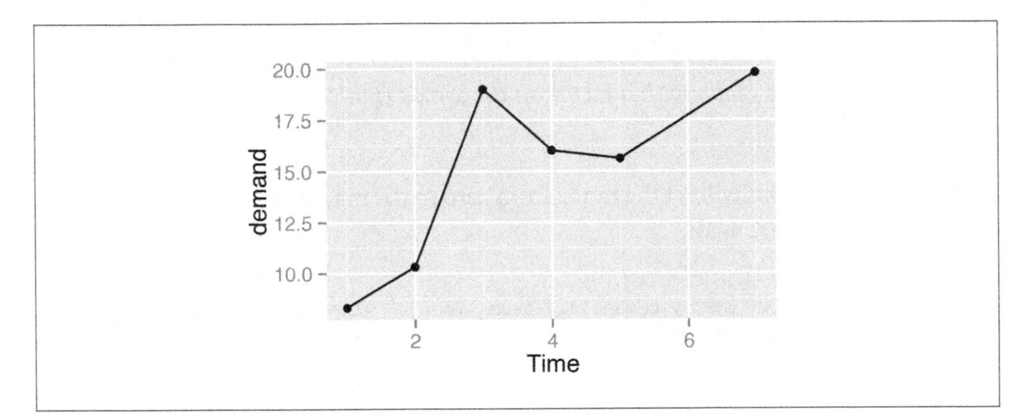

Figure 4-4. Line graph with points

Discussion

Sometimes it is useful to indicate each data point on a line graph. This is helpful when the density of observations is low, or when the observations do not happen at regular intervals. For example, in the `BOD` data set there is no entry for `Time=6`, but this is not apparent from just a bare line graph (compare Figure 4-3 with Figure 4-4).

In the `worldpop` data set, the intervals between each data point are not consistent. In the far past, the estimates were not as frequent as they are in the more recent past. Displaying points on the graph illustrates when each estimate was made (Figure 4-5):

```
library(gcookbook) # For the data set

ggplot(worldpop, aes(x=Year, y=Population)) + geom_line() + geom_point()

# Same with a log y-axis
ggplot(worldpop, aes(x=Year, y=Population)) + geom_line() + geom_point() +
    scale_y_log10()
```

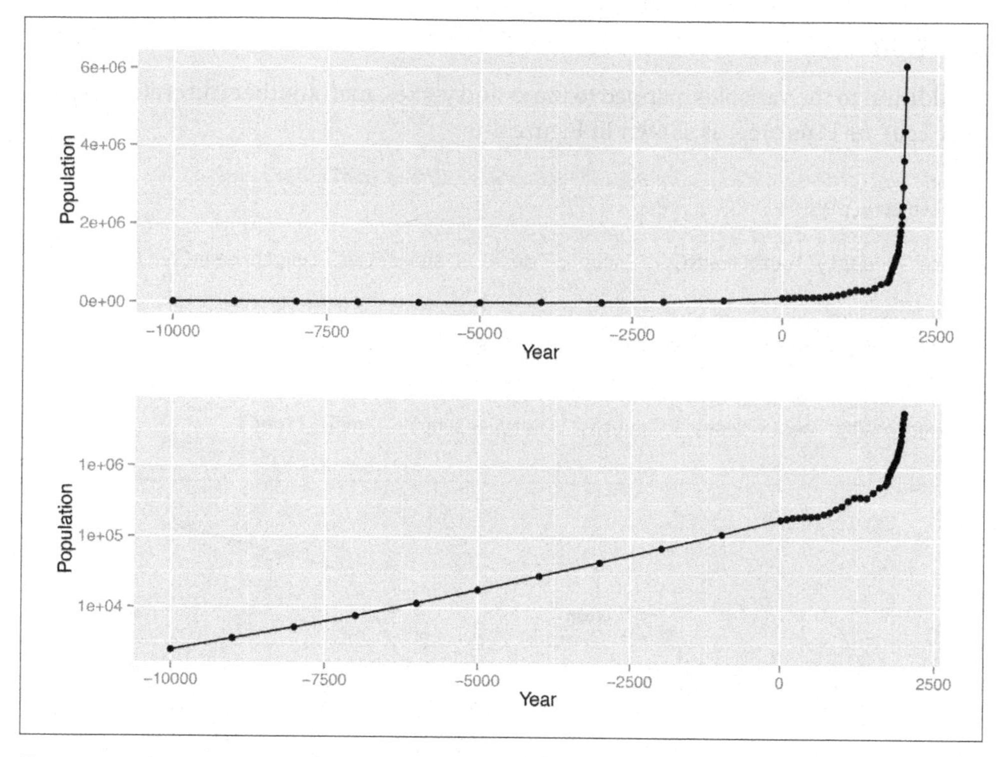

Figure 4-5. Top: points indicate where each data point is; bottom: the same data with a log y-axis

With the log y-axis, you can see that the rate of proportional change has increased in the last thousand years. The estimates for the years before 0 have a roughly constant rate of change of 10 times per 5,000 years. In the most recent 1,000 years, the population has increased at a much faster rate. We can also see that the population estimates are much more frequent in recent times—and probably more accurate!

See Also

To change the appearance of the points, see Recipe 4.5.

4.3. Making a Line Graph with Multiple Lines

Problem

You want to make a line graph with more than one line.

Solution

In addition to the variables mapped to the x- and y-axes, map another (discrete) variable to `colour` or `linetype`, as shown in Figure 4-6:

```
# Load plyr so we can use ddply() to create the example data set
library(plyr)
# Summarize the ToothGrowth data
tg <- ddply(ToothGrowth, c("supp", "dose"), summarise, length=mean(len))

# Map supp to colour
ggplot(tg, aes(x=dose, y=length, colour=supp)) + geom_line()

# Map supp to linetype
ggplot(tg, aes(x=dose, y=length, linetype=supp)) + geom_line()
```

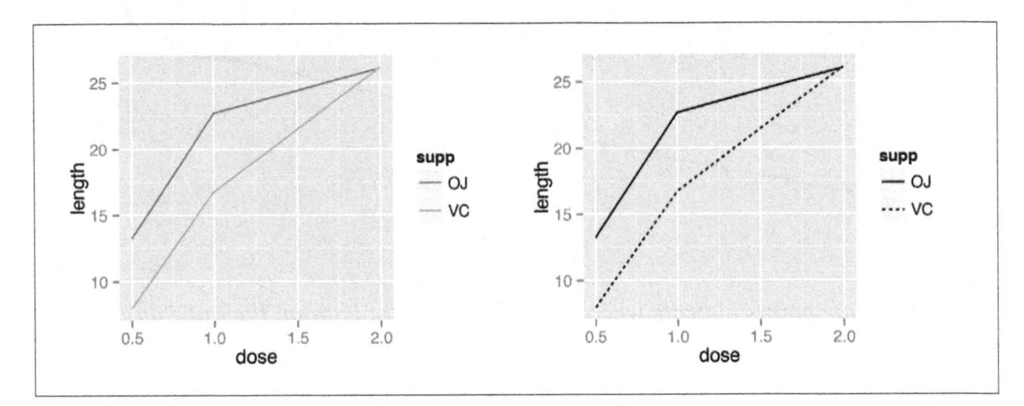

Figure 4-6. Left: a variable mapped to colour; right: a variable mapped to linetype

Discussion

The `tg` data has three columns, including the factor `supp`, which we mapped to `colour` and `linetype`:

```
tg

 supp dose length
   OJ  0.5  13.23
   OJ  1.0  22.70
   OJ  2.0  26.06
   VC  0.5   7.98
   VC  1.0  16.77
   VC  2.0  26.14

str(tg)

'data.frame': 6 obs. of  3 variables:
 $ supp  : Factor w/ 2 levels "OJ","VC": 1 1 1 2 2 2
```

```
 $ dose  : num  0.5 1 2 0.5 1 2
 $ length: num  13.23 22.7 26.06 7.98 16.77 ...
```

If the *x* variable is a factor, you must also tell `ggplot()` to `group` by that same variable, as described momentarily.

Line graphs can be used with a continuous or categorical variable on the x-axis. Sometimes the variable mapped to the x-axis is *conceived* of as being categorical, even when it's stored as a number. In the example here, there are three values of `dose`: 0.5, 1.0, and 2.0. You may want to treat these as categories rather than values on a continuous scale. To do this, convert `dose` to a factor (Figure 4-7):

```
ggplot(tg, aes(x=factor(dose), y=length, colour=supp, group=supp)) + geom_line()
```

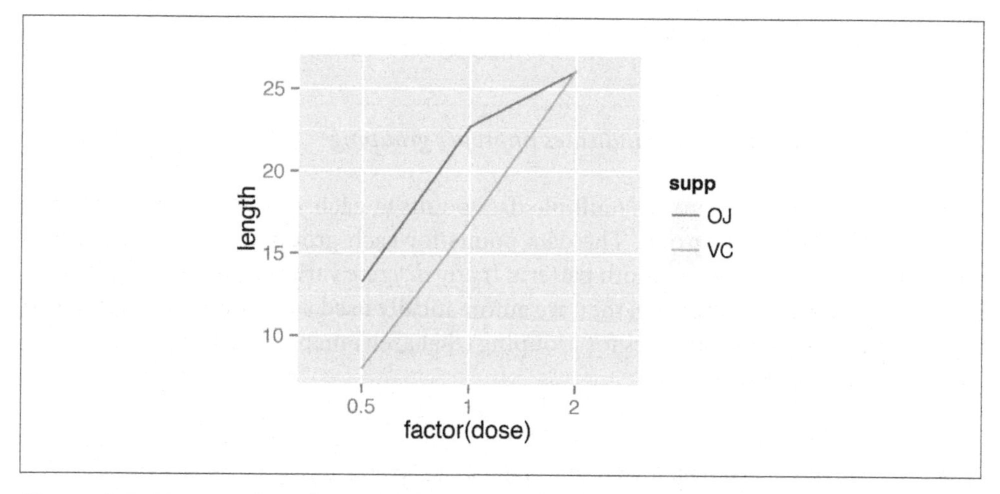

Figure 4-7. Line graph with continuous x variable converted to a factor

Notice the use of `group=supp`. Without this statement, `ggplot()` won't know how to group the data together to draw the lines, and it will give an error:

```
ggplot(tg, aes(x=factor(dose), y=length, colour=supp)) + geom_line()

  geom_path: Each group consists of only one observation. Do you need to adjust the
group aesthetic?
```

Another common problem when the incorrect grouping is used is that you will see a jagged sawtooth pattern, as in Figure 4-8:

```
ggplot(tg, aes(x=dose, y=length)) + geom_line()
```

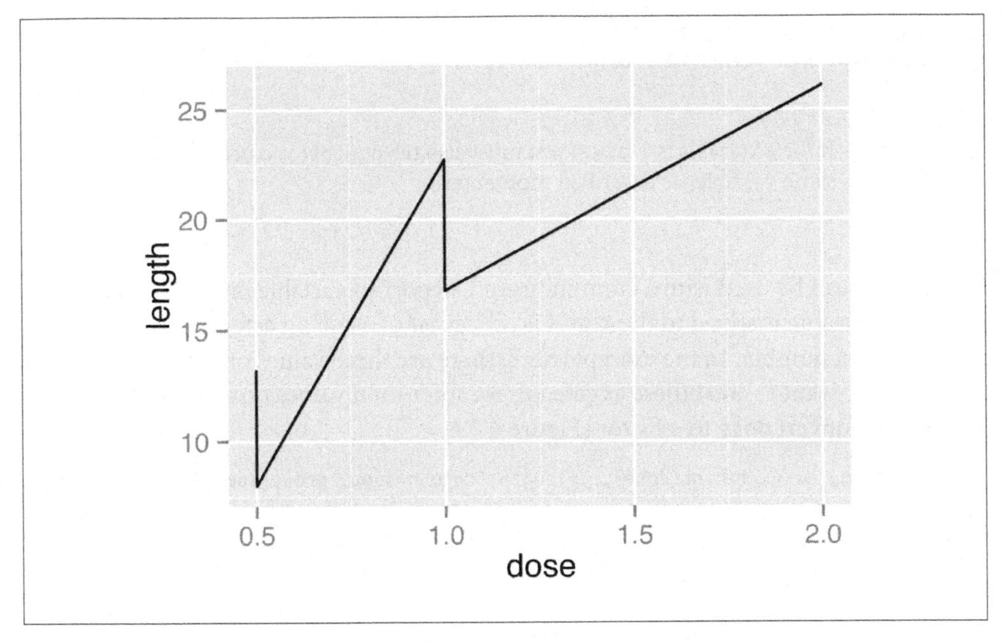

Figure 4-8. A sawtooth pattern indicates improper grouping

This happens because there are multiple data points at each *y* location, and `ggplot()` thinks they're all in one group. The data points for each group are connected with a single line, leading to the sawtooth pattern. If any *discrete* variables are mapped to aesthetics like `colour` or `linetype`, they are automatically used as grouping variables. But if you want to use other variables for grouping (that aren't mapped to an aesthetic), they should be used with `group`.

When in doubt, if your line graph looks wrong, try explicitly specifying the grouping variable with `group`. It's common for problems to occur with line graphs because `ggplot()` is unsure of how the variables should be grouped.

If your plot has points along with the lines, you can also map variables to properties of the points, such as `shape` and `fill` (Figure 4-9):

```
ggplot(tg, aes(x=dose, y=length, shape=supp)) + geom_line() +
    geom_point(size=4)           # Make the points a little larger

ggplot(tg, aes(x=dose, y=length, fill=supp)) + geom_line() +
    geom_point(size=4, shape=21) # Also use a point with a color fill
```

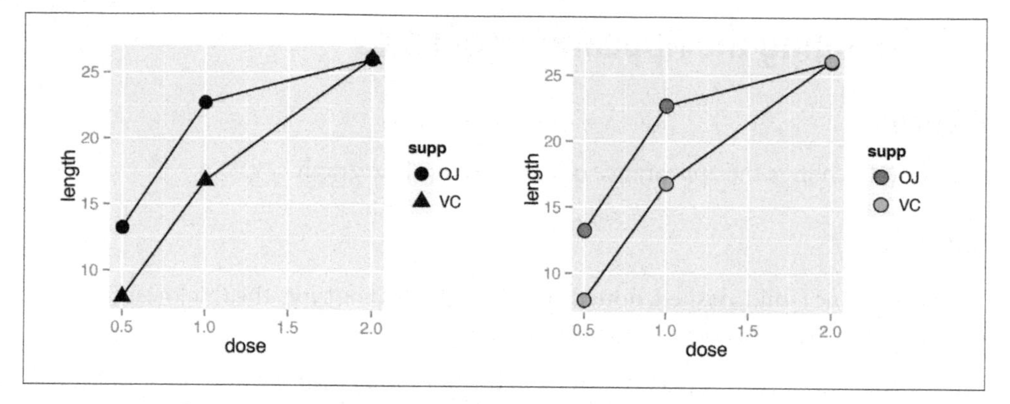

Figure 4-9. Left: line graph with different shapes; right: with different colors

Sometimes points will overlap. In these cases, you may want to *dodge* them, which means their positions will be adjusted left and right (Figure 4-10). When doing so, you must also dodge the lines, or else only the points will move and they will be misaligned. You must also specify how far they should move when dodged:

```
ggplot(tg, aes(x=dose, y=length, shape=supp)) +
    geom_line(position=position_dodge(0.2)) +           # Dodge lines by 0.2
    geom_point(position=position_dodge(0.2), size=4)  # Dodge points by 0.2
```

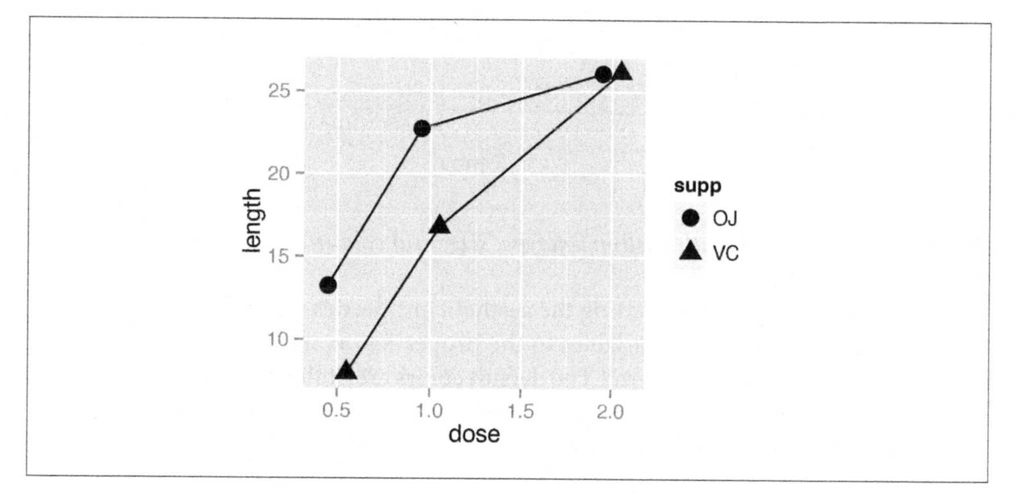

Figure 4-10. Dodging to avoid overlapping points

4.4. Changing the Appearance of Lines

Problem

You want to change the appearance of the lines in a line graph.

Solution

The type of line (solid, dashed, dotted, etc.) is set with `linetype`, the thickness (in mm) with `size`, and the color of the line with `colour`.

These properties can be set (as shown in Figure 4-11) by passing them values in the call to `geom_line()`:

```
ggplot(BOD, aes(x=Time, y=demand)) +
    geom_line(linetype="dashed", size=1, colour="blue")
```

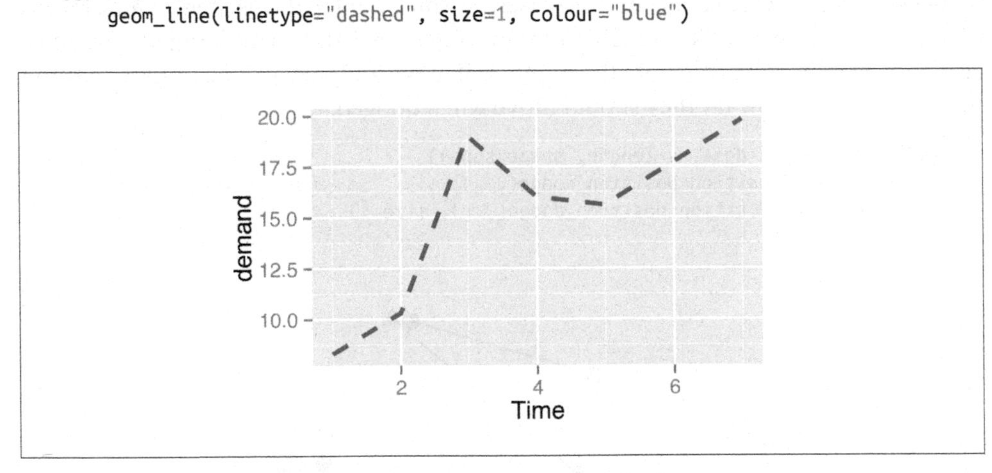

Figure 4-11. Line graph with custom linetype, size, and colour

If there is more than one line, setting the aesthetic properties will affect all of the lines. On the other hand, *mapping* variables to the properties, as we saw in Recipe 4.3, will result in each line looking different. The default colors aren't the most appealing, so you may want to use a different palette, as shown in Figure 4-12, by using `scale_colour_brewer()` or `scale_colour_manual()`:

```
# Load plyr so we can use ddply() to create the example data set
library(plyr)
# Summarize the ToothGrowth data
tg <- ddply(ToothGrowth, c("supp", "dose"), summarise, length=mean(len))

ggplot(tg, aes(x=dose, y=length, colour=supp)) +
    geom_line() +
    scale_colour_brewer(palette="Set1")
```

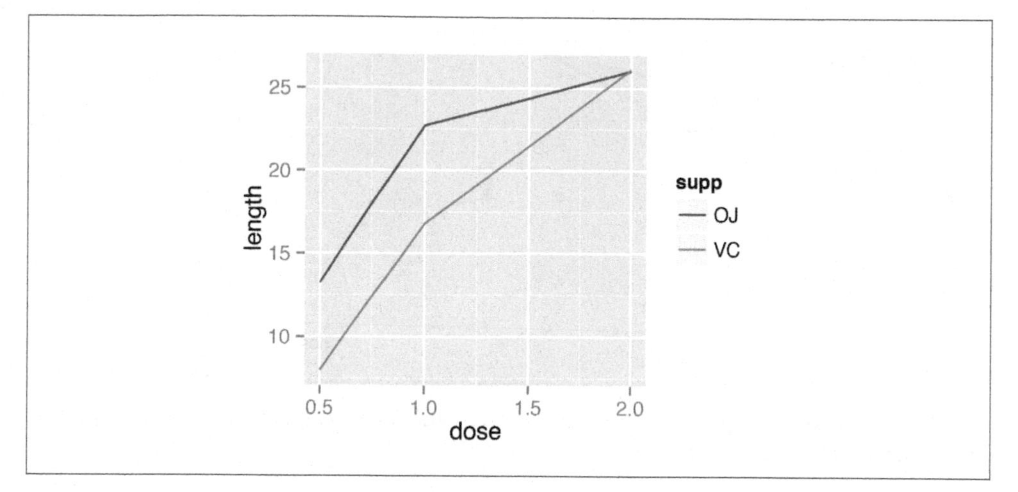

Figure 4-12. Using a palette from RColorBrewer

Discussion

To set a single constant color for all the lines, specify `colour` outside of `aes()`. The same works for `size`, `linetype`, and point `shape` (Figure 4-13). You may also have to specify the grouping variable:

```
# If both lines have the same properties, you need to specify a variable to
# use for grouping
ggplot(tg, aes(x=dose, y=length, group=supp)) +
    geom_line(colour="darkgreen", size=1.5)

# Since supp is mapped to colour, it will automatically be used for grouping
ggplot(tg, aes(x=dose, y=length, colour=supp)) +
    geom_line(linetype="dashed") +
    geom_point(shape=22, size=3, fill="white")
```

See Also

For more information about using colors, see Chapter 12.

4.5. Changing the Appearance of Points

Problem

You want to change the appearance of the points in a line graph.

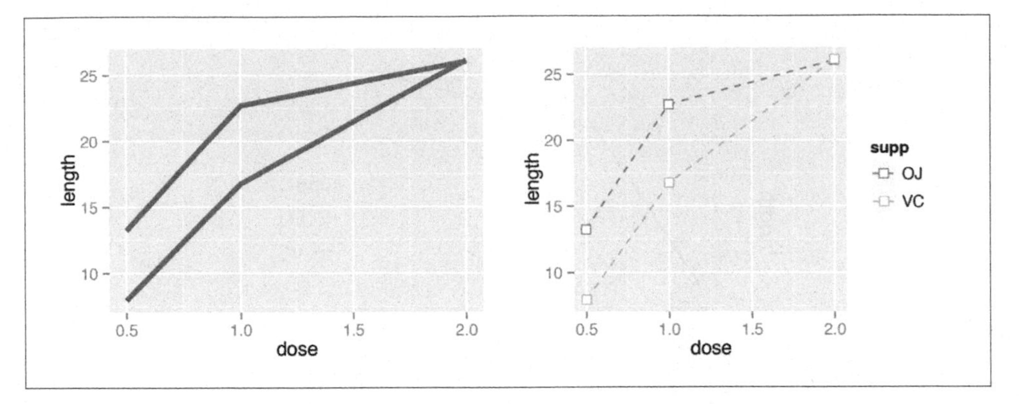

Figure 4-13. Left: line graph with constant size and color; right: with supp mapped to colour, and with points added

Solution

In `geom_point()`, set the `size`, `shape`, `colour`, and/or `fill` outside of `aes()` (the result is shown in Figure 4-14):

```
ggplot(BOD, aes(x=Time, y=demand)) +
    geom_line() +
    geom_point(size=4, shape=22, colour="darkred", fill="pink")
```

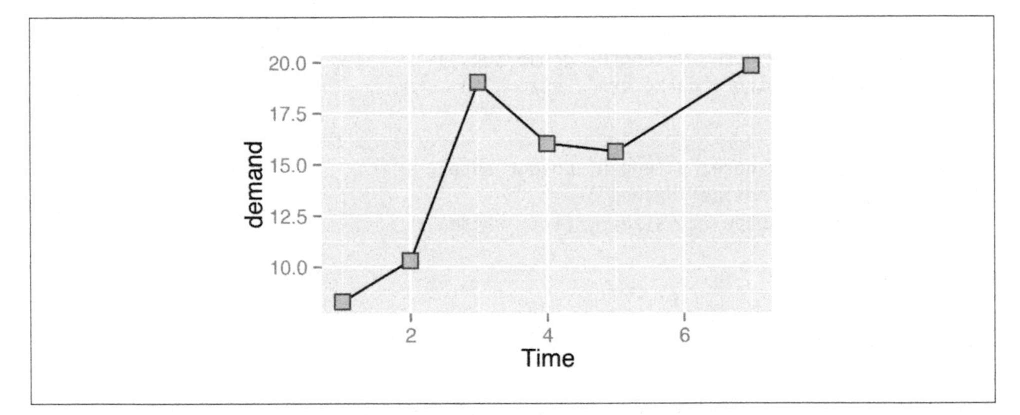

Figure 4-14. Points with custom size, shape, color, and fill

Discussion

The default `shape` for points is a solid circle, the default `size` is 2, and the default `colour` is `"black"`. The `fill` color is relevant only for some point shapes (numbered 21–25), which have separate outline and fill colors (see Recipe 5.3 for a chart of shapes). The fill

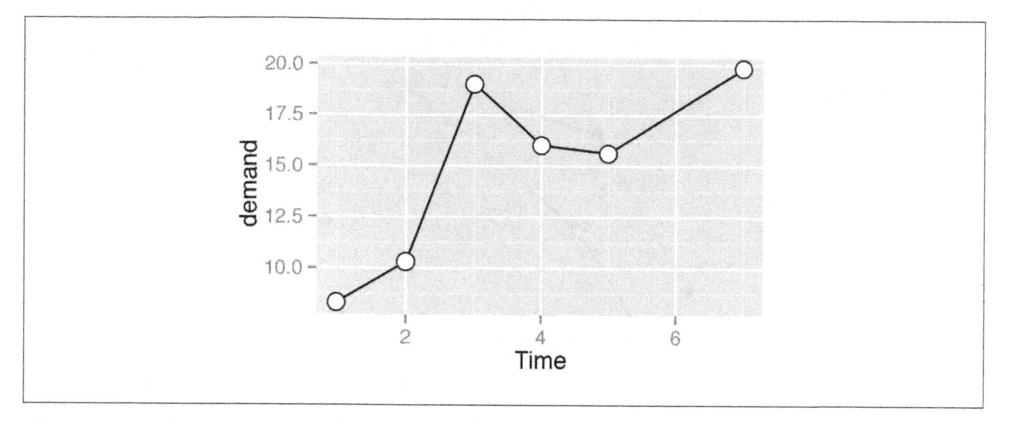

Figure 4-15. Points with a white fill

color is typically NA, or empty; you can fill it with white to get hollow-looking circles, as shown in Figure 4-15:

```
ggplot(BOD, aes(x=Time, y=demand)) +
    geom_line() +
    geom_point(size=4, shape=21, fill="white")
```

If the points and lines have different colors, you should specify the points after the lines, so that they are drawn on top. Otherwise, the lines will be drawn on top of the points.

For multiple lines, we saw in Recipe 4.3 how to draw differently colored points for each group by mapping variables to aesthetic properties of points, inside of `aes()`. The default colors are not very appealing, so you may want to use a different palette, using `scale_colour_brewer()` or `scale_colour_manual()`. To set a single constant shape or size for all the points, as in Figure 4-16, specify `shape` or `size` outside of `aes()`:

```
# Load plyr so we can use ddply() to create the example data set
library(plyr)
# Summarize the ToothGrowth data
tg <- ddply(ToothGrowth, c("supp", "dose"), summarise, length=mean(len))

# Save the position_dodge specification because we'll use it multiple times
pd <- position_dodge(0.2)

ggplot(tg, aes(x=dose, y=length, fill=supp)) +
    geom_line(position=pd) +
    geom_point(shape=21, size=3, position=pd) +
    scale_fill_manual(values=c("black","white"))
```

See Also

See Recipe 5.3 for more on using different shapes, and Chapter 12 for more about colors.

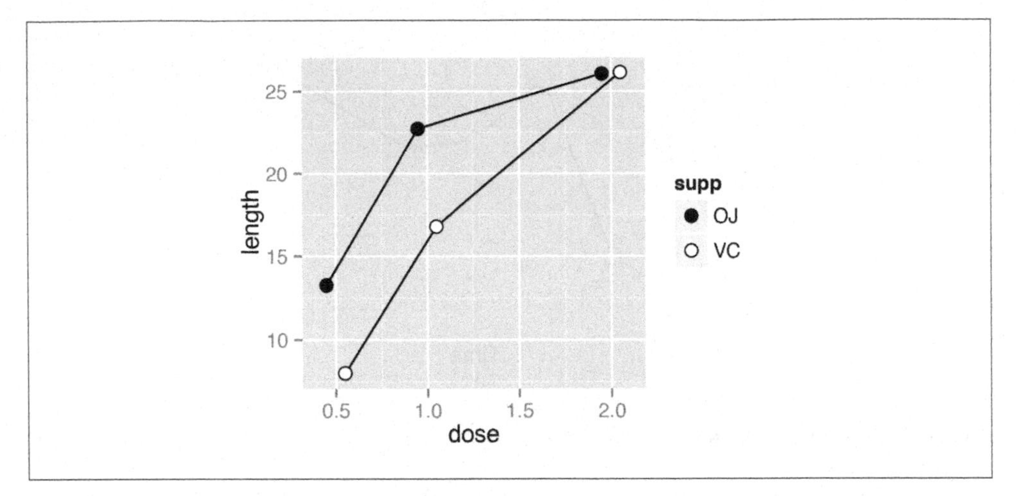

Figure 4-16. Line graph with manually specified fills of black and white, and a slight dodge

4.6. Making a Graph with a Shaded Area

Problem

You want to make a graph with a shaded area.

Solution

Use `geom_area()` to get a shaded area, as in Figure 4-17:

```
# Convert the sunspot.year data set into a data frame for this example
sunspotyear <- data.frame(
    Year     = as.numeric(time(sunspot.year)),
    Sunspots = as.numeric(sunspot.year)
)

ggplot(sunspotyear, aes(x=Year, y=Sunspots)) + geom_area()
```

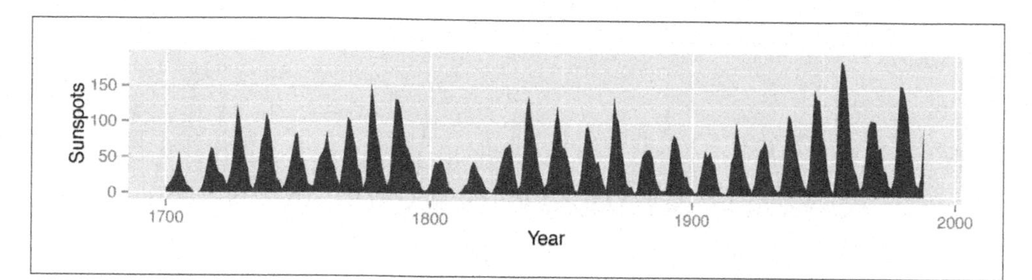

Figure 4-17. Graph with a shaded area

Discussion

By default, the area will be filled with a very dark grey and will have no outline. The color can be changed by setting `fill`. In the following example, we'll set it to `"blue"`, and we'll also make it 80% transparent by setting `alpha` to 0.2. This makes it possible to see the grid lines through the area, as shown in Figure 4-18. We'll also add an outline, by setting `colour`:

```
ggplot(sunspotyear, aes(x=Year, y=Sunspots)) +
    geom_area(colour="black", fill="blue", alpha=.2)
```

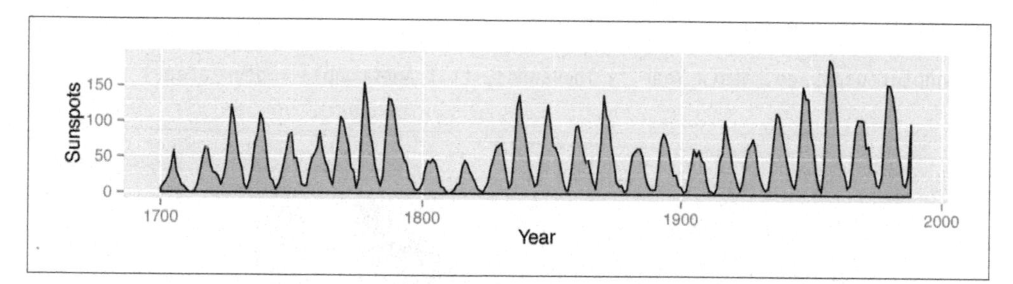

Figure 4-18. Graph with a semitransparent shaded area and an outline

Having an outline around the entire area might not be desirable, because it puts a vertical line at the beginning and end of the shaded area, as well as one along the bottom. To avoid this issue, we can draw the area without an outline (by not specifying `colour`), and then layer a `geom_line()` on top, as shown in Figure 4-19:

```
ggplot(sunspotyear, aes(x=Year, y=Sunspots)) +
    geom_area(fill="blue", alpha=.2) +
    geom_line()
```

See Also

See Chapter 12 for more on choosing colors.

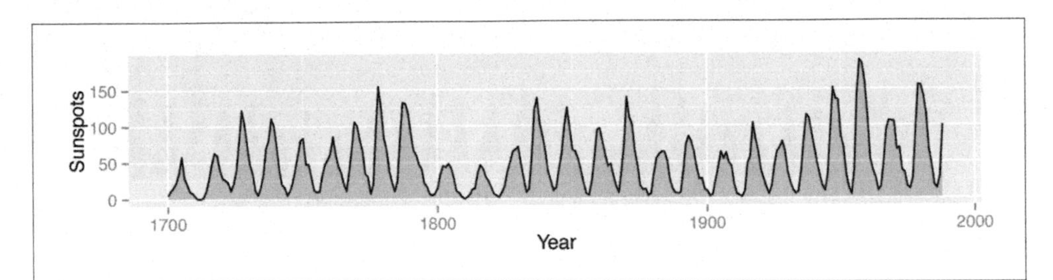

Figure 4-19. Line graph with a line just on top, using geom_line()

4.7. Making a Stacked Area Graph

Problem

You want to make a stacked area graph.

Solution

Use `geom_area()` and map a factor to `fill` (Figure 4-20):

```
library(gcookbook) # For the data set

ggplot(uspopage, aes(x=Year, y=Thousands, fill=AgeGroup)) + geom_area()
```

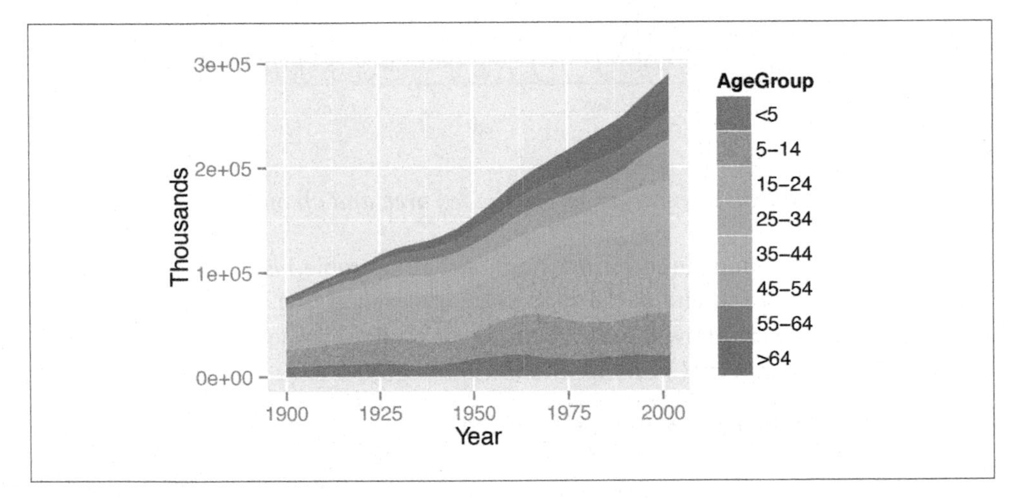

Figure 4-20. Stacked area graph

Discussion

The sort of data that is plotted with a stacked area chart is often provided in a wide format, but `ggplot2()` requires data to be in long format. To convert it, see Recipe 15.19.

In the example here, we used the `uspopage` data set:

```
uspopage

 Year AgeGroup Thousands
 1900       <5      9181
 1900     5-14     16966
 1900    15-24     14951
 1900    25-34     12161
 1900    35-44      9273
 1900    45-54      6437
 1900    55-64      4026
 1900      >64      3099
 1901       <5      9336
 1901     5-14     17158
...
```

The default order of legend items is the opposite of the stacking order. The legend can be reversed by setting the breaks in the scale. This version of the chart (Figure 4-21) reverses the legend order, changes the palette to a range of blues, and adds thin (`size=.2`) lines between each area. It also makes the filled areas semitransparent (`alpha=.4`), so that it is possible to see the grid lines through them:

```
ggplot(uspopage, aes(x=Year, y=Thousands, fill=AgeGroup)) +
    geom_area(colour="black", size=.2, alpha=.4) +
    scale_fill_brewer(palette="Blues", breaks=rev(levels(uspopage$AgeGroup)))
```

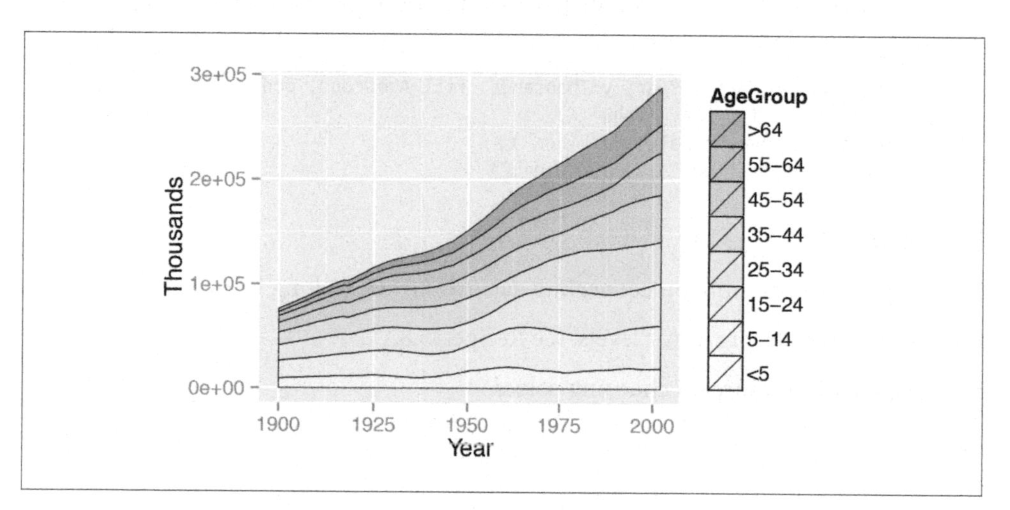

Figure 4-21. Reversed legend order, lines, and a different palette

To reverse the stacking order, we'll put `order=desc(AgeGroup)` inside of `aes()` (Figure 4-22):

```
library(plyr) # For the desc() function
ggplot(uspopage, aes(x=Year, y=Thousands, fill=AgeGroup, order=desc(AgeGroup))) +
    geom_area(colour="black", size=.2, alpha=.4) +
    scale_fill_brewer(palette="Blues")
```

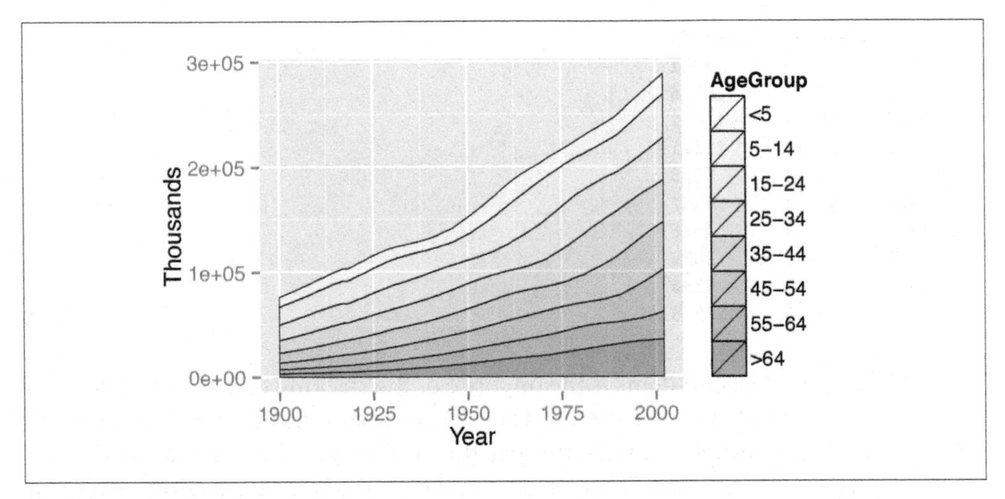

Figure 4-22. Reversed stacking order

Since each filled area is drawn with a polygon, the outline includes the left and right sides. This might be distracting or misleading. To get rid of it (Figure 4-23), first draw the stacked areas *without* an outline (by leaving `colour` as the default `NA` value), and then add a `geom_line()` on top:

```
ggplot(uspopage, aes(x=Year, y=Thousands, fill=AgeGroup, order=desc(AgeGroup))) +
    geom_area(colour=NA, alpha=.4) +
    scale_fill_brewer(palette="Blues") +
    geom_line(position="stack", size=.2)
```

See Also

See Recipe 15.19 for more on converting data from wide to long format.

For more on reordering factor levels, see Recipe 15.8.

See Chapter 12 for more on choosing colors.

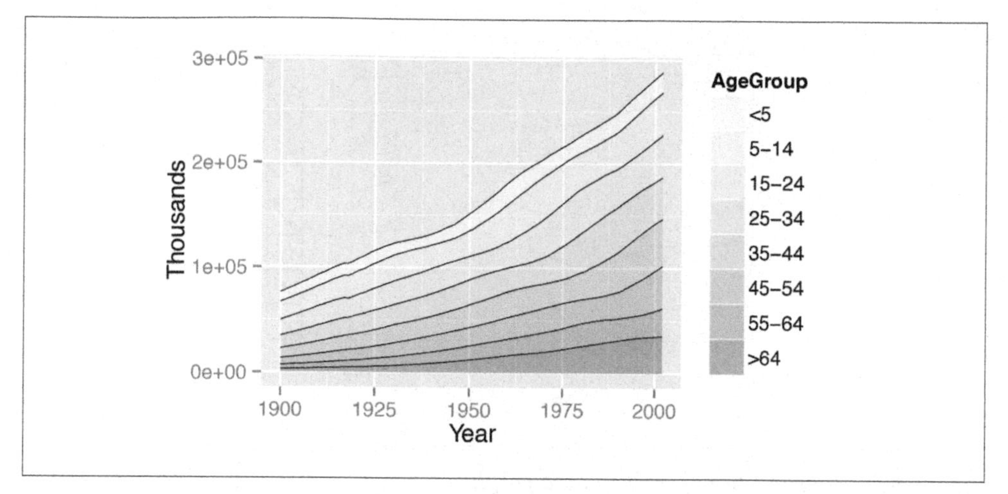

Figure 4-23. No lines on the left and right of the graph

4.8. Making a Proportional Stacked Area Graph

Problem

You want to make a stacked area graph with the overall height scaled to a constant value.

Solution

First, calculate the proportions. In this example, we'll use `ddply()` to break `uspopage` into groups by `Year`, then calculate a new column, `Percent`. This value is the `Thousands` for each row, divided by the sum of `Thousands` for each `Year` group, multiplied by 100 to get a percent value:

```
library(gcookbook) # For the data set
library(plyr)      # For the ddply() function

# Convert Thousands to Percent
uspopage_prop <- ddply(uspopage, "Year", transform,
                       Percent = Thousands / sum(Thousands) * 100)
```

Once we've calculated the proportions, plotting is the same as with a regular stacked area graph (Figure 4-24):

```
ggplot(uspopage_prop, aes(x=Year, y=Percent, fill=AgeGroup)) +
    geom_area(colour="black", size=.2, alpha=.4) +
    scale_fill_brewer(palette="Blues", breaks=rev(levels(uspopage$AgeGroup)))
```

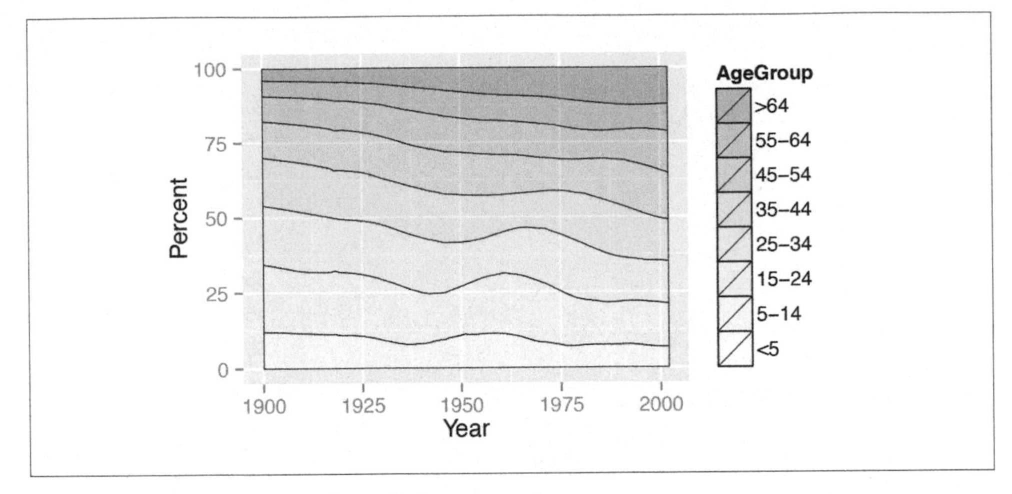

Figure 4-24. A proportional stacked area graph

Discussion

Let's take a closer look at the data and how it was summarized:

```
uspopage

 Year AgeGroup Thousands
 1900       <5      9181
 1900     5-14     16966
 1900    15-24     14951
 1900    25-34     12161
 1900    35-44      9273
 1900    45-54      6437
 1900    55-64      4026
 1900      >64      3099
 1901       <5      9336
 1901     5-14     17158
...
```

We'll use `ddply()` to split it into separate data frames for each value of `Year`, then apply the `transform()` function to each piece and calculate the `Percent` for each piece. Then `ddply()` puts all the data frames back together:

```
uspopage_prop <- ddply(uspopage, "Year", transform,
                       Percent = Thousands / sum(Thousands) * 100)

 Year AgeGroup Thousands   Percent
 1900       <5      9181 12.065340
 1900     5-14     16966 22.296107
 1900    15-24     14951 19.648067
 1900    25-34     12161 15.981549
 1900    35-44      9273 12.186243
```

```
 1900    45-54      6437  8.459274
 1900    55-64      4026  5.290825
 1900      >64      3099  4.072594
 1901       <5      9336 12.033409
 1901     5-14     17158 22.115385
...
```

See Also

For more on summarizing data by groups, see Recipe 15.17.

4.9. Adding a Confidence Region

Problem

You want to add a confidence region to a graph.

Solution

Use `geom_ribbon()` and map values to `ymin` and `ymax`.

In the `climate` data set, `Anomaly10y` is a 10-year running average of the deviation (in Celsius) from the average 1950–1980 temperature, and `Unc10y` is the 95% confidence interval. We'll set `ymax` and `ymin` to `Anomaly10y` plus or minus `Unc10y` (Figure 4-25):

```
library(gcookbook) # For the data set

# Grab a subset of the climate data
clim <- subset(climate, Source == "Berkeley",
                select=c("Year", "Anomaly10y", "Unc10y"))

clim

Year Anomaly10y Unc10y
 1800     -0.435  0.505
 1801     -0.453  0.493
 1802     -0.460  0.486
...
 2003      0.869  0.028
 2004      0.884  0.029

# Shaded region
ggplot(clim, aes(x=Year, y=Anomaly10y)) +
    geom_ribbon(aes(ymin=Anomaly10y-Unc10y, ymax=Anomaly10y+Unc10y),
                alpha=0.2) +
    geom_line()
```

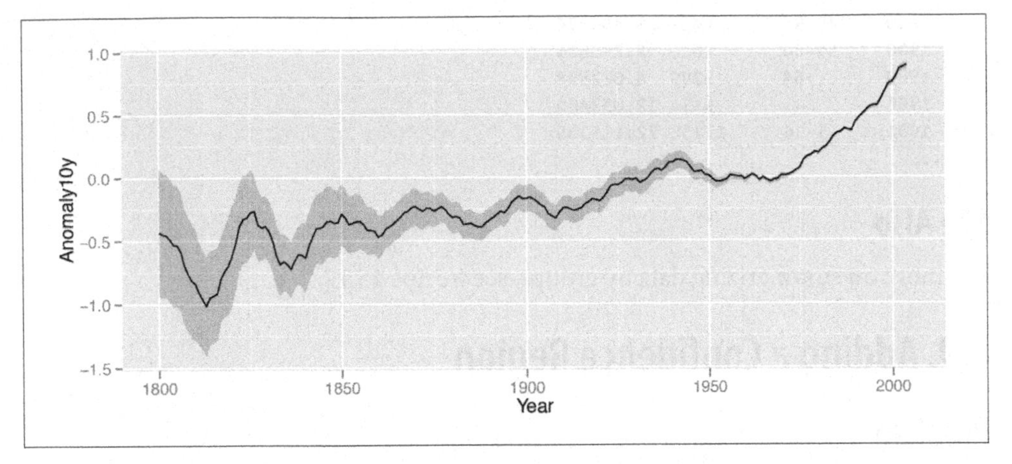

Figure 4-25. A line graph with a shaded confidence region

The shaded region is actually a very dark grey, but it is mostly transparent. The transparency is set with `alpha=0.2`, which makes it 80% transparent.

Discussion

Notice that the `geom_ribbon()` is before `geom_line()`, so that the line is drawn on top of the shaded region. If the reverse order were used, the shaded region could obscure the line. In this particular case that wouldn't be a problem since the shaded region is mostly transparent, but it would be a problem if the shaded region were opaque.

Instead of a shaded region, you can also use dotted lines to represent the upper and lower bounds (Figure 4-26):

```
# With a dotted line for upper and lower bounds
ggplot(clim, aes(x=Year, y=Anomaly10y)) +
    geom_line(aes(y=Anomaly10y-Unc10y), colour="grey50", linetype="dotted") +
    geom_line(aes(y=Anomaly10y+Unc10y), colour="grey50", linetype="dotted") +
    geom_line()
```

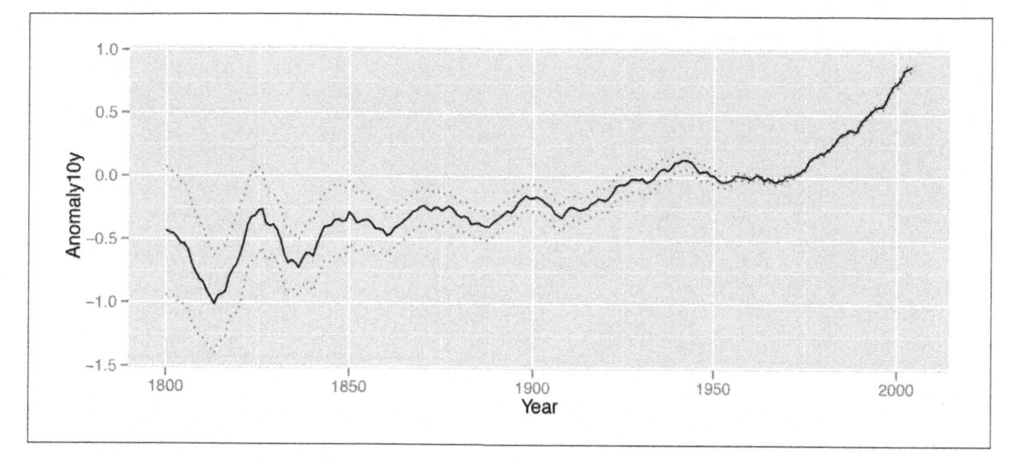

Figure 4-26. A line graph with dotted lines representing a confidence region

Shaded regions can represent things other than confidence regions, such as the difference between two values, for example.

In the area graphs in Recipe 4.7, the *y* range of the shaded area goes from `0` to `y`. Here, it goes from `ymin` to `ymax`.

CHAPTER 5
Scatter Plots

Scatter plots are used to display the relationship between two continuous variables. In a scatter plot, each observation in a data set is represented by a point. Often, a scatter plot will also have a line showing the predicted values based on some statistical model. This is easy to do with R and ggplot2, and can help to make sense of data when the trends aren't immediately obvious just by looking at it.

With large data sets, it can be problematic to plot every single observation because the points will be overplotted, obscuring one another. When this happens, you'll probably want to summarize the data before displaying it. We'll also see how to do that in this chapter.

5.1. Making a Basic Scatter Plot

Problem

You want to make a scatter plot.

Solution

Use `geom_point()`, and map one variable to `x` and one to `y`.

In the `heightweight` data set, there are a number of columns, but we'll only use two in this example (Figure 5-1):

```
library(gcookbook) # For the data set

# List the two columns we'll use
heightweight[, c("ageYear", "heightIn")]

 ageYear heightIn
   11.92     56.3
```

```
   12.92     62.3
   12.75     63.3
...
   13.92     62.0
   12.58     59.3
ggplot(heightweight, aes(x=ageYear, y=heightIn)) + geom_point()
```

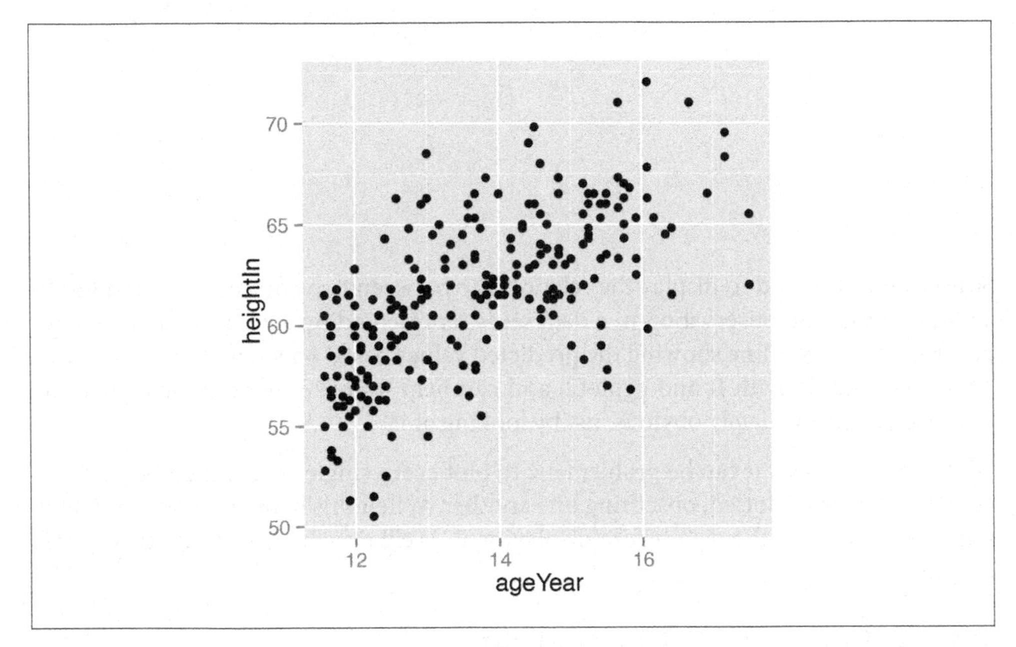

Figure 5-1. A basic scatter plot

Discussion

To use different shapes in a scatter plot, set `shape`. A common alternative to the default solid circles (shape #16) is hollow ones (#21), as seen in Figure 5-2 (left):

```
ggplot(heightweight, aes(x=ageYear, y=heightIn)) + geom_point(shape=21)
```

The size of the points can be controlled with `size`. The default value of `size` is 2. The following will set `size=1.5`, for smaller points (Figure 5-2, right):

```
ggplot(heightweight, aes(x=ageYear, y=heightIn)) + geom_point(size=1.5)
```

When displaying to screen or outputting to bitmap files like PNG, the default solid circle shape (#16) can result in aliased (jagged-looking) edges on some platforms. An alternative is to use shape 19, which is also a solid circle, but comes out smooth in more cases (see Figure 5-3). See Recipe 14.5 for more about anti-aliased output.

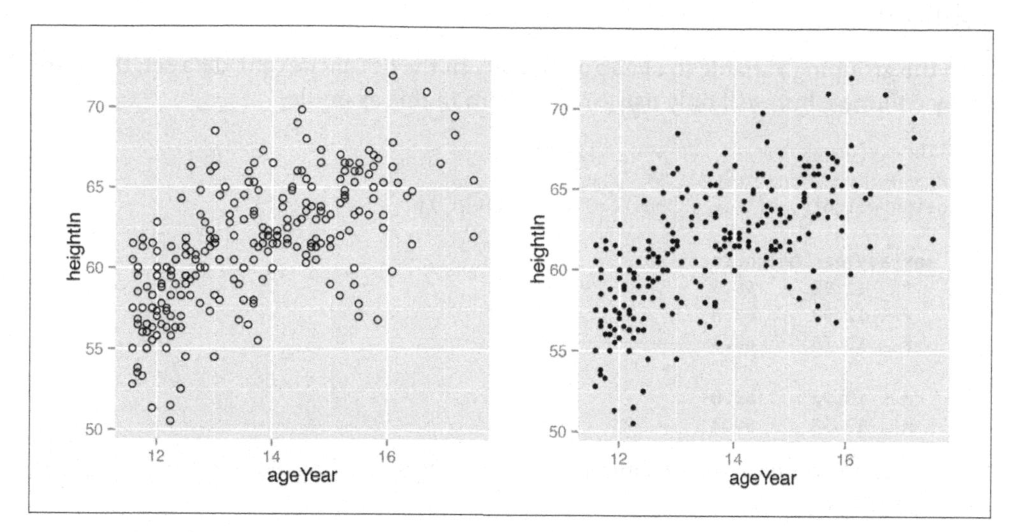

Figure 5-2. Left: scatter plot with hollow circles (shape 21); right: with smaller points

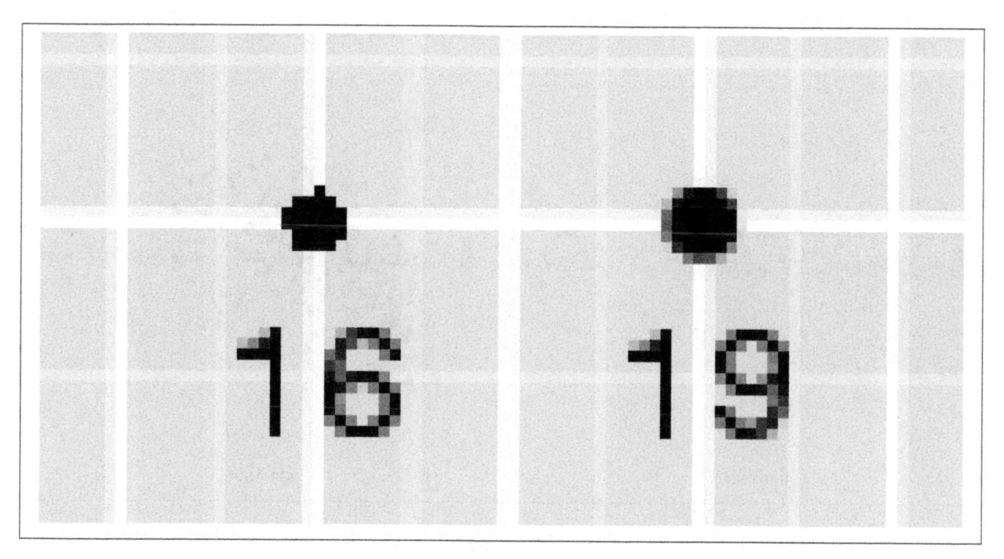

Figure 5-3. Point shapes 16 and 19, as they appear with some bitmap output devices

5.2. Grouping Data Points by a Variable Using Shape or Color

Problem

You want to group points by some variable, using shape or color.

Solution

Map the grouping variable to `shape` or `colour`. In the `heightweight` data set, there are many columns, but we'll only use three of them in this example:

```
library(gcookbook) # For the data set
# Show the three columns we'll use
heightweight[, c("sex", "ageYear", "heightIn")]

 sex ageYear heightIn
   f   11.92     56.3
   f   12.92     62.3
   f   12.75     63.3
...
   m   13.92     62.0
   m   12.58     59.3
```

We can group points on the variable `sex`, by mapping `sex` to one of the aesthetics `colour` or `shape` (Figure 5-4):

```
ggplot(heightweight, aes(x=ageYear, y=heightIn, colour=sex)) + geom_point()

ggplot(heightweight, aes(x=ageYear, y=heightIn, shape=sex)) + geom_point()
```

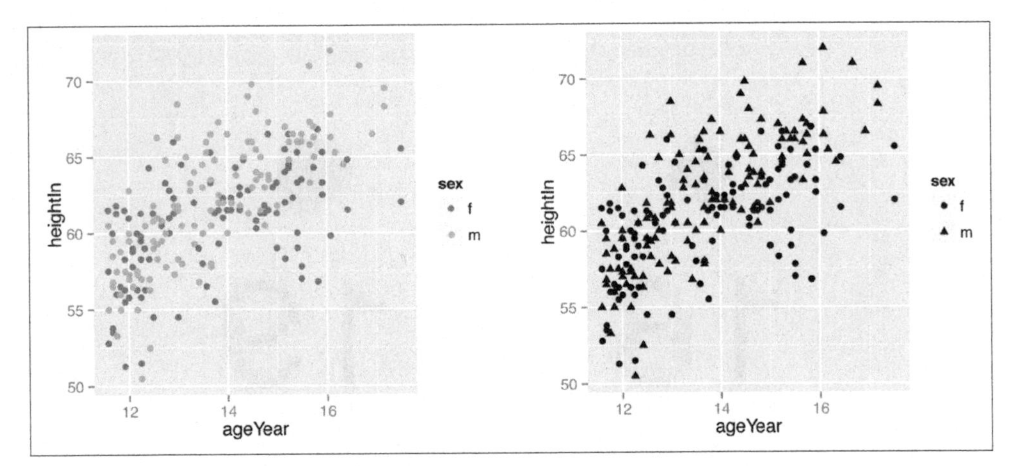

Figure 5-4. Grouping points by a variable mapped to colour (left), and to shape (right)

Discussion

The grouping variable must be categorical—in other words, a factor or character vector. If it is stored as a vector of numeric values, it should be converted to a factor before it is used as a grouping variable.

It is possible to map a variable to both `shape` and `colour`, or, if you have multiple grouping variables, to map different variables to them. Here, we'll map `sex` to `shape` and `colour` (Figure 5-5, left):

```
ggplot(heightweight, aes(x=ageYear, y=heightIn, shape=sex, colour=sex)) +
    geom_point()
```

The default shapes and colors may not be very appealing. Other shapes can be used with `scale_shape_manual()`, and other colors can be used with `scale_colour_brewer()` or `scale_colour_manual()`.

This will set different shapes and colors for the grouping variables (Figure 5-5, right):

```
ggplot(heightweight, aes(x=ageYear, y=heightIn, shape=sex, colour=sex)) +
    geom_point() +
    scale_shape_manual(values=c(1,2)) +
    scale_colour_brewer(palette="Set1")
```

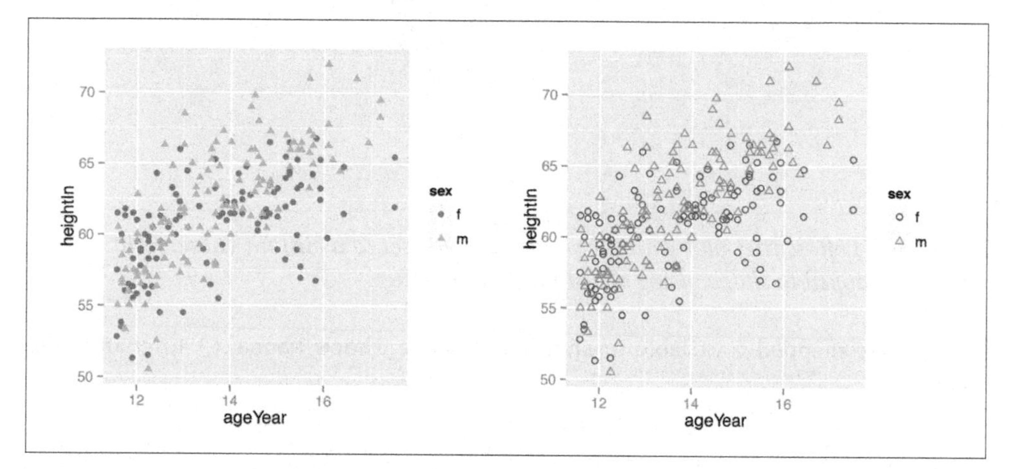

Figure 5-5. Left: mapping to both shape and colour; right: with manually set shapes and colors

See Also

To use different shapes, see Recipe 5.3.

For more on using different colors, see Chapter 12.

5.3. Using Different Point Shapes

Problem

You want to use point shapes that are different from the defaults.

Solution

If you want to set the shape of all the points (Figure 5-6), specify the shape in geom _point():

```
library(gcookbook) # For the data set

ggplot(heightweight, aes(x=ageYear, y=heightIn)) + geom_point(shape=3)
```

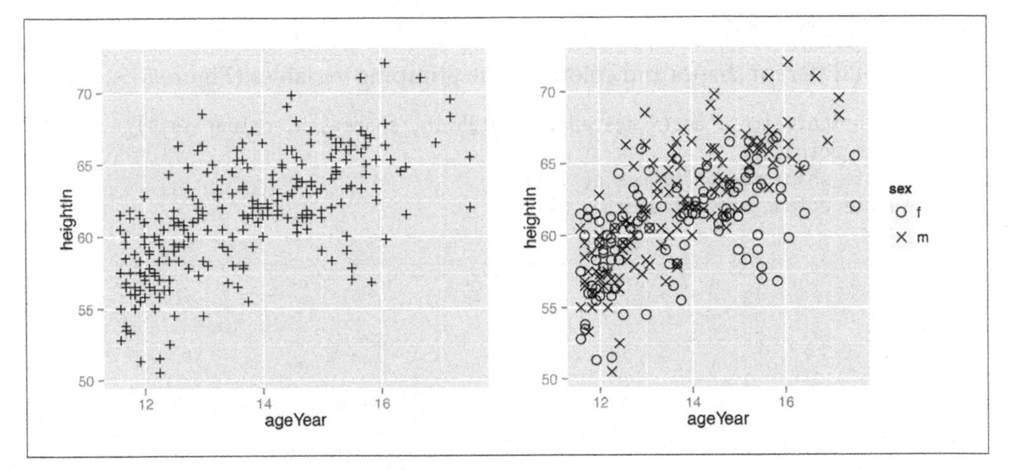

Figure 5-6. Left: scatter plot with the shape aesthetic set to a custom value; right: with a variable mapped to shape, using a custom shape palette

If you have mapped a variable to shape, use scale_shape_manual() to change the shapes:

```
# Use slightly larger points and use a shape scale with custom values
ggplot(heightweight, aes(x=ageYear, y=heightIn, shape=sex)) +
    geom_point(size=3) + scale_shape_manual(values=c(1, 4))
```

Discussion

Figure 5-7 shows the shapes that are available in R graphics. Some of the point shapes (1–14) have just an outline, some (15–20) are solid, and some (21–25) have an outline and fill that can be controlled separately. (You can also use characters for points.)

For shapes 1–20, the color of the entire point—even the points that are solid—is controlled by the colour aesthetic. For shapes 21–25, the outline is controlled by colour and the fill is controlled by fill.

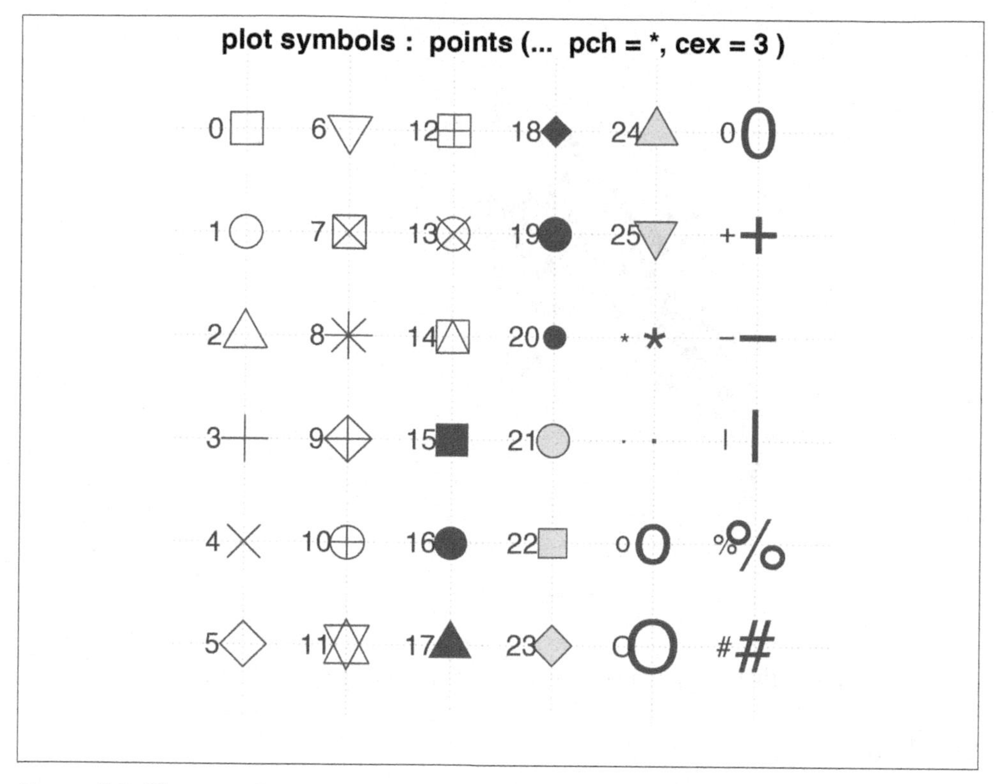

Figure 5-7. Shapes in R

It's possible to have the shape represent one variable and the fill (empty or solid) represent another variable. This is done a little indirectly, by choosing shapes that have both `colour` and `fill`, and a color palette that includes `NA` and another color (the `NA` will result in a hollow shape). For example, we'll take the `heightweight` data set and add another column that indicates whether the child weighed 100 pounds or more (Figure 5-8):

```
# Make a copy of the data
hw <- heightweight
# Categorize into <100 and >=100 groups
hw$weightGroup <- cut(hw$weightLb, breaks=c(-Inf, 100, Inf),
                      labels=c("< 100", ">= 100"))

# Use shapes with fill and color, and use colors that are empty (NA) and
# filled
ggplot(hw, aes(x=ageYear, y=heightIn, shape=sex, fill=weightGroup)) +
    geom_point(size=2.5) +
    scale_shape_manual(values=c(21, 24)) +
    scale_fill_manual(values=c(NA, "black"),
      guide=guide_legend(override.aes=list(shape=21)))
```

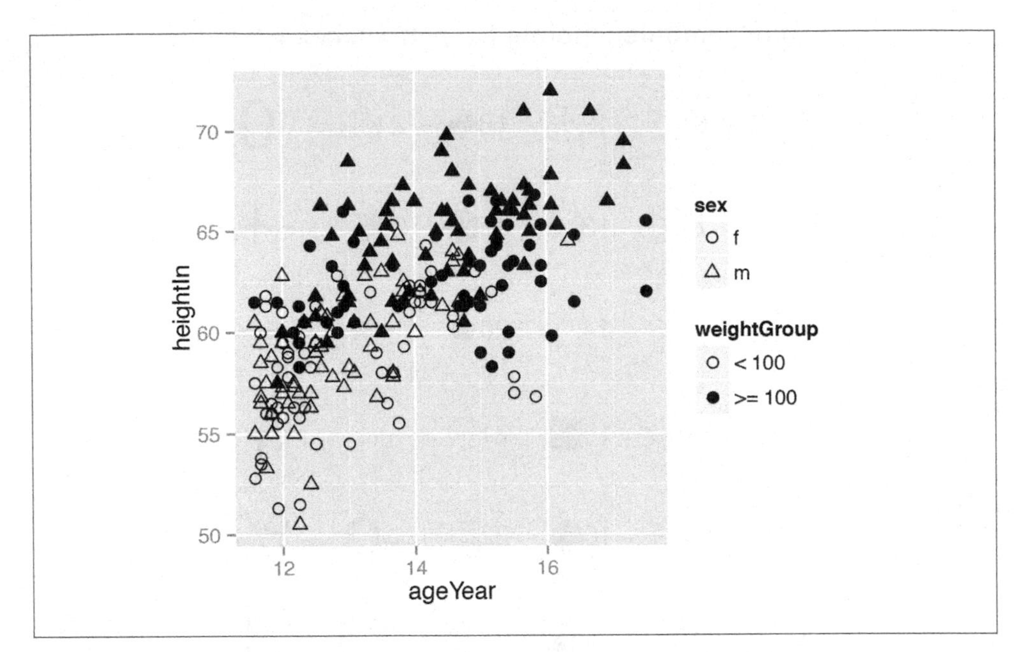

Figure 5-8. A variable mapped to shape and another mapped to fill

See Also

For more on using different colors, see Chapter 12.

For more information about recoding a continuous variable to a categorical one, see Recipe 15.14.

5.4. Mapping a Continuous Variable to Color or Size

Problem

You want to represent a third continuous variable using color or size.

Solution

Map the continuous variable to `size` or `colour`. In the `heightweight` data set, there are many columns, but we'll only use four of them in this example:

```
library(gcookbook) # For the data set

# List the four columns we'll use
heightweight[, c("sex", "ageYear", "heightIn", "weightLb")]

 sex ageYear heightIn weightLb
```

```
    f   11.92      56.3     85.0
    f   12.92      62.3    105.0
    f   12.75      63.3    108.0
...
    m   13.92      62.0    107.5
    m   12.58      59.3     87.0
```

The basic scatter plot in Recipe 5.1 shows the relationship between the continuous variables `ageYear` and `heightIn`. To represent a third continuous variable, `weightLb`, we must map it to another aesthetic property. We can map it to `colour` or `size`, as shown in Figure 5-9:

```
ggplot(heightweight, aes(x=ageYear, y=heightIn, colour=weightLb)) + geom_point()

ggplot(heightweight, aes(x=ageYear, y=heightIn, size=weightLb)) + geom_point()
```

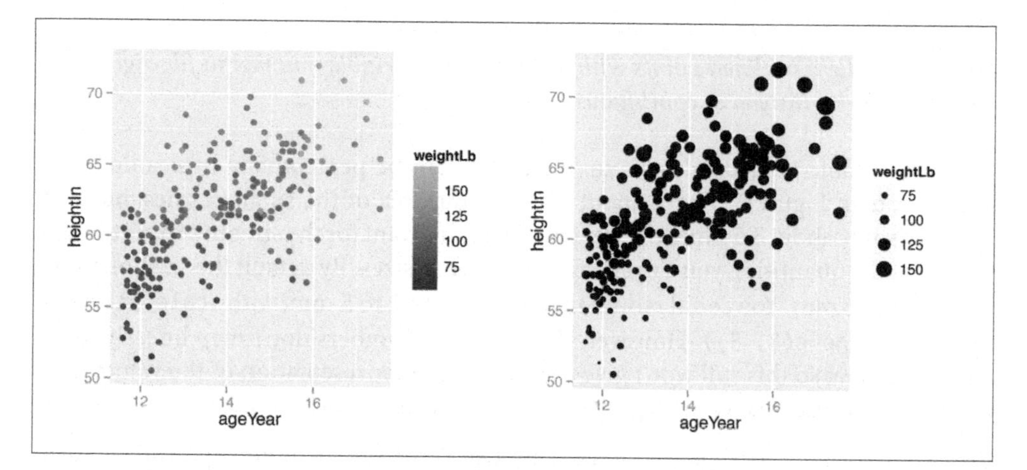

Figure 5-9. Left: a continuous variable mapped to colour; right: mapped to size

Discussion

A basic scatter plot shows the relationship between two continuous variables: one mapped to the x-axis, and one to the y-axis. When there are more than two continuous variables, they must be mapped to other aesthetics: size and/or color.

We can easily perceive small differences in spatial position, so we can interpret the variables mapped to *x* and *y* coordinates with high accuracy. We aren't very good at perceiving small differences in size and color, though, so we will interpret variables mapped to these aesthetic attributes with a much lower accuracy. When you map a variable to one of these properties, it should be one where accuracy is not very important for interpretation.

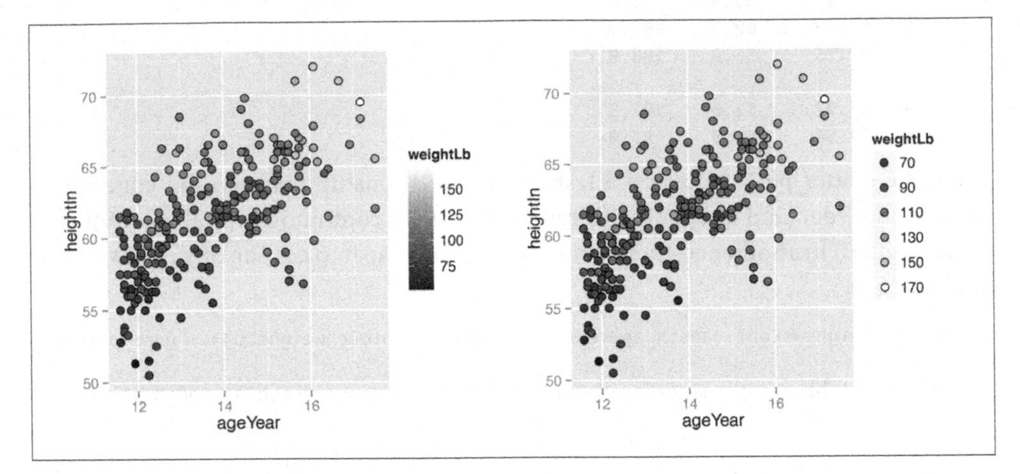

Figure 5-10. Left: outlined points with a continuous variable mapped to fill; right: with a discrete legend instead of continuous colorbar

When a variable is mapped to `size`, the results can be perceptually misleading. The largest dots in Figure 5-9 have about 36 times the area of the smallest ones, but they represent only about 3.5 times the weight. If it is important for the sizes to proportionally represent the quantities, you can change the range of sizes. By default the sizes of points go from 1 to 6 mm. You could reduce the range to, say, 2 to 5 mm, with `scale_size_continuous(range=c(2, 5))`. However, the point size numbers don't map linearly to diameter or area, so this still won't give a very accurate representation of the values. (See Recipe 5.12 for details on making the area of dots proportional to the value.)

When it comes to color, there are actually two aesthetic attributes that can be used: `colour` and `fill`. For most point shapes, you use `colour`. However, shapes 21–25 have an outline with a solid region in the middle where the color is controlled by `fill`. These outlined shapes can be useful when using a color scale with light colors, as in Figure 5-10, because the outline sets them off from the background. In this example, we also set the fill gradient to go from black to white and make the points larger so that the fill is easier to see:

```
ggplot(heightweight, aes(x=ageYear, y=heightIn, fill=weightLb)) +
    geom_point(shape=21, size=2.5) +
    scale_fill_gradient(low="black", high="white")

# Using guide_legend() will result in a discrete legend instead of a colorbar
ggplot(heightweight, aes(x=ageYear, y=heightIn, fill=weightLb)) +
    geom_point(shape=21, size=2.5) +
    scale_fill_gradient(low="black", high="white", breaks=seq(70, 170, by=20),
                        guide=guide_legend())
```

When we map a continuous variable to an aesthetic, that doesn't prevent us from mapping a categorical variable to other aesthetics. In Figure 5-11, we'll map `weightLb` to `size`, and also map `sex` to `colour`. Because there is a fair amount of overplotting, we'll make the points 50% transparent by setting `alpha=.5`. We'll also use `scale_size_area()` to make the area of the points proportional to the value (see Recipe 5.12), and change the color palette to one that is a little more appealing:

```
ggplot(heightweight, aes(x=ageYear, y=heightIn, size=weightLb, colour=sex)) +
    geom_point(alpha=.5) +
    scale_size_area() +     # Make area proportional to numeric value
    scale_colour_brewer(palette="Set1")
```

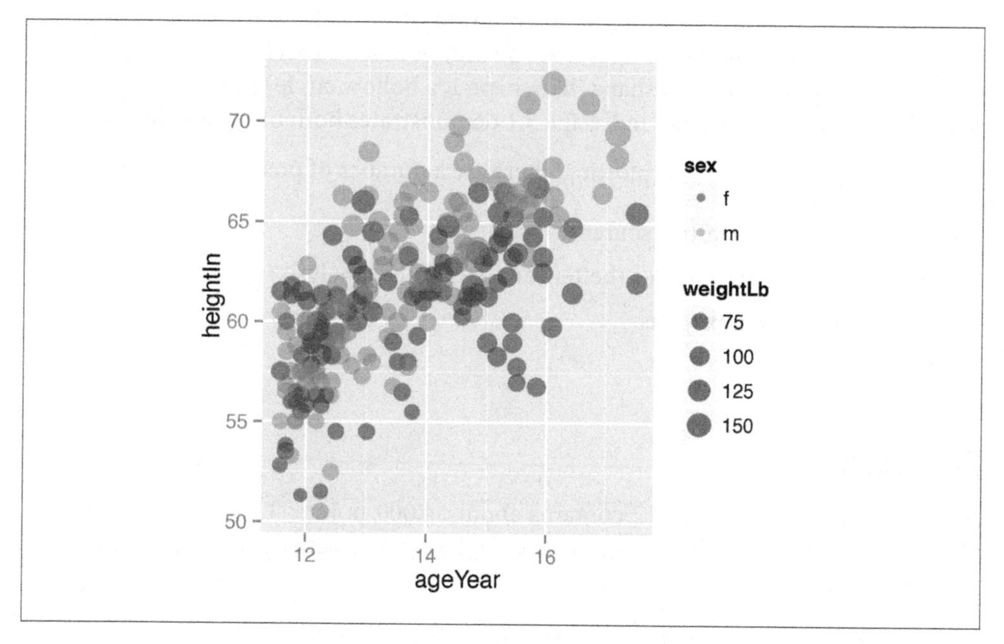

Figure 5-11. Continuous variable mapped to size and categorical variable mapped to colour

When a variable is mapped to `size`, it's a good idea to *not* map a variable to `shape`. This is because it is difficult to compare the sizes of different shapes; for example, a size 4 triangle could appear larger than a size 3.5 circle. Also, some of the shapes really are different sizes: shapes 16 and 19 are both circles, but at any given numeric size, shape 19 circles are visually larger than shape 16 circles.

See Also

To use different colors from the default, see Recipe 12.6.

See Recipe 5.12 for creating a balloon plot.

5.5. Dealing with Overplotting

Problem

You have many points and they obscure each other.

Solution

With large data sets, the points in a scatter plot may obscure each other and prevent the viewer from accurately assessing the distribution of the data. This is called *overplotting*. If the amount of overplotting is low, you may be able to alleviate it by using smaller points, or by using a different shape (like shape 1, a hollow circle) through which other points can be seen. Figure 5-2 in Recipe 5.1 demonstrates both of these solutions.

If there's a high degree of overplotting, there are a number of possible solutions:

- Make the points semitransparent
- Bin the data into rectangles (better for quantitative analysis)
- Bin the data into hexagons
- Use box plots

Discussion

The scatter plot in Figure 5-12 contains about 54,000 points. They are heavily overplotted, making it impossible to get a sense of the relative density of points in different areas of the graph:

```
sp <- ggplot(diamonds, aes(x=carat, y=price))

sp + geom_point()
```

We can make the points semitransparent using `alpha`, as in Figure 5-13. Here, we'll make them 90% transparent and then 99% transparent, by setting `alpha=.1` and `alpha=.01`:

```
sp + geom_point(alpha=.1)

sp + geom_point(alpha=.01)
```

Now we can see that there are vertical bands at nice round values of `carats`, indicating that diamonds tend to be cut to those sizes. Still, the data is so dense that even when the points are 99% transparent, much of the graph appears solid black, and the data distribution is still somewhat obscured.

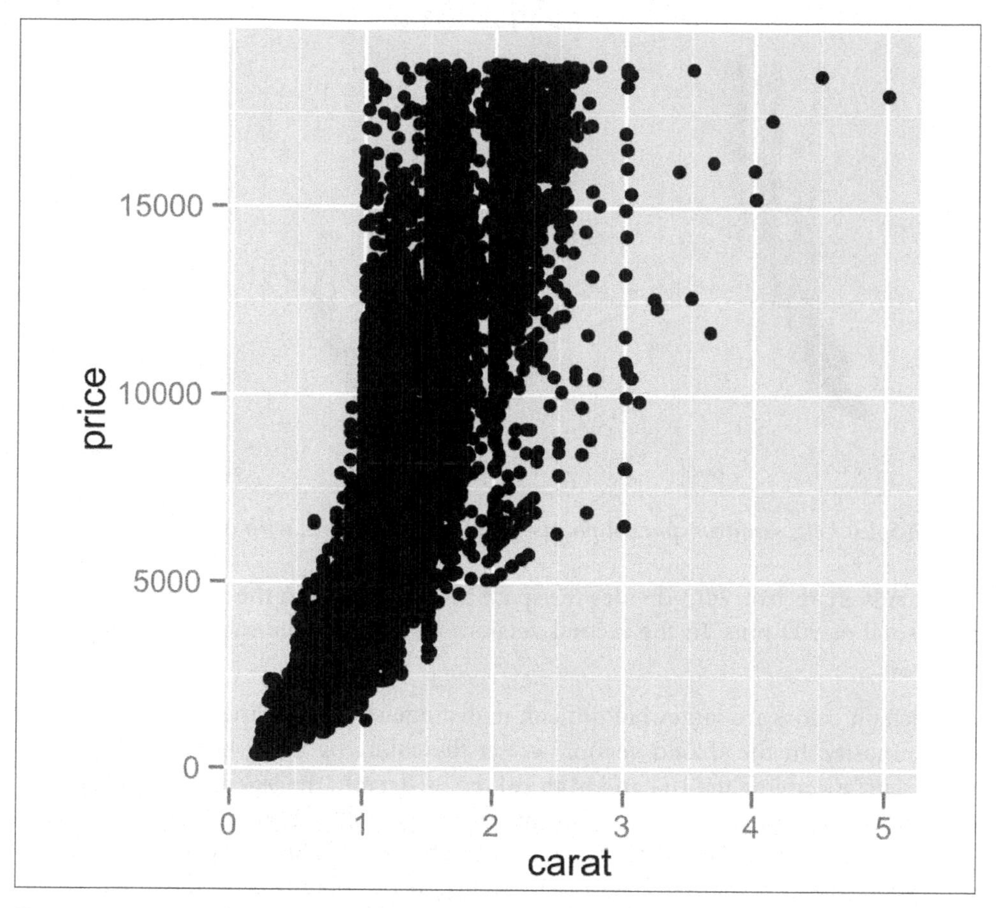

Figure 5-12. Overplotting, with about 54,000 points

For most graphs, vector formats (such as PDF, EPS, and SVG) result in smaller output files than bitmap formats (such as TIFF and PNG). But in cases where there are tens of thousands of points, vector output files can be very large and slow to render—the scatter plot here with 99% transparent points is 1.5 MB! In these cases, high-resolution bitmaps will be smaller and faster to display on computer screens. See Chapter 14 for more information.

Another solution is to *bin* the points into rectangles and map the density of the points to the fill color of the rectangles, as shown in Figure 5-14. With the binned visualization, the vertical bands are barely visible. The density of points in the lower-left corner is much greater, which tells us that the vast majority of diamonds are small and inexpensive.

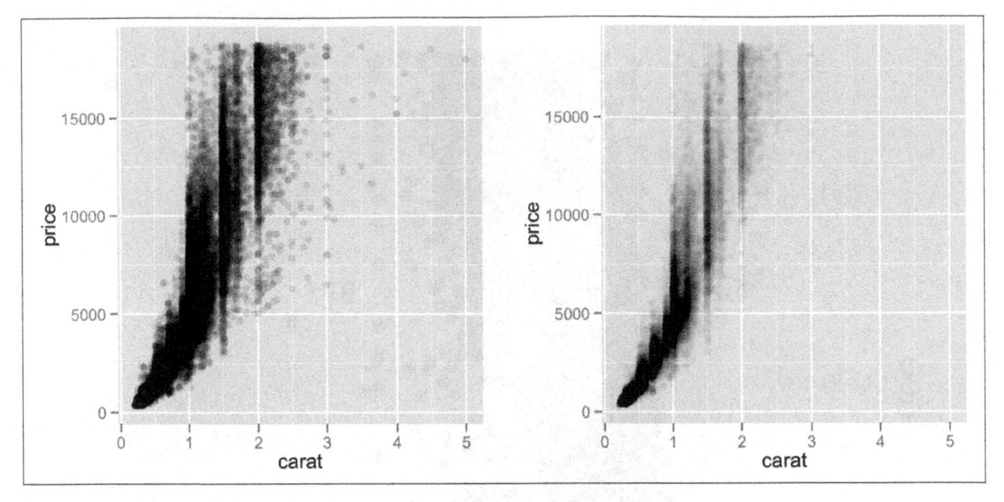

Figure 5-13. Left: semitransparent points with alpha=.1; right: with alpha=.01

By default, `stat_bin_2d()` divides the space into 30 groups in the *x* and *y* directions, for a total of 900 bins. In the second version, we increase the number of bins with `bins=50`.

The default colors are somewhat difficult to distinguish because they don't vary much in luminosity. In the second version we set the colors by using `scale_fill_gradient()` and specifying the `low` and `high` colors. By default, the legend doesn't show an entry for the lowest values. This is because the range of the color scale starts not from zero, but from the smallest nonzero quantity in a bin—probably 1, in this case. To make the legend show a zero (as in Figure 5-14, right), we can manually set the range from 0 to the maximum, 6000, using `limits` (Figure 5-14, left):

```
sp + stat_bin2d()

sp + stat_bin2d(bins=50) +
    scale_fill_gradient(low="lightblue", high="red", limits=c(0, 6000))
```

Another alternative is to bin the data into hexagons instead of rectangles, with `stat_binhex()` (Figure 5-15). It works just like `stat_bin2d()`. To use it, you must first install the hexbin package, with `install.packages("hexbin")`:

```
library(hexbin)

sp + stat_binhex() +
    scale_fill_gradient(low="lightblue", high="red",
                        limits=c(0, 8000))

sp + stat_binhex() +
    scale_fill_gradient(low="lightblue", high="red",
```

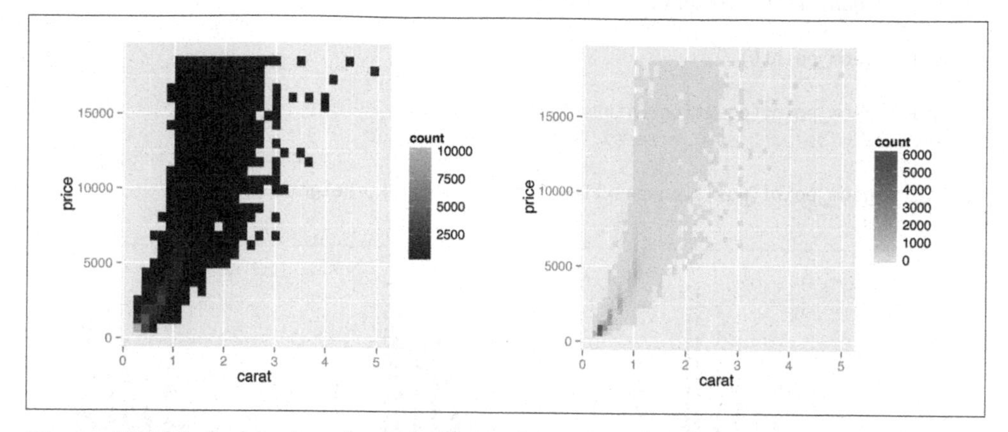

Figure 5-14. Left: binning data with stat_bin2d(); right: with more bins, manually specified colors, and legend breaks

```
breaks=c(0, 250, 500, 1000, 2000, 4000, 6000),
limits=c(0, 6000))
```

For both of these methods, if you manually specify the range, and there is a bin that falls outside that range because it has too many or too few points, that bin will show up as grey rather than the color at the high or low end of the range, as seen in the graph on the right in Figure 5-15.

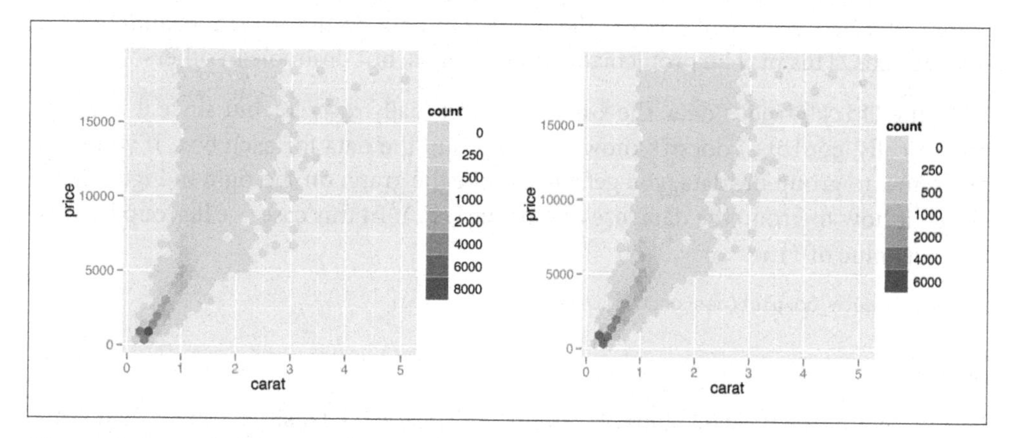

Figure 5-15. Left: binning data with stat_binhex(); right: cells outside of the range shown in grey

Overplotting can also occur when the data is *discrete* on one or both axes, as shown in Figure 5-16. In these cases, you can randomly *jitter* the points with `position_jitter()`. By default the amount of jitter is 40% of the resolution of the data in each direction, but these amounts can be controlled with `width` and `height`:

```
sp1 <- ggplot(ChickWeight, aes(x=Time, y=weight))

sp1 + geom_point()

sp1 + geom_point(position="jitter")
# Could also use geom_jitter(), which is equivalent

sp1 + geom_point(position=position_jitter(width=.5, height=0))
```

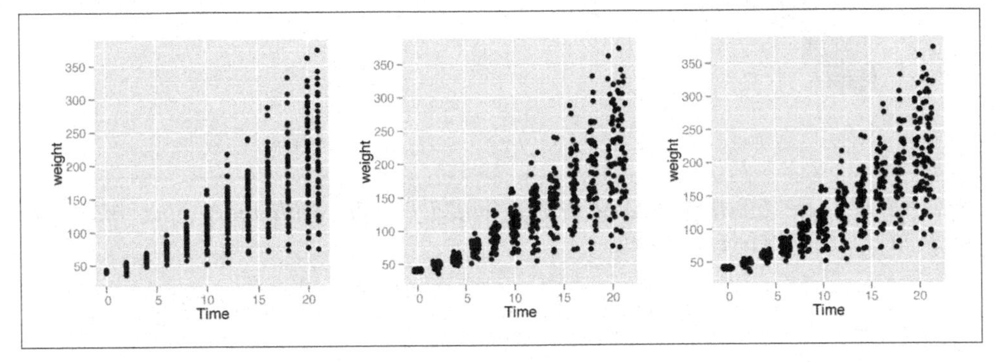

Figure 5-16. Left: data with a discrete x variable; middle: jittered; right: jittered horizontally only

When the data has one discrete axis and one continuous axis, it might make sense to use box plots, as shown in Figure 5-17. This will convey a different story than a standard scatter plot because it will obscure the *number* of data points at each location on the discrete axis. This may be problematic in some cases, but desirable in others.

With the `ChickWeights` data, the x-axis is conceptually discrete, but since it is stored numerically, `ggplot()` doesn't know how to group the data for each box. If you don't tell it how to group the data, you get a result like the graph on the right in Figure 5-17. To tell it how to group the data, use `aes(group=...)`. In this case, we'll group by each distinct value of `Time`:

```
sp1 + geom_boxplot(aes(group=Time))
```

See Also

Instead of binning the data, it may be useful to display a 2D density estimate. To do this, see Recipe 6.12.

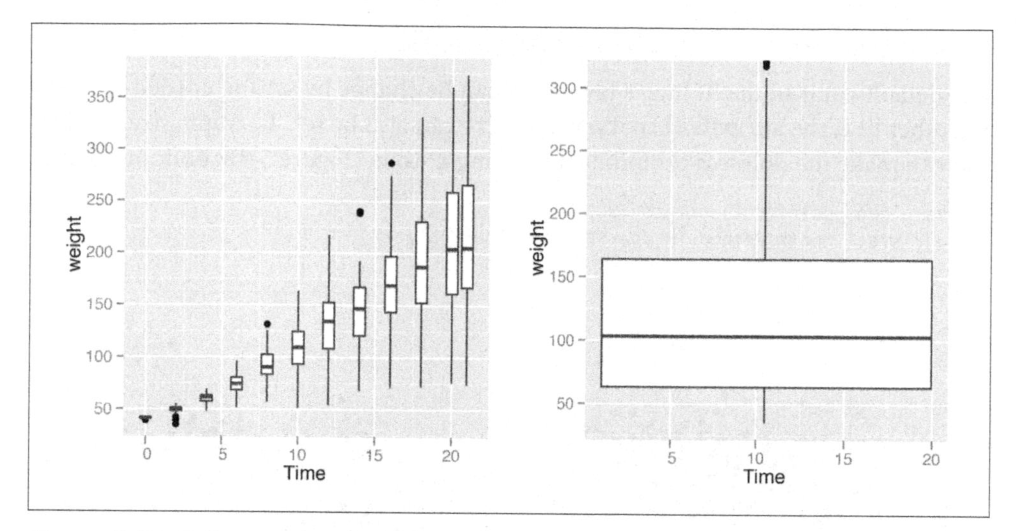

Figure 5-17. Left: grouping into box plots; right: what happens if you don't specify groups

5.6. Adding Fitted Regression Model Lines

Problem

You want to add lines from a fitted regression model to a scatter plot.

Solution

To add a linear regression line to a scatter plot, add `stat_smooth()` and tell it to use `method=lm`. This instructs it to fit the data with the `lm()` (linear model) function. First we'll save the base plot object in `sp`, then we'll add different components to it:

```
library(gcookbook) # For the data set

# The base plot
sp <- ggplot(heightweight, aes(x=ageYear, y=heightIn))

sp + geom_point() + stat_smooth(method=lm)
```

By default, `stat_smooth()` also adds a 95% confidence region for the regression fit. The confidence interval can be changed by setting `level`, or it can be disabled with `se=FALSE` (Figure 5-18):

```
# 99% confidence region
sp + geom_point() + stat_smooth(method=lm, level=0.99)
```

```
# No confidence region
sp + geom_point() + stat_smooth(method=lm, se=FALSE)
```

The default color of the fit line is blue. This can be change by setting `colour`. As with any other line, the attributes `linetype` and `size` can also be set. To emphasize the line, you can make the dots less prominent by setting `colour` (Figure 5-18, bottom right):

```
sp + geom_point(colour="grey60") +
    stat_smooth(method=lm, se=FALSE, colour="black")
```

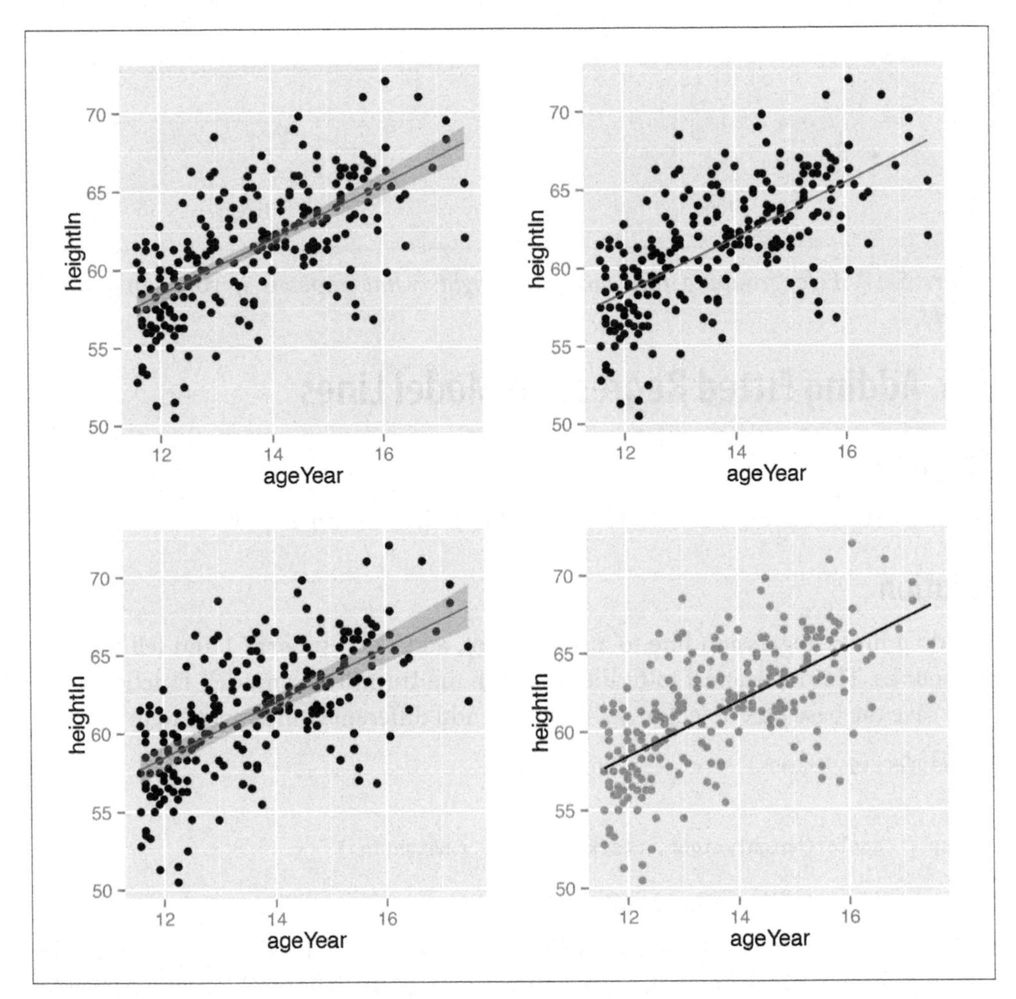

Figure 5-18. Top left: an lm fit with the default 95% confidence region; bottom left: a 99% confidence region; top right: no confidence region; bottom right: in black with grey points

Discussion

The linear regression line is not the only way of fitting a model to the data—in fact, it's not even the default. If you add `stat_smooth()` without specifying the method, it will use a `loess` (locally weighted polynomial) curve, as shown in Figure 5-19. Both of these will have the same result:

```
sp + geom_point(colour="grey60") + stat_smooth()
sp + geom_point(colour="grey60") + stat_smooth(method=loess)
```

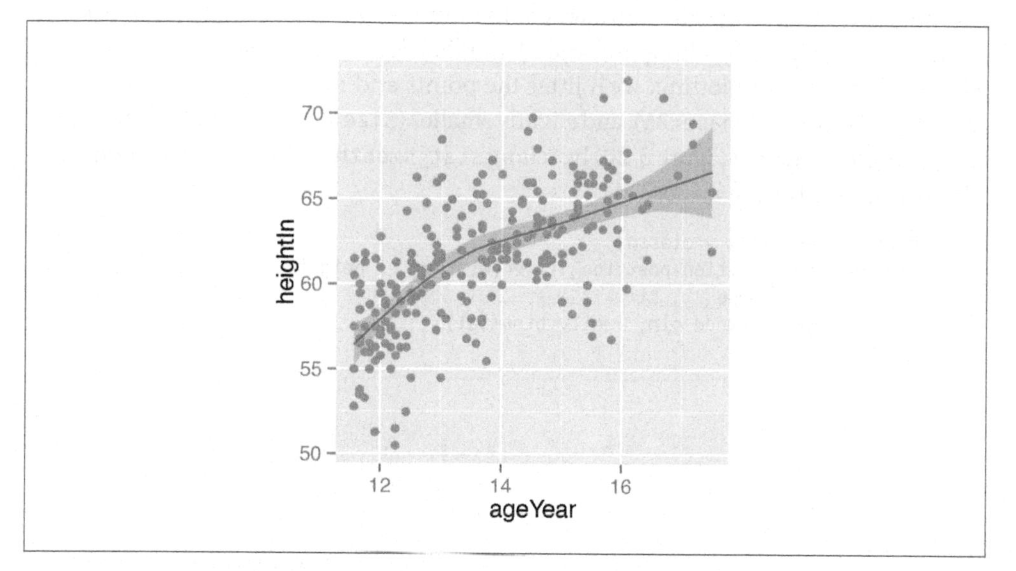

Figure 5-19. A LOESS fit

Additional parameters can be passed along to the `loess()` function by just passing them to `stat_smooth()`.

Another common type of model fit is a logistic regression. Logistic regression isn't appropriate for the `heightweight` data set, but it's perfect for the `biopsy` data set in the `MASS` library. In this data set, there are nine different measured attributes of breast cancer biopsies, as well as the class of the tumor, which is either `benign` or `malignant`. To prepare the data for logistic regression, we must convert the factor `class`, with the levels `benign` and `malignant`, to a vector with numeric values of 0 and 1. We'll make a copy of the `biopsy` data frame, then store the numeric coded class in a column called `classn`:

```
library(MASS) # For the data set

b <- biopsy

b$classn[b$class=="benign"]    <- 0
b$classn[b$class=="malignant"] <- 1
```

```
b

     ID V1 V2 V3 V4 V5 V6 V7 V8 V9     class classn
1000025  5  1  1  1  2  1  3  1  1    benign      0
1002945  5  4  4  5  7 10  3  2  1    benign      0
1015425  3  1  1  1  2  2  3  1  1    benign      0
...
 897471  4  8  6  4  3  4 10  6  1 malignant      1
 897471  4  8  8  5  4  5 10  4  1 malignant      1
```

Although there are many attributes we could examine, for this example we'll just look at the relationship of `V1` (clump thickness) and the class of the tumor. Because there is a large degree of overplotting, we'll jitter the points and make them semitransparent (`alpha=0.4`), hollow (`shape=21`), and slightly smaller (`size=1.5`). Then we'll add a fitted logistic regression line (Figure 5-20) by telling `stat_smooth()` to use the `glm()` function with the option `family=binomial`:

```
ggplot(b, aes(x=V1, y=classn)) +
    geom_point(position=position_jitter(width=0.3, height=0.06), alpha=0.4,
               shape=21, size=1.5) +
    stat_smooth(method=glm, family=binomial)
```

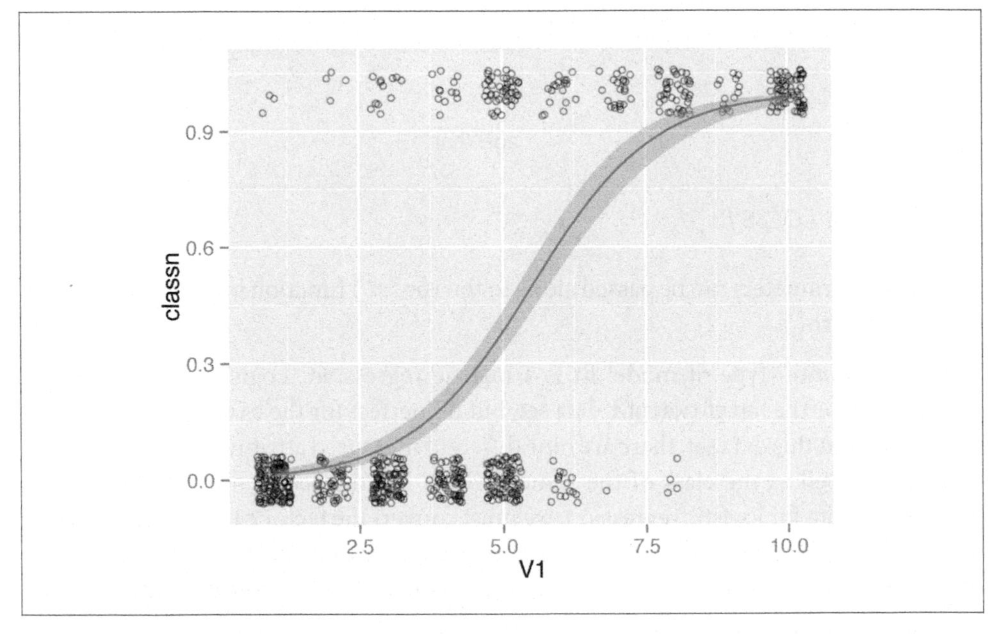

Figure 5-20. A logistic model

If your scatter plot has points grouped by a factor, using `colour` or `shape`, one fit line will be drawn for each group. First we'll make the base plot object `sps`, then we'll add the loess lines to it. We'll also make the points less prominent by making them semi-transparent, using alpha=.4 (Figure 5-21):

```
sps <- ggplot(heightweight, aes(x=ageYear, y=heightIn, colour=sex)) +
       geom_point() +
       scale_colour_brewer(palette="Set1")

sps + geom_smooth()
```

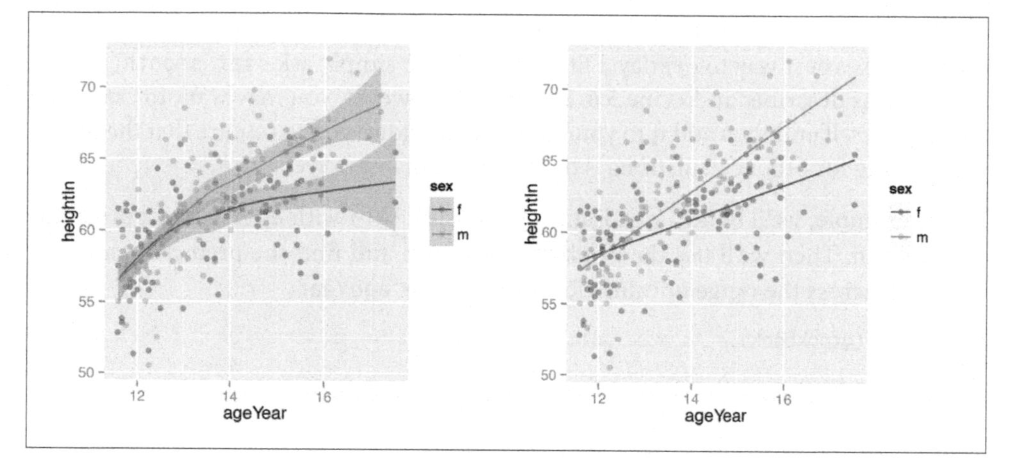

Figure 5-21. Left: LOESS fit lines for each group; right: extrapolated linear fit lines

Notice that the blue line, for males, doesn't run all the way to the right side of the graph. There are two reasons for this. The first is that, by default, stat_smooth() limits the prediction to within the range of the predictor data (on the x-axis). The second is that even if it extrapolates, the loess() function only offers prediction within the *x* range of the data.

If you want the lines to extrapolate from the data, as shown in the right-hand image of Figure 5-21, you must use a model method that allows extrapolation, like lm(), and pass stat_smooth() the option fullrange=TRUE:

```
sps + geom_smooth(method=lm, se=FALSE, fullrange=TRUE)
```

In this example with the heightweight data set, the default settings for stat_smooth() (with LOESS and no extrapolation) make more sense than the extrapolated linear predictions, because we don't grow linearly and we don't grow forever.

5.7. Adding Fitted Lines from an Existing Model

Problem

You have already created a fitted regression model object for a data set, and you want to plot the lines for that model.

Solution

Usually the easiest way to overlay a fitted model is to simply ask `stat_smooth()` to do it for you, as described in Recipe 5.6. Sometimes, however, you may want to create the model yourself and then add it to your graph. This allows you to be sure that the model you're using for other calculations is the same one that you see.

In this example, we'll build a quadratic model using `lm()` with `ageYear` as a predictor of `heightIn`. Then we'll use the `predict()` function and find the predicted values of `heightIn` across the range of values for the predictor, `ageYear`:

```
library(gcookbook) # For the data set

model <- lm(heightIn ~ ageYear + I(ageYear^2), heightweight)
model

Call:
lm(formula = heightIn ~ ageYear + I(ageYear^2), data = heightweight)

Coefficients:
 (Intercept)      ageYear  I(ageYear^2)
    -10.3136       8.6673       -0.2478


# Create a data frame with ageYear column, interpolating across range
xmin <- min(heightweight$ageYear)
xmax <- max(heightweight$ageYear)
predicted <- data.frame(ageYear=seq(xmin, xmax, length.out=100))

# Calculate predicted values of heightIn
predicted$heightIn <- predict(model, predicted)
predicted

  ageYear heightIn
 11.58000 56.82624
 11.63980 57.00047
 ...
 17.44020 65.47875
 17.50000 65.47933
```

We can now plot the data points along with the values predicted from the model (as you'll see in Figure 5-22):

```
sp <- ggplot(heightweight, aes(x=ageYear, y=heightIn)) +
      geom_point(colour="grey40")

sp + geom_line(data=predicted, size=1)
```

Discussion

Any model object can be used, so long as it has a corresponding `predict()` method. For example, `lm` has `predict.lm()`, `loess` has `predict.loess()`, and so on.

Adding lines from a model can be simplified by using the function `predictvals()`, defined next. If you simply pass in a model, it will do the work of finding the variable names and range of the predictor, and will return a data frame with predictor and predicted values. That data frame can then be passed to `geom_line()` to draw the fitted line, as we did earlier:

```
# Given a model, predict values of yvar from xvar
# This supports one predictor and one predicted variable
# xrange: If NULL, determine the x range from the model object. If a vector with
#   two numbers, use those as the min and max of the prediction range.
# samples: Number of samples across the x range.
# ...: Further arguments to be passed to predict()
predictvals <- function(model, xvar, yvar, xrange=NULL, samples=100, ...) {

  # If xrange isn't passed in, determine xrange from the models.
  # Different ways of extracting the x range, depending on model type
  if (is.null(xrange)) {
    if (any(class(model) %in% c("lm", "glm")))
      xrange <- range(model$model[[xvar]])
    else if (any(class(model) %in% "loess"))
      xrange <- range(model$x)
  }

  newdata <- data.frame(x = seq(xrange[1], xrange[2], length.out = samples))
  names(newdata) <- xvar
  newdata[[yvar]] <- predict(model, newdata = newdata, ...)
  newdata
}
```

With the `heightweight` data set, we'll make a linear model with `lm()` and a LOESS model with `loess()` (Figure 5-22):

```
modlinear <- lm(heightIn ~ ageYear, heightweight)

modloess  <- loess(heightIn ~ ageYear, heightweight)
```

Then we can call `predictvals()` on each model, and pass the resulting data frames to `geom_line()`:

```
lm_predicted    <- predictvals(modlinear, "ageYear", "heightIn")
loess_predicted <- predictvals(modloess, "ageYear", "heightIn")
```

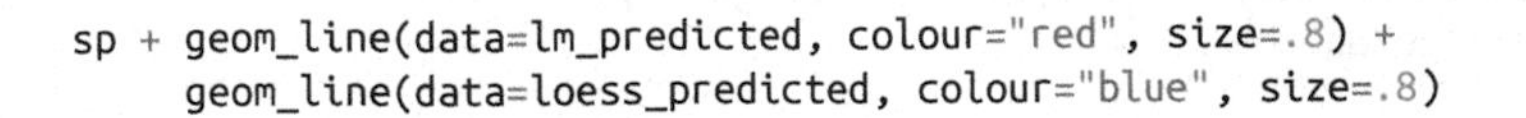

```
sp + geom_line(data=lm_predicted, colour="red", size=.8) +
     geom_line(data=loess_predicted, colour="blue", size=.8)
```

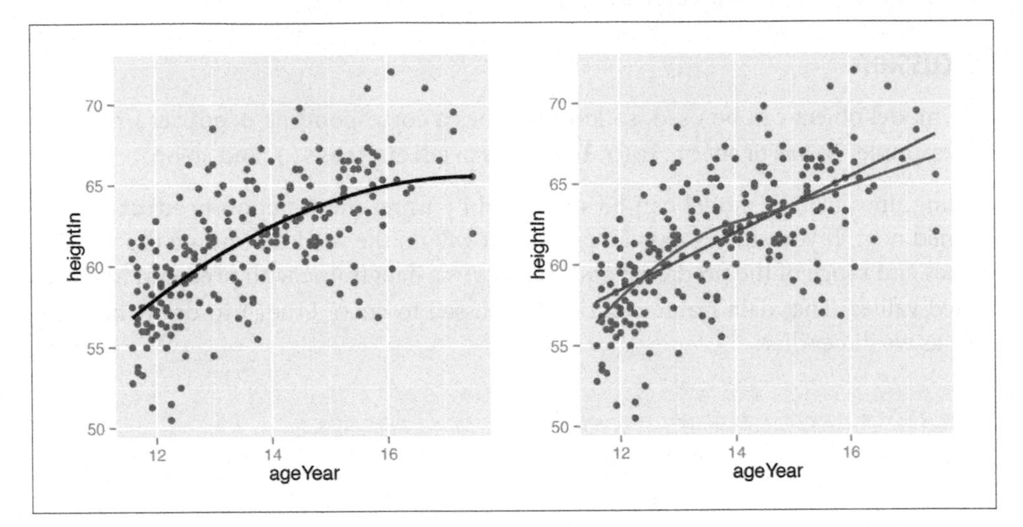

Figure 5-22. Left: a quadratic prediction line from an lm object; right: prediction lines from linear (red) and LOESS (blue) models

For `glm` models that use a nonlinear link function, you need to specify `type="response"` to the `predictvals()` function. This is because the default behavior is to return predicted values in the scale of the linear predictors, instead of in the scale of the response (*y*) variable.

To illustrate this, we'll use the `biopsy` data set from the `MASS` library. As we did in Recipe 5.6, we'll use `V1` to predict `class`. Since logistic regression uses values from 0 to 1, while `class` is a factor, we'll first have to convert `class` to 0s and 1s:

```
library(MASS) # For the data set
b <- biopsy

b$classn[b$class=="benign"]    <- 0
b$classn[b$class=="malignant"] <- 1
```

Next, we'll perform the logistic regression:

```
fitlogistic <- glm(classn ~ V1, b, family=binomial)
```

Finally, we'll make the graph with jittered points and the `fitlogistic` line. We'll make the line in a shade of blue by specifying a color in RGB values, and slightly thicker, with `size=1` (Figure 5-23):

```
# Get predicted values
glm_predicted <- predictvals(fitlogistic, "V1", "classn", type="response")
```

```
ggplot(b, aes(x=V1, y=classn)) +
    geom_point(position=position_jitter(width=.3, height=.08), alpha=0.4,
               shape=21, size=1.5) +
    geom_line(data=glm_predicted, colour="#1177FF", size=1)
```

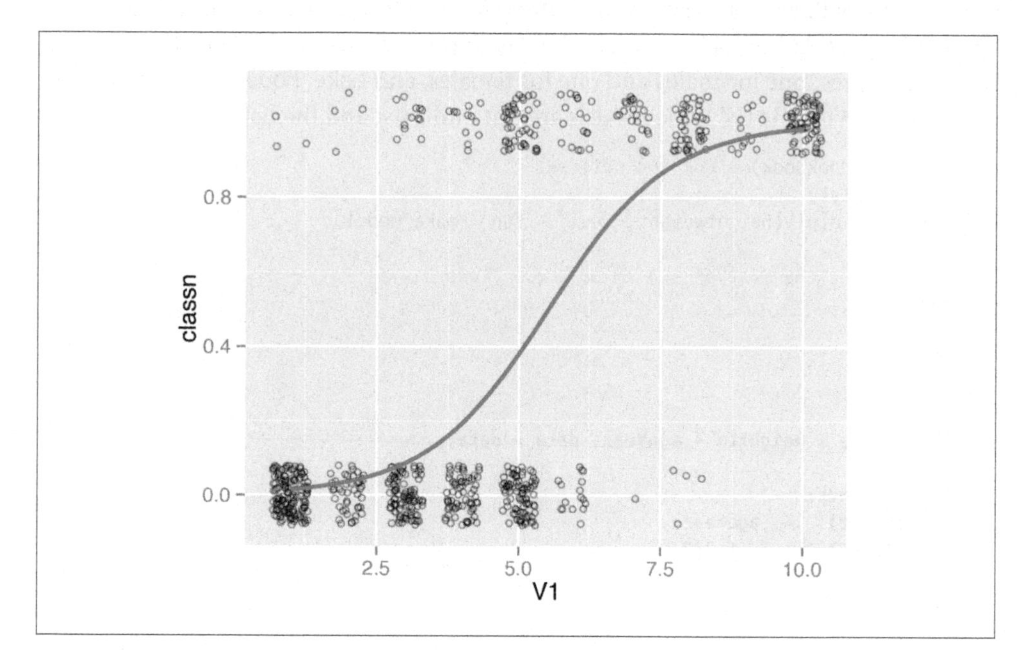

Figure 5-23. A fitted logistic model

5.8. Adding Fitted Lines from Multiple Existing Models

Problem

You have already created a fitted regression model object for a data set, and you want to plot the lines for that model.

Solution

Use the `predictvals()` function from the previous recipe along with `dlply()` and `ldply()` from the plyr package.

With the `heightweight` data set, we'll make a linear model with `lm()` for each of the levels of `sex`, and put those model objects in a list. The model building is done with a function, `make_model()`, defined here. If you pass it a data frame, it simply returns an `lm` object. The model can be customized for your data:

```
make_model <- function(data) {
    lm(heightIn ~ ageYear, data)
}
```

With this function, we can use the `dlply()` function to build a model for each subset of data. This will split the data frame into subsets by the grouping variable `sex`, and apply `make_model()` to each subset. In this case, the `heightweight` data will be split into two data frames, one for males and one for females, and `make_model()` will be run on each subset. With `dlply()`, the models are put into a list and the list is returned:

```
library(gcookbook) # For the data set
library(plyr)
models <- dlply(heightweight, "sex", .fun = make_model)

# Print out the list of two lm objects, f and m
models

$f

Call:
lm(formula = heightIn ~ ageYear, data = data)

Coefficients:
(Intercept)      ageYear
     43.963        1.209

$m

Call:
lm(formula = heightIn ~ ageYear, data = data)

Coefficients:
(Intercept)      ageYear
     30.658        2.301

attr(,"split_type")
[1] "data.frame"
attr(,"split_labels")
  sex
1   f
2   m
```

Now that we have the list of model objects, we can run `predictvals()` to get predicted values from each model, using the `ldply()` function:

```
predvals <- ldply(models, .fun=predictvals, xvar="ageYear", yvar="heightIn")
predvals

 sex  ageYear heightIn
   f 11.58000 57.96250
   f 11.63980 58.03478
   f 11.69960 58.10707
```

```
...
m 17.38040 70.64912
m 17.44020 70.78671
m 17.50000 70.92430
```

Finally, we can plot the data with the predicted values (Figure 5-24):

```
ggplot(heightweight, aes(x=ageYear, y=heightIn, colour=sex)) +
    geom_point() + geom_line(data=predvals)
```

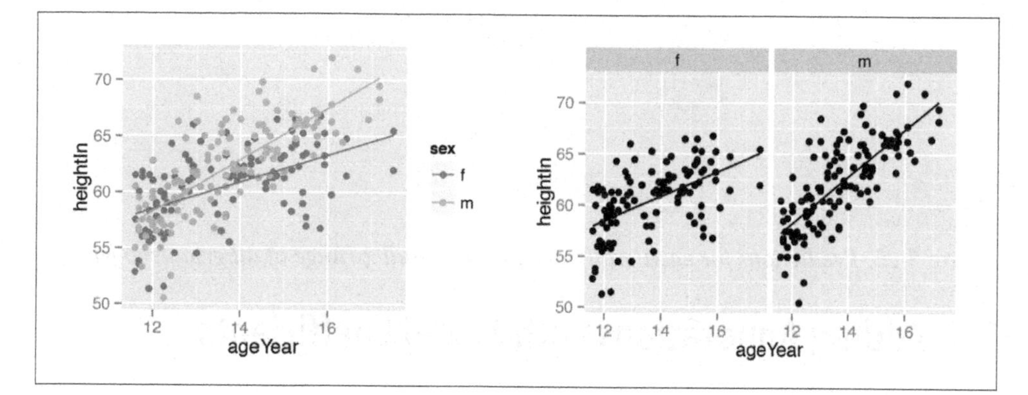

Figure 5-24. Left: predictions from two separate lm objects, one for each subset of data; right: with facets

Discussion

The `dlply()` and `ldply()` calls are used for splitting the data into parts, running functions on those parts, and then reassembling the output.

With the preceding code, the *x* range of the predicted values for each group spans the *x* range of each group, and no further; for the males, the prediction line stops at the oldest male, while for females, the prediction line continues further right, to the oldest female. To form prediction lines that have the same *x* range across all groups, we can simply pass in `xrange`, like this:

```
predvals <- ldply(models, .fun=predictvals, xvar="ageYear", yvar="heightIn",
                  xrange=range(heightweight$ageYear))
```

Then we can plot it, the same as we did before:

```
ggplot(heightweight, aes(x=ageYear, y=heightIn, colour=sex)) +
    geom_point() + geom_line(data=predvals)
```

As you can see in Figure 5-25, the line for males now extends as far to the right as the line for females. Keep in mind that extrapolating past the data isn't always appropriate, though; whether or not it's justified will depend on the nature of your data and the assumptions you bring to the table.

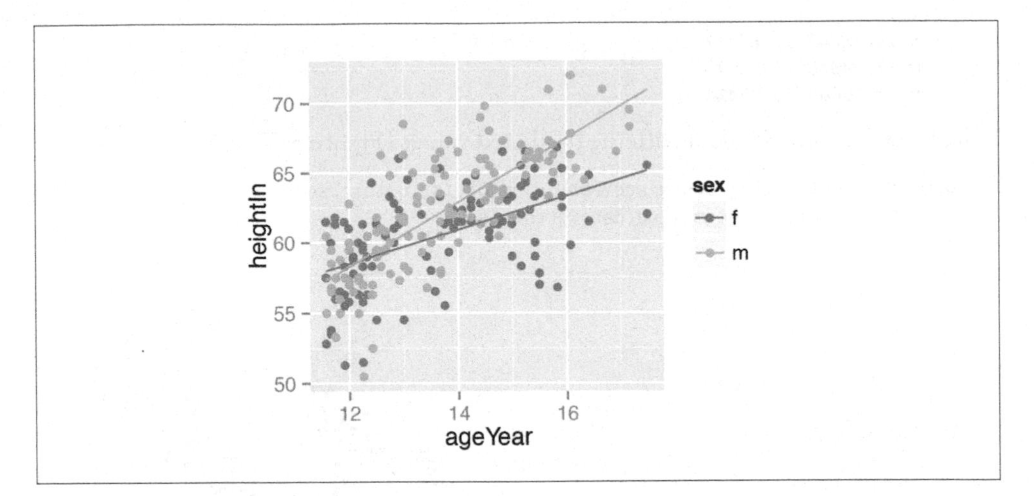

Figure 5-25. Predictions for each group extend to the full x range of all groups together

5.9. Adding Annotations with Model Coefficients

Problem

You want to add numerical information about a model to a plot.

Solution

To add simple text to a plot, simply add an annotation. In this example, we'll create a linear model and use the `predictvals()` function defined in Recipe 5.7 to create a prediction line from the model. Then we'll add an annotation:

```
library(gcookbook) # For the data set

model <- lm(heightIn ~ ageYear, heightweight)
summary(model)

Call:
lm(formula = heightIn ~ ageYear, data = heightweight)

Residuals:
    Min      1Q  Median      3Q     Max
-8.3517 -1.9006  0.1378  1.9071  8.3371

Coefficients:
            Estimate Std. Error t value Pr(>|t|)
(Intercept)  37.4356     1.8281   20.48   <2e-16 ***
ageYear       1.7483     0.1329   13.15   <2e-16 ***
---
```

```
Signif. codes:  0 '***' 0.001 '**' 0.01 '*' 0.05 '.' 0.1 ' ' 1

Residual standard error: 2.989 on 234 degrees of freedom
Multiple R-squared: 0.4249, Adjusted R-squared: 0.4225
F-statistic: 172.9 on 1 and 234 DF,  p-value: < 2.2e-16
```

This shows that the r^2 value is 0.4249. We'll create a graph and manually add the text using `annotate()` (Figure 5-26):

```
# First generate prediction data
pred <- predictvals(model, "ageYear", "heightIn")
sp <- ggplot(heightweight, aes(x=ageYear, y=heightIn)) + geom_point() +
    geom_line(data=pred)

sp + annotate("text", label="r^2=0.42", x=16.5, y=52)
```

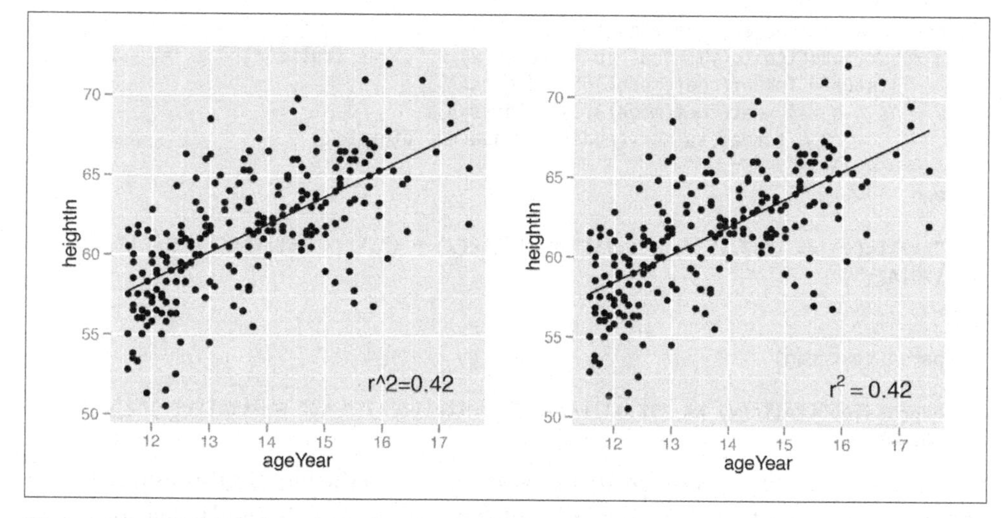

Figure 5-26. Left: plain text; right: math expression

Instead of using a plain text string, it's also possible to enter formulas using R's math expression syntax, by setting `parse=TRUE`:

```
sp + annotate("text", label="r^2 == 0.42", parse = TRUE, x=16.5, y=52)
```

Discussion

Text geoms in ggplot2 do not take expression objects directly; instead, they take character strings that are turned into expressions with `parse(text="a + b")`.

If you use a math expression, the syntax must be correct for it to be a valid R expression object. You can test validity by wrapping it in `expression()` and seeing if it throws an error (make sure *not* to use quotes around the expression). In the example here, == is a valid construct in an expression to express equality, but = is not:

```
expression(r^2 == 0.42)  # Valid

expression(r^2 == 0.42)


expression(r^2 = 0.42)    # Not valid

Error: unexpected '=' in "expression(r^2 ="
```

It's possible to automatically extract values from the model object and build an expression using those values. In this example, we'll create a string that, when parsed, returns a valid expression:

```
eqn <- as.character(as.expression(
  substitute(italic(y) == a + b * italic(x) * "," ~~ italic(r)^2 ~ "=" ~ r2,
    list(a = format(coef(model)[1], digits=3),
         b = format(coef(model)[2], digits=3),
         r2 = format(summary(model)$r.squared, digits=2)
        ))))
eqn

"italic(y) == \"37.4\" + \"1.75\" * italic(x) * \",\" ~ ~italic(r)^2 ~ \"=\" ~
\"0.42\""


parse(text=eqn)  # Parsing turns it into an expression

expression(italic(y) == "37.4" + "1.75" * italic(x) * "," ~ ~italic(r)^2 ~ "=" ~
"0.42")
```

Now that we have the expression string, we can add it to the plot. In this example we'll put the text in the bottom-right corner, by setting `x=Inf` and `y=-Inf` and using horizontal and vertical adjustments so that the text all fits inside the plotting area (Figure 5-27):

```
sp + annotate("text", label=eqn, parse=TRUE, x=Inf, y=-Inf, hjust=1.1, vjust=-.5)
```

See Also

The math expression syntax in R can be a bit tricky. See Recipe 7.2 for more information.

5.10. Adding Marginal Rugs to a Scatter Plot

Problem

You want to add marginal rugs to a scatter plot.

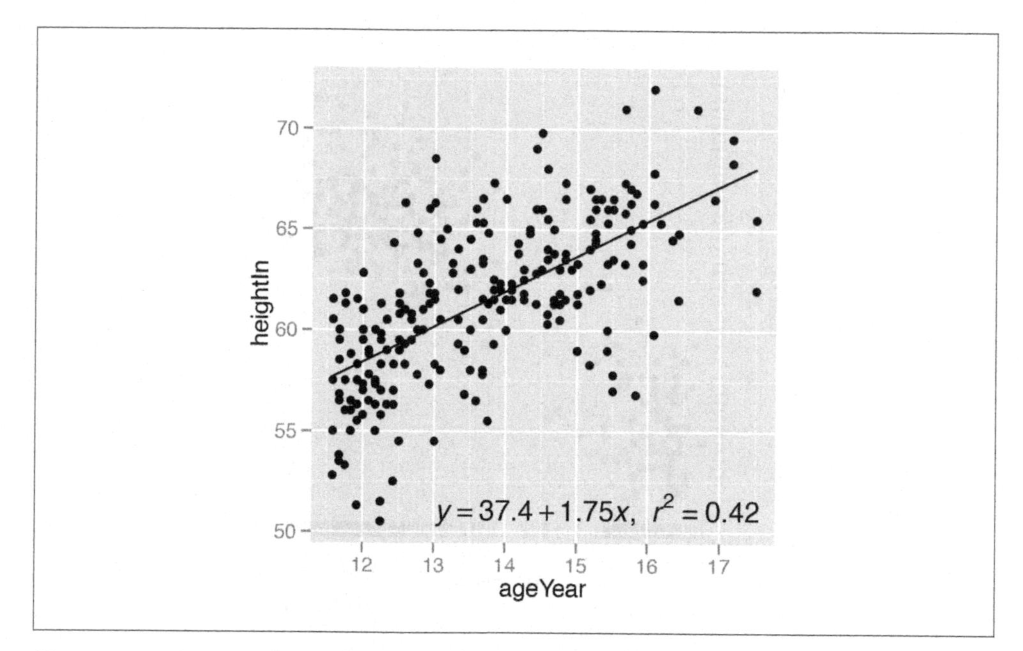

Figure 5-27. Scatter plot with automatically generated expression

Solution

Use `geom_rug()`. For this example (Figure 5-28), we'll use the `faithful` data set, which contains data about the Old Faithful geyser in two columns—`eruptions`, which is the length of each eruption, and `waiting`, which is the length of time to the next eruption:

```
ggplot(faithful, aes(x=eruptions, y=waiting)) + geom_point() + geom_rug()
```

Discussion

A marginal rug plot is essentially a one-dimensional scatter plot that can be used to visualize the distribution of data on each axis.

In this particular data set, the marginal rug is not as informative as it could be. The resolution of the `waiting` variable is in whole minutes, and because of this, the rug lines have a lot of overplotting. To reduce the overplotting, we can jitter the line positions and make them slightly thinner by specifying `size` (Figure 5-29). This helps the viewer see the distribution more clearly:

```
ggplot(faithful, aes(x=eruptions, y=waiting)) + geom_point() +
    geom_rug(position="jitter", size=.2)
```

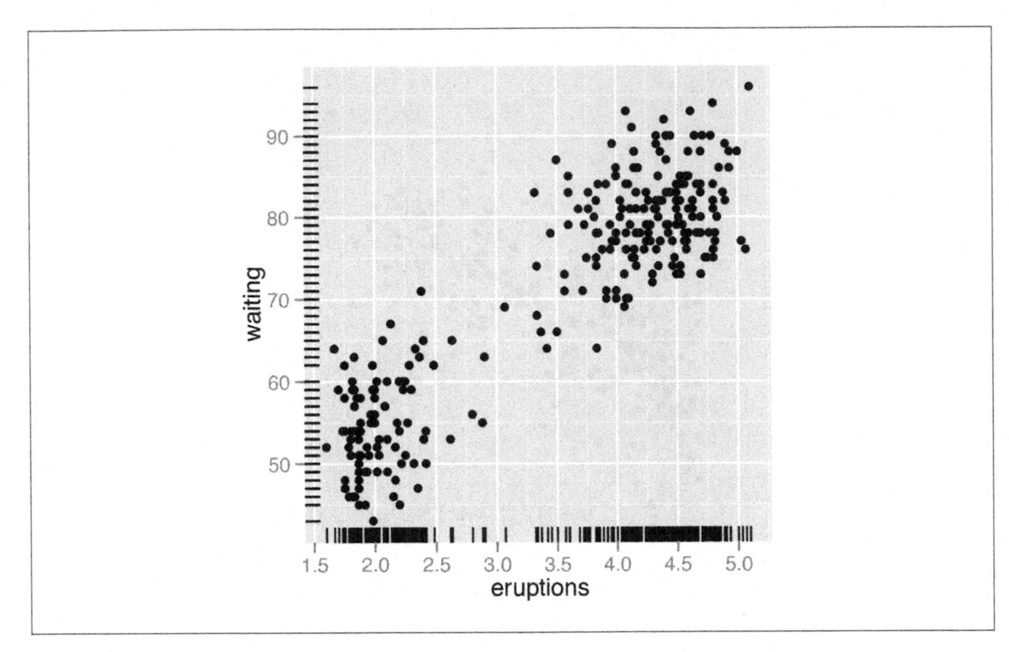

Figure 5-28. Marginal rug added to a scatter plot

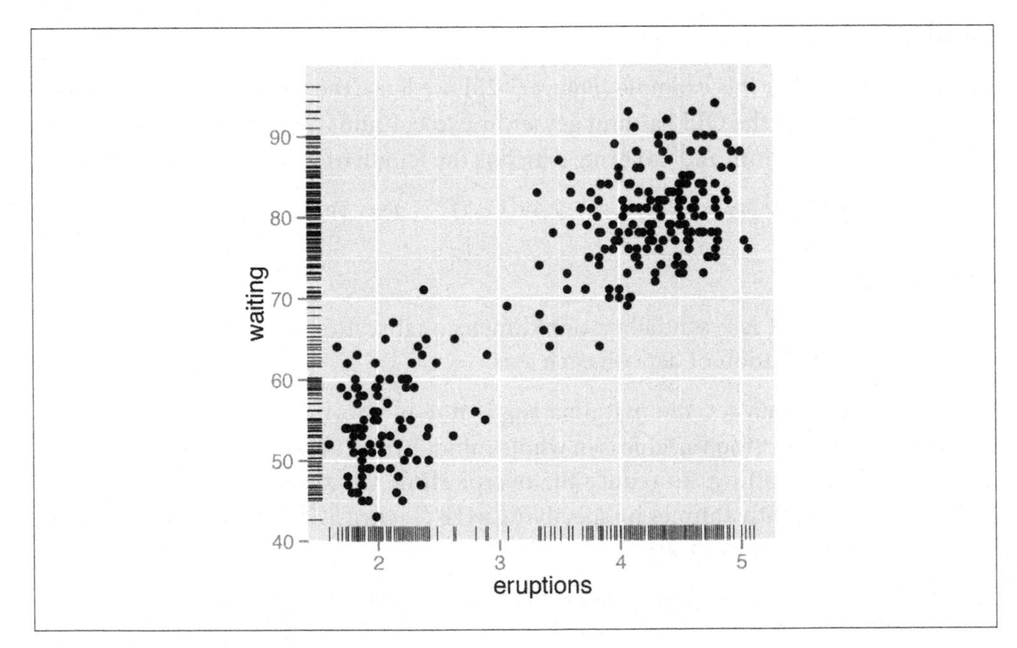

Figure 5-29. Marginal rug with thinner, jittered lines

See Also

For more about overplotting, see Recipe 5.5.

5.11. Labeling Points in a Scatter Plot

Problem

You want to add labels to points in a scatter plot.

Solution

For annotating just one or a few points, you can use `annotate()` or `geom_text()`. For this example, we'll use the `countries` data set and visualize the relationship between health expenditures and infant mortality rate per 1,000 live births. To keep things manageable, we'll just take the subset of countries that spent more than $2000 USD per capita:

```
library(gcookbook) # For the data set
subset(countries, Year==2009 & healthexp>2000)
```

```
           Name Code Year      GDP laborrate healthexp infmortality
        Andorra  AND 2009       NA        NA  3089.636          3.1
      Australia  AUS 2009 42130.82      65.2  3867.429          4.2
        Austria  AUT 2009 45555.43      60.4  5037.311          3.6
...
 United Kingdom  GBR 2009 35163.41      62.2  3285.050          4.7
  United States  USA 2009 45744.56      65.0  7410.163          6.6
```

We'll save the basic scatter plot object in `sp` and add then add things to it. To manually add annotations, use `annotate()`, and specify the coordinates and label (Figure 5-30, left). It may require some trial-and-error tweaking to get them positioned just right:

```
sp <- ggplot(subset(countries, Year==2009 & healthexp>2000),
             aes(x=healthexp, y=infmortality)) +
      geom_point()

sp + annotate("text", x=4350, y=5.4, label="Canada") +
     annotate("text", x=7400, y=6.8, label="USA")
```

To automatically add the labels from your data (Figure 5-30, right), use `geom_text()` and map a column that is a factor or character vector to the `label` aesthetic. In this case, we'll use `Name`, and we'll make the font slightly smaller to reduce crowding. The default value for `size` is 5, which doesn't correspond directly to a point size:

```
sp + geom_text(aes(label=Name), size=4)
```

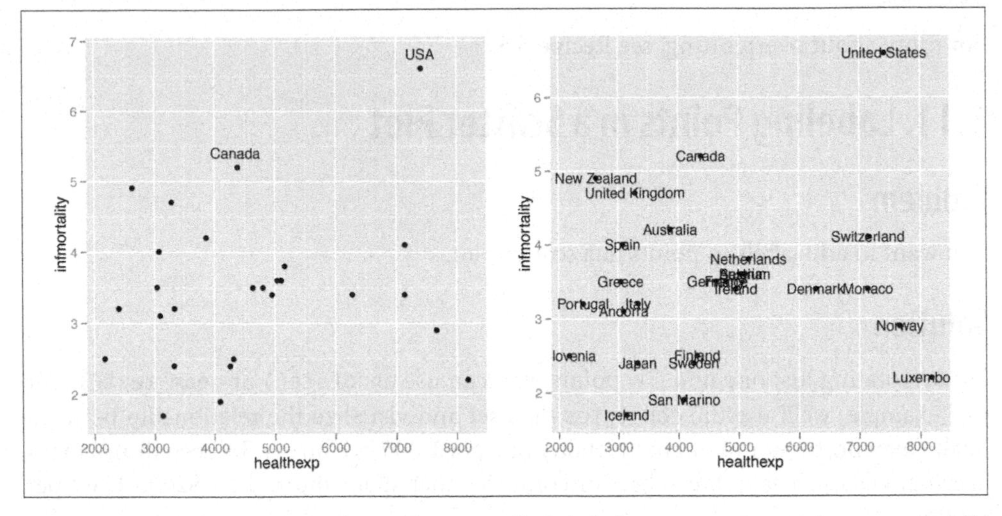

Figure 5-30. Left: a scatter plot with manually labeled points; right: with automatically labeled points and a smaller font

Discussion

The automatic method for placing annotations centers each annotation on the *x* and *y* coordinates. You'll probably want to shift the text vertically, horizontally, or both.

Setting `vjust=0` will make the baseline of the text on the same level as the point (Figure 5-31, left), and setting `vjust=1` will make the top of the text level with the point. This usually isn't enough, though—you can increase or decrease `vjust` to shift the labels higher or lower, or you can add or subtract a bit to or from the `y` mapping to get the same effect (Figure 5-31, right):

```
sp + geom_text(aes(label=Name), size=4, vjust=0)

# Add a little extra to y
sp + geom_text(aes(y=infmortality+.1, label=Name), size=4, vjust=0)
```

It often makes sense to right- or left-justify the labels relative to the points. To left-justify, set `hjust=0` (Figure 5-32, left), and to right-justify, set `hjust=1`. As was the case with `vjust`, the labels will still slightly overlap with the points. This time, though, it's not a good idea to try to fix it by increasing or decreasing `hjust`. Doing so will shift the labels a distance proportional to the length of the label, making longer labels move further than shorter ones. It's better to just set `hjust` to 0 or 1, and then add or subtract a bit to or from `x` (Figure 5-32, right):

```
sp + geom_text(aes(label=Name), size=4, hjust=0)

sp + geom_text(aes(x=healthexp+100, label=Name), size=4, hjust=0)
```

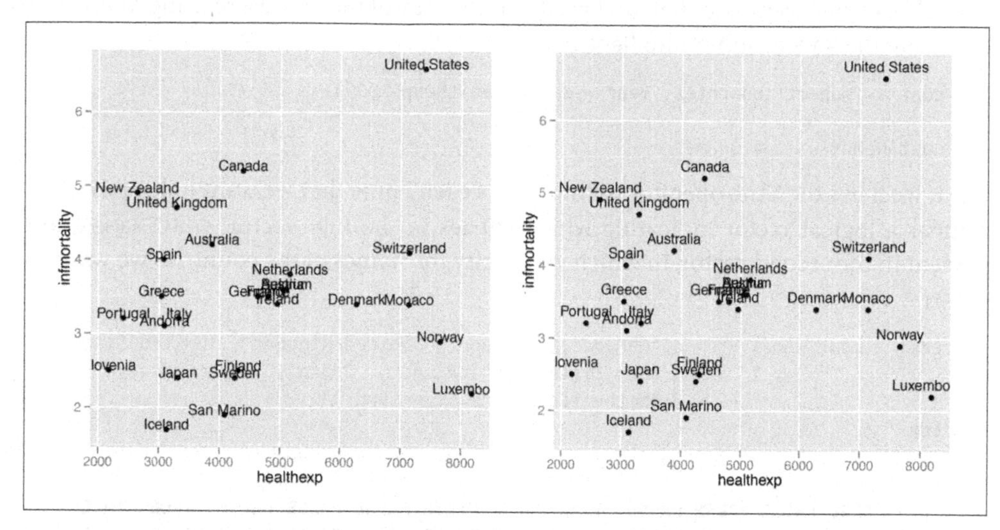

Figure 5-31. Left: a scatter plot with vjust=0; right: with a little extra added to y

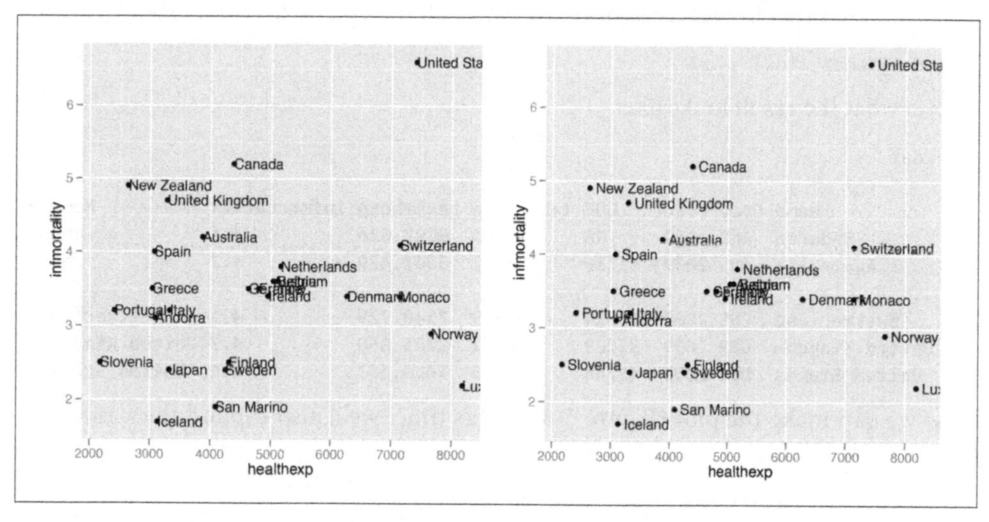

Figure 5-32. Left: a scatter plot with hjust=0; right: with a little extra added to x

If you are using a logarithmic axis, instead of adding to x or y, you'll need to *multiply* the x or y value by a number to shift the labels a consistent amount.

If you want to label just some of the points but want the placement to be handled automatically, you can add a new column to your data frame containing just the labels you want. Here's one way to do that: first we'll make a copy of the data we're using, then we'll duplicate the `Name` column into `Name1`:

```
cdat <- subset(countries, Year==2009 & healthexp>2000)

cdat$Name1 <- cdat$Name
```

Next, we'll use the `%in%` operator to find *where* each name that we want to keep is. This returns a logical vector indicating which entries in the first vector, `cdat$Name1`, are present in the second vector, in which we specify the names of the countries we want to show:

```
idx <- cdat$Name1 %in% c("Canada", "Ireland", "United Kingdom", "United States",
                         "New Zealand", "Iceland", "Japan", "Luxembourg",
                         "Netherlands", "Switzerland")
idx


 [1] FALSE FALSE FALSE FALSE  TRUE FALSE FALSE FALSE FALSE FALSE  TRUE  TRUE
[13] FALSE  TRUE  TRUE FALSE  TRUE  TRUE FALSE FALSE FALSE FALSE FALSE FALSE
[25]  TRUE  TRUE  TRUE
```

Then we'll use that Boolean vector to overwrite all the *other* entries in `Name1` with `NA`:

```
cdat$Name1[!idx] <- NA
```

This is what the result looks like:

```
cdat

           Name Code Year   GDP laborrate healthexp infmortality          Name1
        Andorra  AND 2009    NA        NA  3089.636          3.1           <NA>
      Australia  AUS 2009 42130      65.2  3867.429          4.2           <NA>
...
    Switzerland  CHE 2009 63524      66.9  7140.729          4.1    Switzerland
 United Kingdom  GBR 2009 35163      62.2  3285.050          4.7 United Kingdom
  United States  USA 2009 45744      65.0  7410.163          6.6  United States
```

Now we can make the plot (Figure 5-33). This time, we'll also expand the *x* range so that the text will fit:

```
ggplot(cdat, aes(x=healthexp, y=infmortality)) +
    geom_point() +
    geom_text(aes(x=healthexp+100, label=Name1), size=4, hjust=0) +
    xlim(2000, 10000)
```

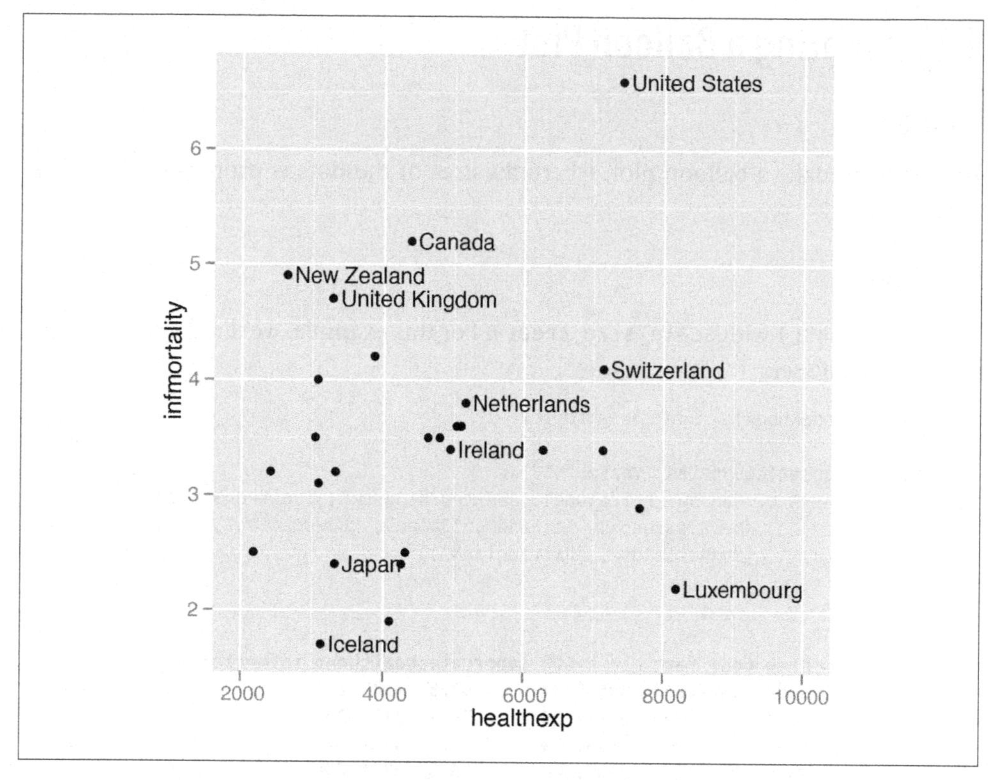

Figure 5-33. Scatter plot with selected labels and expanded x range

If any individual position adjustments are needed, you have a couple of options. One option is to copy the columns used for the *x* and *y* coordinates and modify the numbers for the individual items to move the text around. Make sure to use the original numbers for the coordinates of the points, of course! Another option is to save the output to a vector format such as PDF or SVG (see Recipes 14.1 and 14.2), then edit it in a program like Illustrator or Inkscape.

See Also

For more on controlling the appearance of the text, see Recipe 9.2.

If you want to manually edit a PDF or SVG file, see Recipe 14.4.

5.12. Creating a Balloon Plot

Problem

You want to make a balloon plot, where the area of the dots is proportional to their numerical value.

Solution

Use `geom_point()` with `scale_size_area()`. For this example, we'll use a subset of the `countries` data set:

```
library(gcookbook) # For the data set

cdat <- subset(countries, Year==2009 &
    Name %in% c("Canada", "Ireland", "United Kingdom", "United States",
                "New Zealand", "Iceland", "Japan", "Luxembourg",
                "Netherlands", "Switzerland"))

cdat
```

```
           Name Code Year       GDP laborrate healthexp infmortality
         Canada  CAN 2009  39599.04      67.8  4379.761          5.2
        Iceland  ISL 2009  37972.24      77.5  3130.391          1.7
        Ireland  IRL 2009  49737.93      63.6  4951.845          3.4
          Japan  JPN 2009  39456.44      59.5  3321.466          2.4
     Luxembourg  LUX 2009 106252.24      55.5  8182.855          2.2
    Netherlands  NLD 2009  48068.35      66.1  5163.740          3.8
    New Zealand  NZL 2009  29352.45      68.6  2633.625          4.9
    Switzerland  CHE 2009  63524.65      66.9  7140.729          4.1
 United Kingdom  GBR 2009  35163.41      62.2  3285.050          4.7
  United States  USA 2009  45744.56      65.0  7410.163          6.6
```

If we just map `GDP` to `size`, the value of `GDP` gets mapped to the *radius* of the dots (Figure 5-34, left), which is not what we want; a doubling of value results in a quadrupling of area, and this will distort the interpretation of the data. We instead want to map it to the *area*, and we can do this using `scale_size_area()` (Figure 5-34, right):

```
p <- ggplot(cdat, aes(x=healthexp, y=infmortality, size=GDP)) +
    geom_point(shape=21, colour="black", fill="cornsilk")

# GDP mapped to radius (default with scale_size_continuous)
p

# GDP mapped to area instead, and larger circles
p + scale_size_area(max_size=15)
```

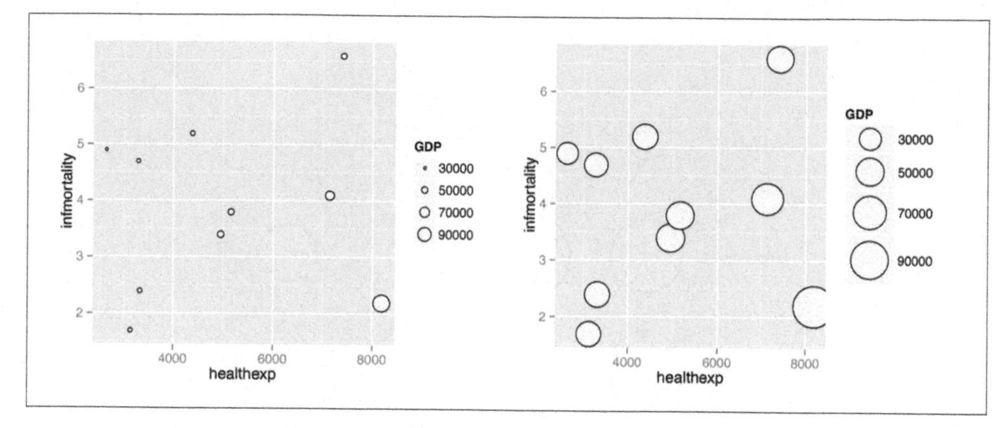

Figure 5-34. Left: balloon plot with value mapped to radius; right: with value mapped to area

Discussion

The example here is a scatter plot, but that is not the only way to use balloon plots. It may also be useful to use them to represent values on a grid, where the x- and y-axes are categorical, as in Figure 5-35:

```
# Add up counts for male and female
hec <- HairEyeColor[,,"Male"] + HairEyeColor[,,"Female"]

# Convert to long format
library(reshape2)
hec <- melt(hec, value.name="count")

ggplot(hec, aes(x=Eye, y=Hair)) +
    geom_point(aes(size=count), shape=21, colour="black", fill="cornsilk") +
    scale_size_area(max_size=20, guide=FALSE) +
    geom_text(aes(y=as.numeric(Hair)-sqrt(count)/22, label=count), vjust=1,
              colour="grey60", size=4)
```

In this example we've used a few tricks to add the text labels under the circles. First, we used `vjust=1` to top-justify the text to the *y* coordinate. Next, we wanted to set the *y* coordinate so that it is just underneath the bottom of each circle. This requires a little arithmetic: take the *numeric* value of `Hair` and subtract a small value from it, where the value depends in some way on `count`. This actually requires taking the square root of `count`, since the radius has a linear relationship with the square root of `count`. The number that this value divided by (22 in this case) is found by trial and error; it depends on the particular data values, radius, and text size.

The text under the circles is in a shade of grey. This is so that it doesn't jump out at the viewer and overwhelm the perceptual impact of the circles, but is still available if the viewer wants to know the exact values.

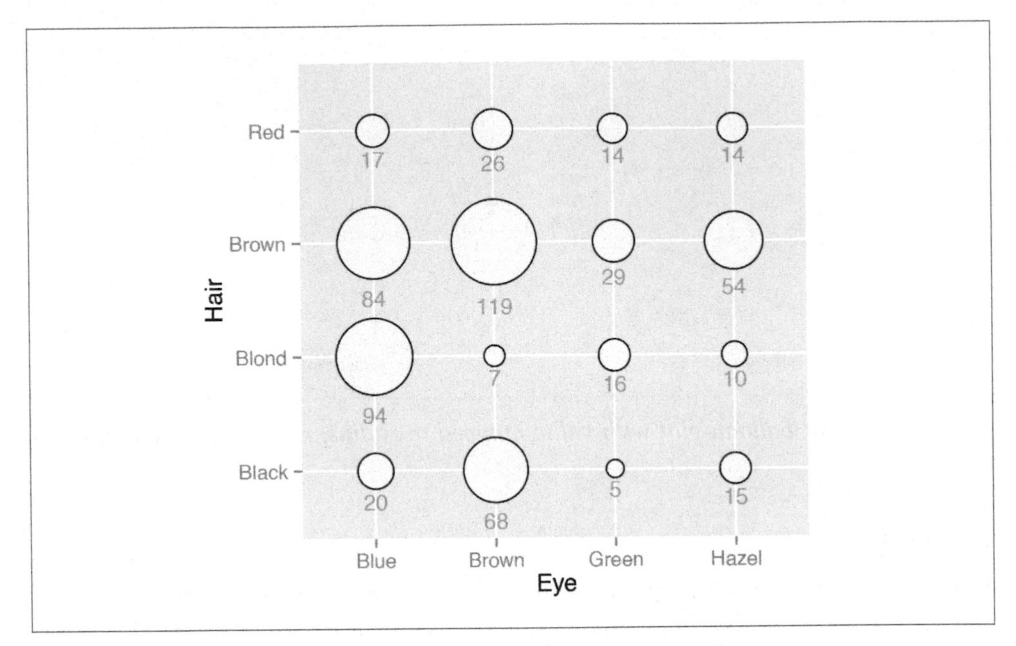

Figure 5-35. Balloon plot with categorical axes and text labels

See Also

To add labels to the circles, see Recipes 5.11 and 7.1.

See Recipe 5.4 for ways of mapping variables to other aesthetics in a scatter plot.

5.13. Making a Scatter Plot Matrix

Problem

You want to make a scatter plot matrix.

Solution

A scatter plot matrix is an excellent way of visualizing the pairwise relationships among several variables. To make one, use the `pairs()` function from R's base graphics.

For this example, we'll use a subset of the `countries` data set. We'll pull out the data for the year 2009, and keep only the columns that are relevant:

```
library(gcookbook) # For the data set
c2009 <- subset(countries, Year==2009,
                select=c(Name, GDP, laborrate, healthexp, infmortality))
```

```
c2009

         Name       GDP laborrate  healthexp infmortality
Afghanistan        NA      59.8   50.88597        103.2
    Albania 3772.6047      59.5  264.60406         17.2
    Algeria 4022.1989      58.5  267.94653         32.0
...
     Zambia 1006.3882      69.2   47.05637         71.5
   Zimbabwe  467.8534      66.8         NA         52.2
```

To make the scatter plot matrix (Figure 5-36), we'll use columns 2 through 5—using the `Name` column wouldn't make sense, and it would produce strange-looking results:

```
pairs(c2009[,2:5])
```

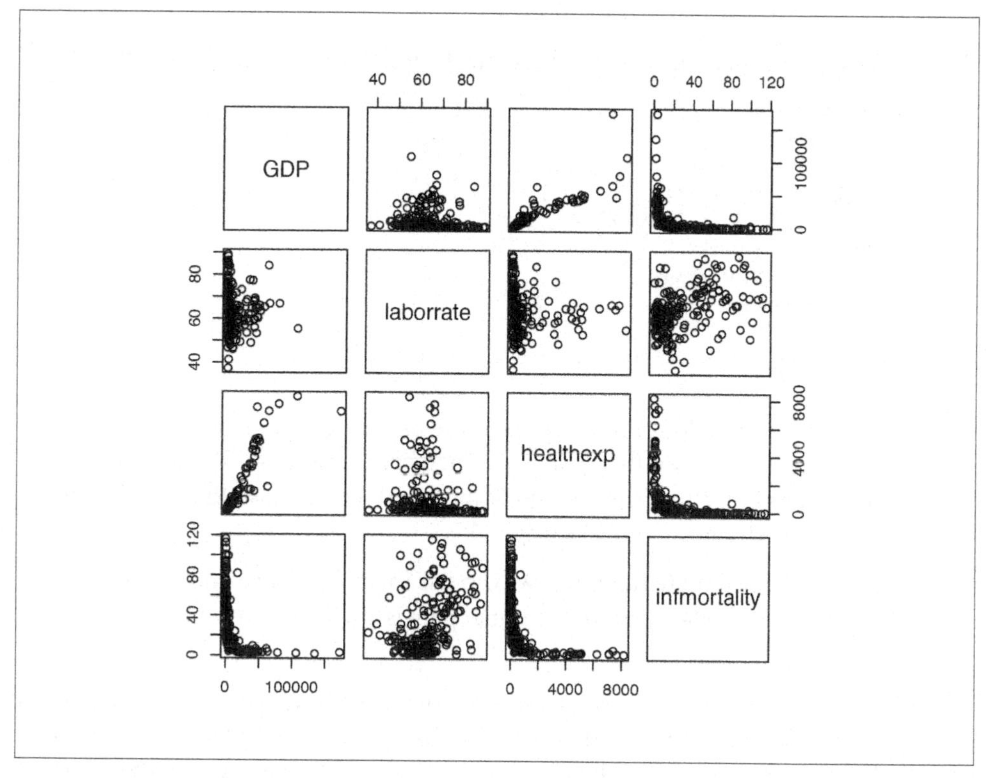

Figure 5-36. A scatter plot matrix

Discussion

We didn't use ggplot2 here because it doesn't make scatter plot matrices (at least, not well).

You can also use customized functions for the panels. To show the correlation coefficient of each pair of variables instead of a scatter plot, we'll define the function `panel.cor`. This will also show higher correlations in a larger font. Don't worry about the details for now—just paste this code into your R session or script:

```
panel.cor <- function(x, y, digits=2, prefix="", cex.cor, ...) {
    usr <- par("usr")
    on.exit(par(usr))
    par(usr = c(0, 1, 0, 1))
    r <- abs(cor(x, y, use="complete.obs"))
    txt <- format(c(r, 0.123456789), digits=digits)[1]
    txt <- paste(prefix, txt, sep="")
    if(missing(cex.cor)) cex.cor <- 0.8/strwidth(txt)
    text(0.5, 0.5, txt, cex =  cex.cor * (1 + r) / 2)
}
```

To show histograms of each variable along the diagonal, we'll define `panel.hist`:

```
panel.hist <- function(x, ...) {
    usr <- par("usr")
    on.exit(par(usr))
    par(usr = c(usr[1:2], 0, 1.5) )
    h <- hist(x, plot = FALSE)
    breaks <- h$breaks
    nB <- length(breaks)
    y <- h$counts
    y <- y/max(y)
    rect(breaks[-nB], 0, breaks[-1], y, col="white", ...)
}
```

Both of these panel functions are taken from the `pairs` help page, so if it's more convenient, you can simply open that help page, then copy and paste. The last line of this version of the `panel.cor` function is slightly modified, however, so that the changes in font size aren't as extreme as with the original.

Now that we've defined these functions we can use them for our scatter plot matrix, by telling `pairs()` to use `panel.cor` for the upper panels and `panel.hist` for the diagonal panels.

We'll also throw in one more thing: `panel.smooth` for the lower panels, which makes a scatter plot and adds a LOWESS smoothed line, as shown in Figure 5-37. (LOWESS is slightly different from LOESS, which we saw in Recipe 5.6, but the differences aren't important for this sort of rough exploratory visualization):

```
pairs(c2009[,2:5], upper.panel = panel.cor,
                   diag.panel  = panel.hist,
                   lower.panel = panel.smooth)
```

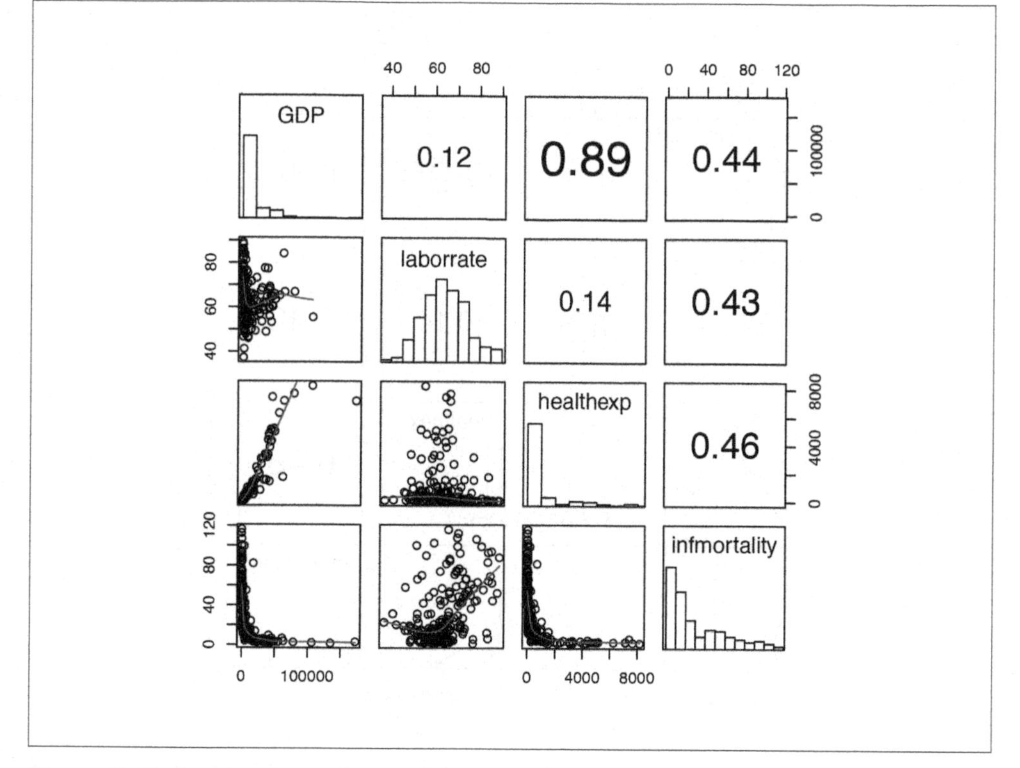

Figure 5-37. Scatter plot with correlations in the upper triangle, smoothing lines in the lower triangle, and histograms on the diagonal

It may be more desirable to use linear regression lines instead of LOWESS lines. The `panel.lm` function will do the trick (unlike the previous panel functions, this one isn't in the `pairs` help page):

```
panel.lm <- function (x, y, col = par("col"), bg = NA, pch = par("pch"),
                      cex = 1, col.smooth = "black", ...) {
    points(x, y, pch = pch, col = col, bg = bg, cex = cex)
    abline(stats::lm(y ~ x),  col = col.smooth, ...)
}
```

This time the default line color is black instead of red, though you can change it here (and with `panel.smooth`) by setting `col.smooth` when you call `pairs()`.

We'll also use small points in the visualization, so that we can distinguish them a bit better (Figure 5-38). This is done by setting `pch="."`:

```
pairs(c2009[,2:5], pch=".",
                   upper.panel = panel.cor,
                   diag.panel  = panel.hist,
                   lower.panel = panel.lm)
```

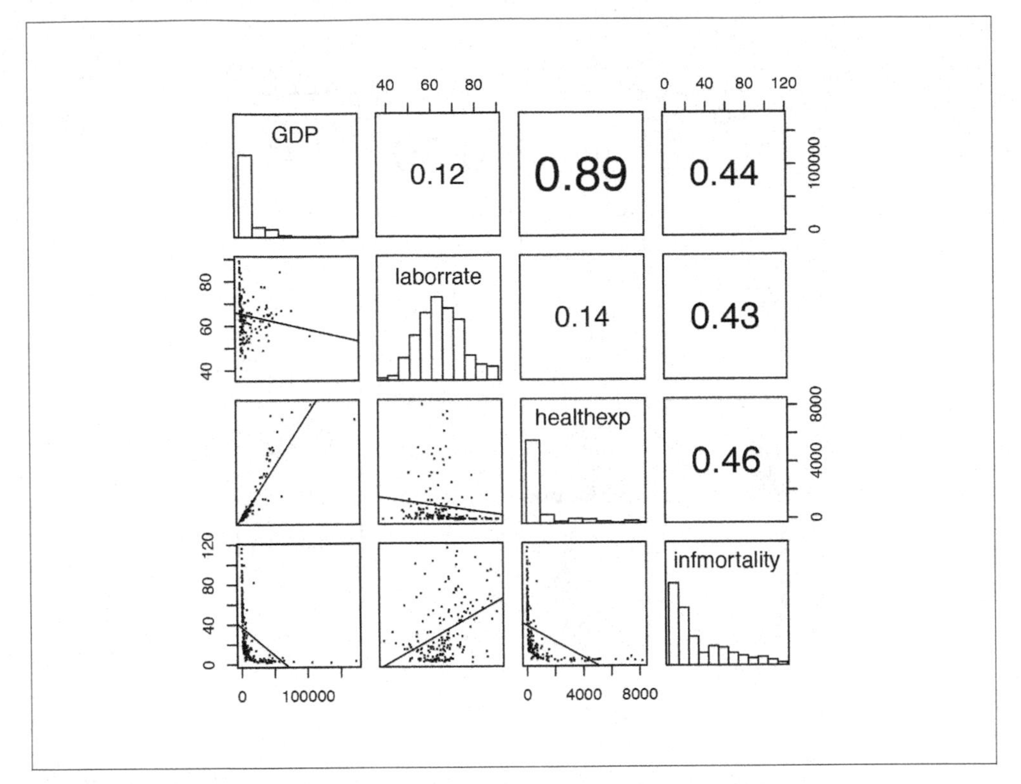

Figure 5-38. Scatter plot matrix with smaller points and linear fit lines

The size of the points can also be controlled using the `cex` parameter. The default value for `cex` is 1; make it smaller for smaller points and larger for larger points. Values below .5 might not render properly with PDF output.

See Also

To create a correlation matrix, see Recipe 13.1.

The `ggpairs()` function from the GGally package can also make scatter plot matrices.

CHAPTER 6
Summarized Data Distributions

This chapter explores how to visualize summarized distributions of data.

6.1. Making a Basic Histogram

Problem

You want to make a histogram.

Solution

Use `geom_histogram()` and map a continuous variable to `x` (Figure 6-1):

```
ggplot(faithful, aes(x=waiting)) + geom_histogram()
```

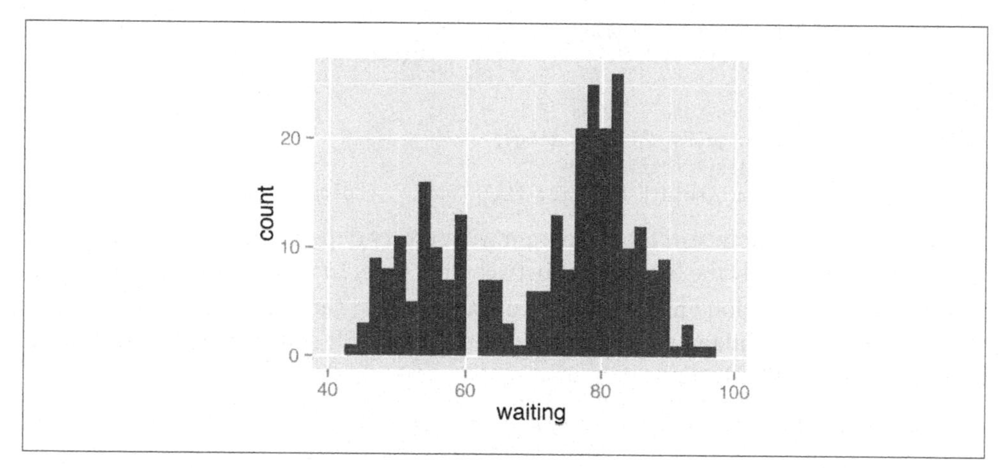

Figure 6-1. A basic histogram

Discussion

All geom_histogram() requires is one column from a data frame or a single vector of data. For this example we'll use the faithful data set, which contains data about the Old Faithful geyser in two columns: eruptions, which is the length of each eruption, and waiting, which is the length of time to the next eruption. We'll only use the waiting column in this example:

```
faithful

eruptions waiting
    3.600      79
    1.800      54
    3.333      74
...
```

If you just want to get a quick look at some data that isn't in a data frame, you can get the same result by passing in NULL for the data frame and giving ggplot() a vector of values. This would have the same result as the previous code:

```
# Store the values in a simple vector
w <- faithful$waiting

ggplot(NULL, aes(x=w)) + geom_histogram()
```

By default, the data is grouped into 30 bins. This may be too fine or too coarse for your data. You can change the size of the bins by using binwidth, or you can divide the range of the data into a specific number of bins. The default colors—a dark fill without an outline—can make it difficult to see which bar corresponds to which value, so we'll also change the colors, as shown in Figure 6-2.

```
# Set the width of each bin to 5
ggplot(faithful, aes(x=waiting)) +
    geom_histogram(binwidth=5, fill="white", colour="black")

# Divide the x range into 15 bins
binsize <- diff(range(faithful$waiting))/15
ggplot(faithful, aes(x=waiting)) +
    geom_histogram(binwidth=binsize, fill="white", colour="black")
```

Sometimes the appearance of the histogram will be very dependent on the width of the bins and where exactly the boundaries between bins are. In Figure 6-3, we'll use a bin width of 8. In the version on the left, we'll use the origin parameter to put boundaries at 31, 39, 47, etc., while in the version on the right, we'll shift it over by 4, putting boundaries at 35, 43, 51, etc.:

```
h <- ggplot(faithful, aes(x=waiting))  # Save the base object for reuse

h + geom_histogram(binwidth=8, fill="white", colour="black", origin=31)
```

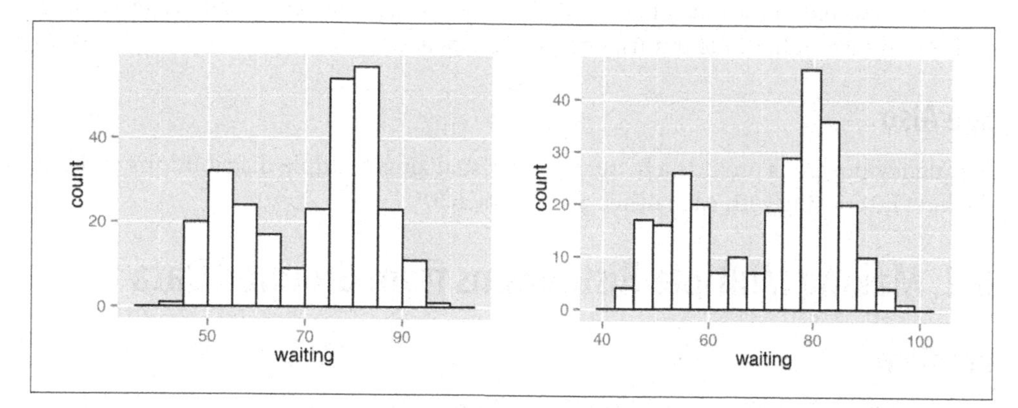

Figure 6-2. Left: histogram with binwidth=5, and with different colors; right: with 15 bins

```
h + geom_histogram(binwidth=8, fill="white", colour="black", origin=35)
```

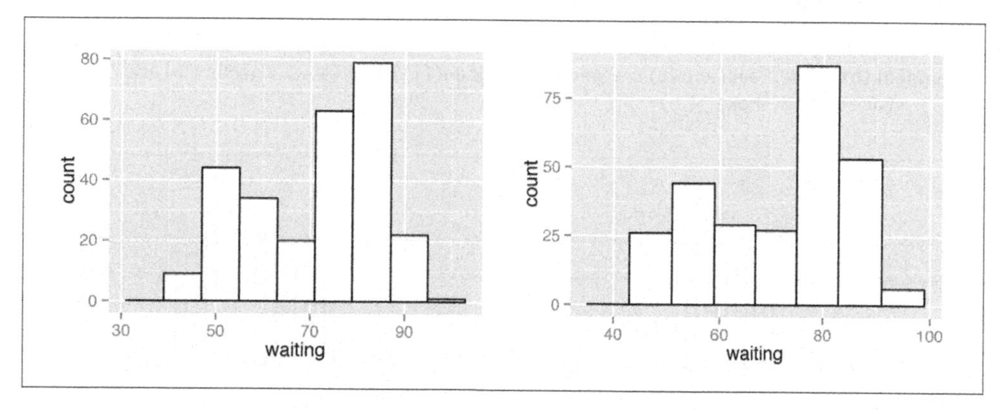

Figure 6-3. Different appearance of histograms with the origin at 31 and 35

The results look quite different, even though they have the same bin size. The `faithful` data set is not particularly small, with 272 observations; with smaller data sets, this is even more of an issue. When visualizing your data, it's a good idea to experiment with different bin sizes and boundary points.

Also, if your data has discrete values, it may matter that the histogram bins are asymmetrical. They are *closed* on the lower bound and *open* on the upper bound. If you have bin boundaries at 1, 2, 3, etc., then the bins will be [1, 2), [2, 3), and so on. In other words, the first bin contains 1 but not 2, and the second bin contains 2 but not 3.

It is also possible to use `geom_bar(stat="bin")` for the same effect, although I find it easier to interpret the code if it uses `geom_histogram()`.

See Also

Frequency polygons provide a better way of visualizing multiple distributions without the bars interfering with each other. See Recipe 6.5.

6.2. Making Multiple Histograms from Grouped Data

Problem

You want to make histograms of multiple groups of data.

Solution

Use `geom_histogram()` and use facets for each group, as shown in Figure 6-4:

```
library(MASS) # For the data set

# Use smoke as the faceting variable
ggplot(birthwt, aes(x=bwt)) + geom_histogram(fill="white", colour="black") +
    facet_grid(smoke ~ .)
```

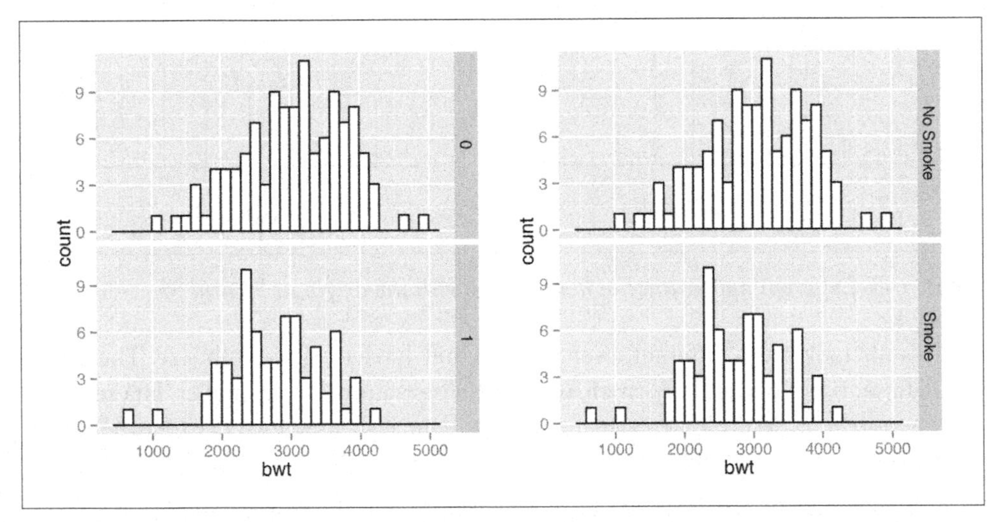

Figure 6-4. Left: two histograms with facets; right: with different facet labels

Discussion

To make these plots, the data must all be in one data frame, with one column containing a categorical variable used for grouping.

For this example, we used the `birthwt` data set. It contains data about birth weights and a number of risk factors for low birth weight:

```
birthwt

low age lwt race smoke ptl ht ui ftv  bwt
  0  19 182    2     0   0  0  1   0 2523
  0  33 155    3     0   0  0  0   3 2551
  0  20 105    1     1   0  0  0   1 2557
...
```

One problem with the faceted graph is that the facet labels are just 0 and 1, and there's no label indicating that those values are for `smoke`. To change the labels, we need to change the names of the factor levels. First we'll take a look at the factor levels, then we'll assign new factor level names, in the same order:

```
birthwt1 <- birthwt # Make a copy of the data

# Convert smoke to a factor
birthwt1$smoke <- factor(birthwt1$smoke)
levels(birthwt1$smoke)

 "0" "1"


library(plyr) # For the revalue() function
birthwt1$smoke <- revalue(birthwt1$smoke, c("0"="No Smoke", "1"="Smoke"))
```

Now when we plot it again, it shows the new labels (Figure 6-4, right).

```
ggplot(birthwt1, aes(x=bwt)) + geom_histogram(fill="white", colour="black") +
    facet_grid(smoke ~ .)
```

With facets, the axes have the same *y* scaling in each facet. If your groups have different sizes, it might be hard to compare the *shapes* of the distributions of each one. For example, see what happens when we facet the birth weights by `race` (Figure 6-5, left):

```
ggplot(birthwt, aes(x=bwt)) + geom_histogram(fill="white", colour="black") +
    facet_grid(race ~ .)
```

To allow the *y* scales to be resized independently (Figure 6-5, right), use `scales="free"`. Note that this will only allow the *y* scales to be free—the *x* scales will still be fixed because the histograms are aligned with respect to that axis:

```
ggplot(birthwt, aes(x=bwt)) + geom_histogram(fill="white", colour="black") +
    facet_grid(race ~ ., scales="free")
```

Another approach is to map the grouping variable to `fill`, as shown in Figure 6-6. The grouping variable must be a factor or character vector. In the `birthwt` data set, the desired grouping variable, `smoke`, is stored as a number, so we'll use the `birthwt1` data set we created above, in which smoke is a factor:

```
# Convert smoke to a factor
birthwt1$smoke <- factor(birthwt1$smoke)

# Map smoke to fill, make the bars NOT stacked, and make them semitransparent
ggplot(birthwt1, aes(x=bwt, fill=smoke)) +
    geom_histogram(position="identity", alpha=0.4)
```

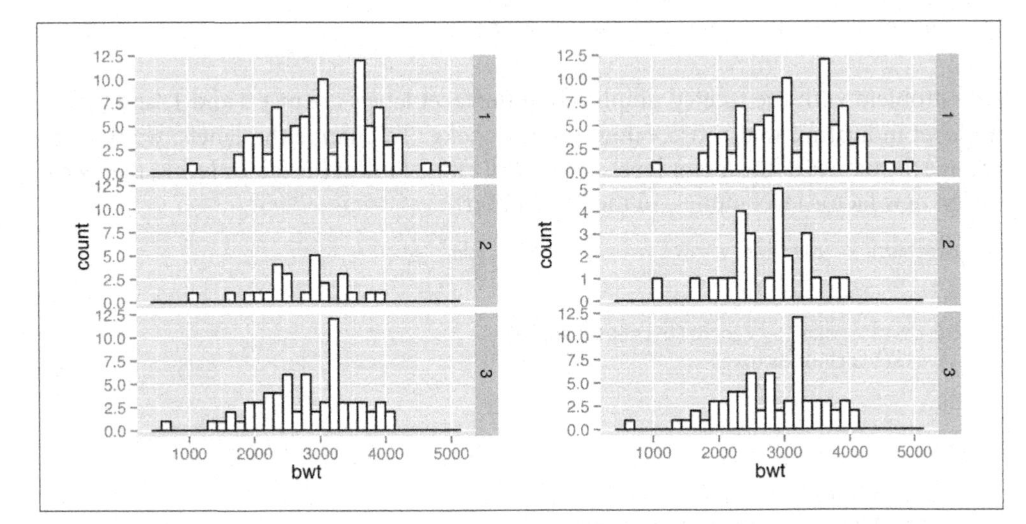

Figure 6-5. Left: histograms with the default fixed scales; right: with scales="free"

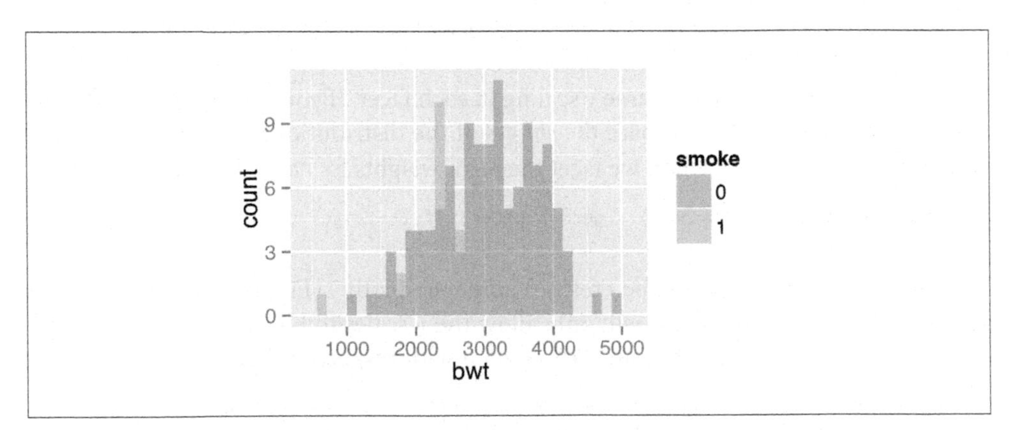

Figure 6-6. Multiple histograms with different fill colors

The position="identity" is important. Without it, ggplot() will stack the histogram bars on top of each other vertically, making it much more difficult to see the distribution of each group.

6.3. Making a Density Curve

Problem

You want to make a kernel density curve.

Solution

Use geom_density() and map a continuous variable to x (Figure 6-7):

```
ggplot(faithful, aes(x=waiting)) + geom_density()
```

If you don't like the lines along the side and bottom, you can use geom_line(stat="density") (see Figure 6-7, right):

```
# The expand_limits() increases the y range to include the value 0
ggplot(faithful, aes(x=waiting)) + geom_line(stat="density") +
    expand_limits(y=0)
```

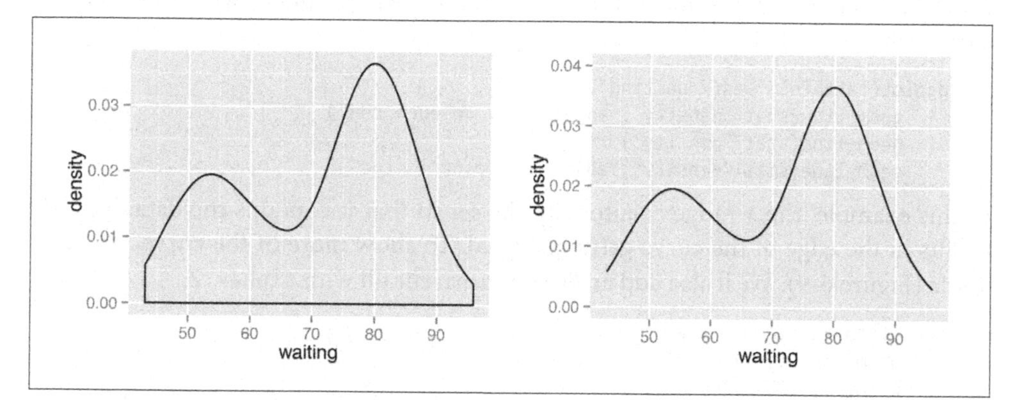

Figure 6-7. Left: a kernel density estimate curve with geom_density(); right: with geom_line()

Discussion

Like geom_histogram(), geom_density() requires just one column from a data frame. For this example, we'll use the faithful data set, which contains data about the Old Faithful geyser in two columns: eruptions, which is the length of each eruption, and waiting, which is the length of time to the next eruption. We'll only use the waiting column in this example:

```
faithful

eruptions waiting
    3.600      79
    1.800      54
    3.333      74
...
```

The second method mentioned earlier uses `geom_line()` and tells it to use the `"density"` statistical transformation. This is essentially the same as the first method, using `geom_density()`, except the former draws it with a closed polygon.

As with `geom_histogram()`, if you just want to get a quick look at data that isn't in a data frame, you can get the same result by passing in `NULL` for the data frame and giving `ggplot()` a vector of values. This would have the same result as the first solution:

```
# Store the values in a simple vector
w <- faithful$waiting

ggplot(NULL, aes(x=w)) + geom_density()
```

A kernel density curve is an estimate of the population distribution, based on the sample data. The amount of smoothing depends on the *kernel bandwidth*: the larger the bandwidth, the more smoothing there is. The bandwidth can be set with the `adjust` parameter, which has a default value of 1. Figure 6-8 shows what happens with a smaller and larger value of `adjust`:

```
ggplot(faithful, aes(x=waiting)) +
    geom_line(stat="density", adjust=.25, colour="red") +
    geom_line(stat="density") +
    geom_line(stat="density", adjust=2, colour="blue")
```

In this example, the *x* range is automatically set so that it contains the data, but this results in the edge of the curve getting clipped. To show more of the curve, set the *x* limits (Figure 6-9). We'll also add an 80% transparent fill with `alpha=.2`:

```
ggplot(faithful, aes(x=waiting)) +
    geom_density(fill="blue", alpha=.2) +
    xlim(35, 105)

# This draws a blue polygon with geom_density(), then adds a line on top
ggplot(faithful, aes(x=waiting)) +
    geom_density(fill="blue", colour=NA, alpha=.2) +
    geom_line(stat="density") +
    xlim(35, 105)
```

If this edge-clipping happens with your data, it might mean that your curve is too smooth—if the curve is wider than your data, it might not be the best model of your data. Or it could be because you have a small data set.

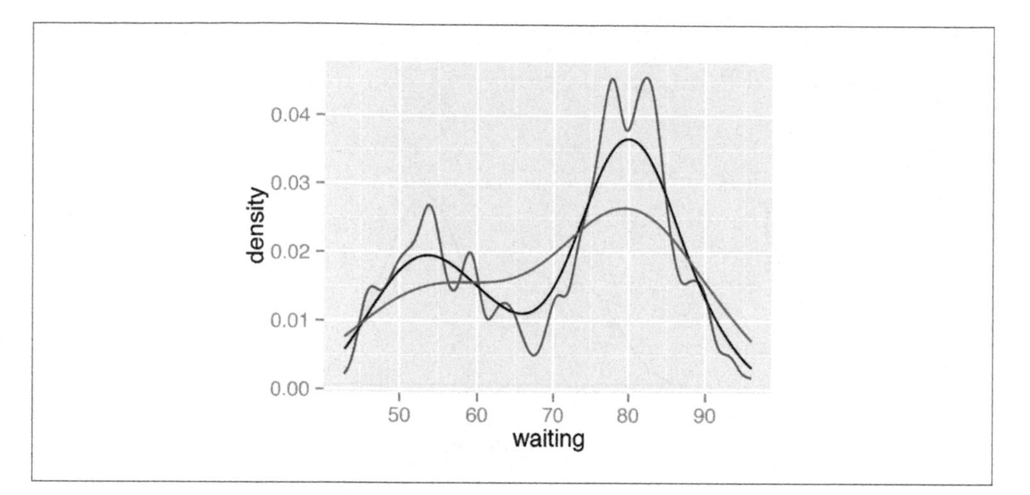

Figure 6-8. Density curves with adjust set to .25 (red), default value of 1 (black), and 2 (blue)

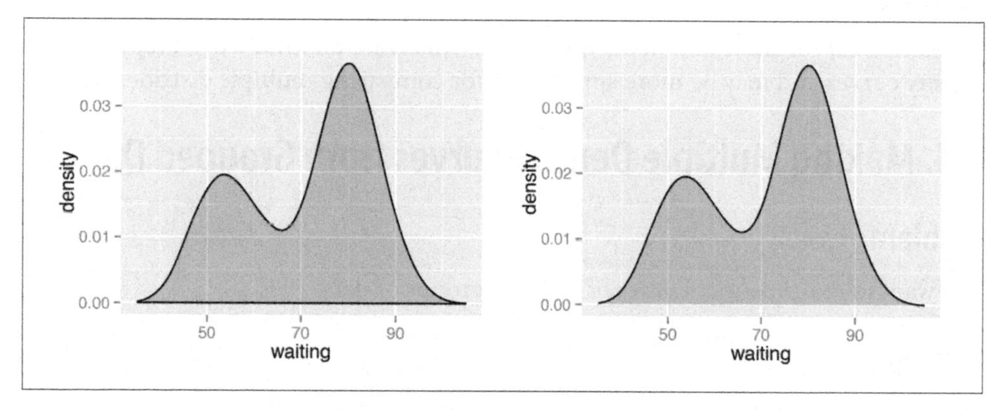

Figure 6-9. Left: density curve with wider x limits and a semitransparent fill; right: in two parts, with geom_density() and geom_line()

To compare the theoretical and observed distributions, you can overlay the density curve with the histogram. Since the *y* values for the density curve are small (the area under the curve always sums to 1), it would be barely visible if you overlaid it on a histogram without any transformation. To solve this problem, you can scale down the histogram to match the density curve with the mapping `y=..density..`. Here we'll add `geom_histogram()` first, and then layer `geom_density()` on top (Figure 6-10):

```
ggplot(faithful, aes(x=waiting, y=..density..)) +
    geom_histogram(fill="cornsilk", colour="grey60", size=.2) +
    geom_density() +
    xlim(35, 105)
```

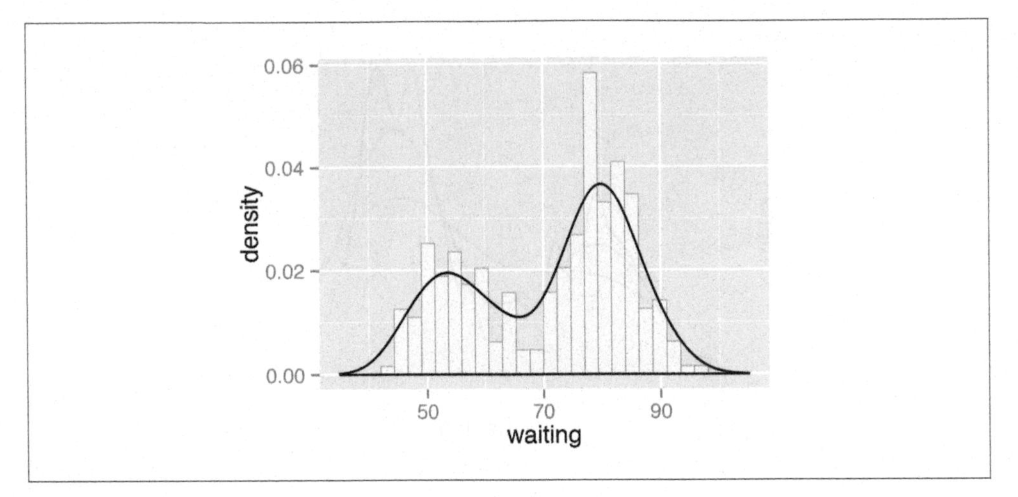

Figure 6-10. Density curve overlaid on a histogram

See Also

See Recipe 6.9 for information on violin plots, which are another way of representing density curves and may be more appropriate for comparing multiple distributions.

6.4. Making Multiple Density Curves from Grouped Data

Problem

You want to make density curves of multiple groups of data.

Solution

Use `geom_density()`, and map the grouping variable to an aesthetic like `colour` or `fill`, as shown in Figure 6-11. The grouping variable must be a factor or character vector. In the `birthwt` data set, the desired grouping variable, `smoke`, is stored as a number, so we have to convert it to a factor first:

```
library(MASS) # For the data set
# Make a copy of the data
birthwt1 <- birthwt

# Convert smoke to a factor
birthwt1$smoke <- factor(birthwt1$smoke)

# Map smoke to colour
ggplot(birthwt1, aes(x=bwt, colour=smoke)) + geom_density()
```

```
# Map smoke to fill and make the fill semitransparent by setting alpha
ggplot(birthwt1, aes(x=bwt, fill=smoke)) + geom_density(alpha=.3)
```

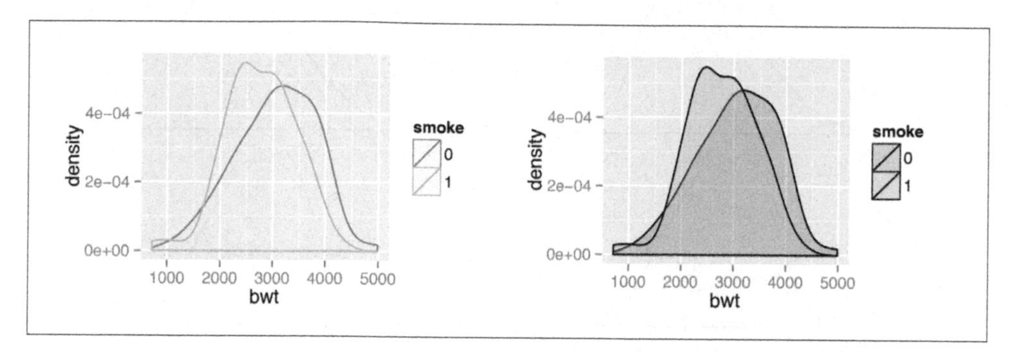

Figure 6-11. Left: different line colors for each group; right: different semitransparent fill colors for each group

Discussion

To make these plots, the data must all be in one data frame, with one column containing a categorical variable used for grouping.

For this example, we used the `birthwt` data set. It contains data about birth weights and a number of risk factors for low birth weight:

```
birthwt

low age lwt race smoke ptl ht ui ftv  bwt
  0  19 182    2     0   0  0  1   0 2523
  0  33 155    3     0   0  0  0   3 2551
  0  20 105    1     1   0  0  0   1 2557
...
```

We looked at the relationship between `smoke` (smoking) and `bwt` (birth weight in grams). The value of `smoke` is either 0 or 1, but since it's stored as a numeric vector, `ggplot()` doesn't know that it should be treated as a categorical variable. To make it so `ggplot()` knows to treat `smoke` as categorical, we can either convert that column of the data frame to a factor, or tell `ggplot()` to treat it as a factor by using `factor(smoke)` inside of the `aes()` statement. For these examples, we converted it to a factor in the data.

Another method for visualizing the distributions is to use facets, as shown in Figure 6-12. We can align the facets vertically or horizontally. Here we'll align them vertically so that it's easy to compare the two distributions:

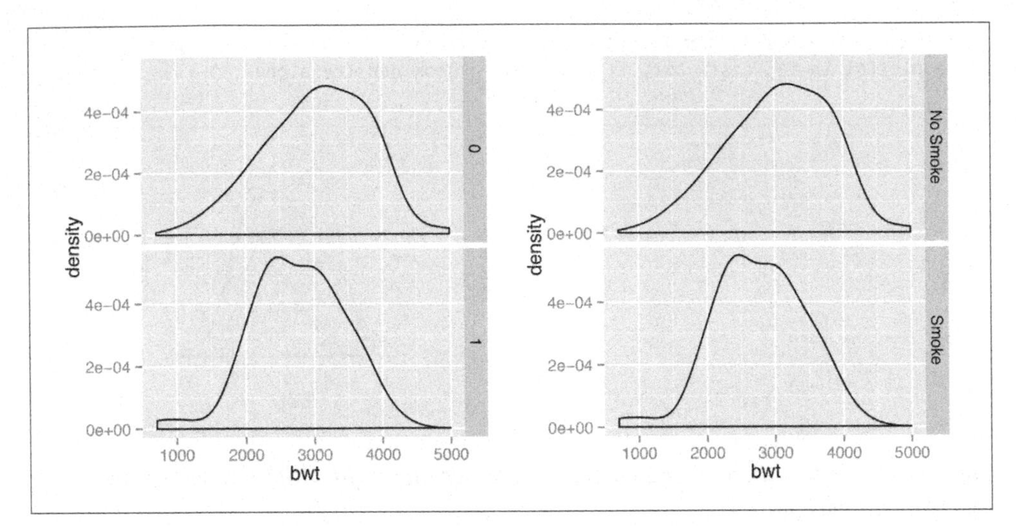

Figure 6-12. Left: density curves with facets; right: with different facet labels

```
ggplot(birthwt1, aes(x=bwt)) + geom_density() + facet_grid(smoke ~ .)
```

One problem with the faceted graph is that the facet labels are just 0 and 1, and there's no label indicating that those values are for `smoke`. To change the labels, we need to change the names of the factor levels. First we'll take a look at the factor levels, then we'll assign new factor level names, in the same order:

```
levels(birthwt1$smoke)

 "0" "1"

library(plyr) # For the revalue function
birthwt1$smoke <- revalue(birthwt1$smoke, c("0"="No Smoke", "1"="Smoke"))
```

Now when we plot it again, it shows the new labels (Figure 6-12, right):

```
ggplot(birthwt1, aes(x=bwt)) + geom_density() + facet_grid(smoke ~ .)
```

If you want to see the histograms along with the density curves, the best option is to use facets, since other methods of visualizing both histograms in a single graph can be difficult to interpret. To do this, map `y=..density..`, so that the histogram is scaled down to the height of the density curves. In this example, we'll also make the histogram bars a little less prominent by changing the colors (Figure 6-13):

```
ggplot(birthwt1, aes(x=bwt, y=..density..)) +
    geom_histogram(binwidth=200, fill="cornsilk", colour="grey60", size=.2) +
    geom_density() +
    facet_grid(smoke ~ .)
```

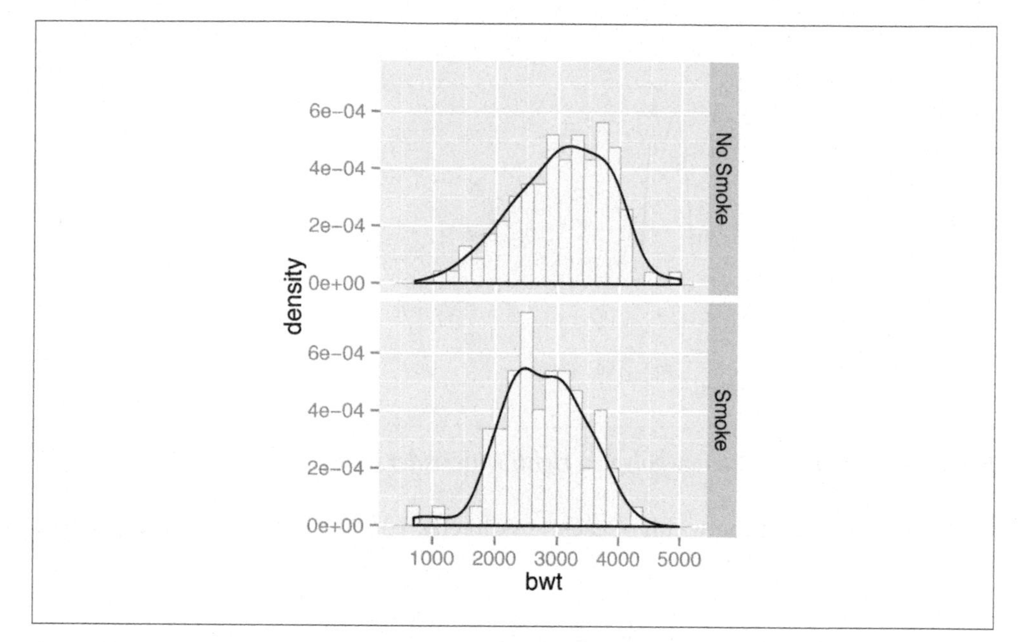

Figure 6-13. Density curves overlaid on histograms

6.5. Making a Frequency Polygon

Problem

You want to make a frequency polygon.

Solution

Use `geom_freqpoly()` (Figure 6-14):

```
ggplot(faithful, aes(x=waiting)) + geom_freqpoly()
```

Discussion

A frequency polygon appears similar to a kernel density estimate curve, but it shows the same information as a histogram. That is, like a histogram, it shows what is in the data, whereas a kernel density estimate is just that—an estimate—and requires you to pick some value for the bandwidth.

Also like a histogram, you can control the bin width for the frequency polygon (Figure 6-14, right):

```
ggplot(faithful, aes(x=waiting)) + geom_freqpoly(binwidth=4)
```

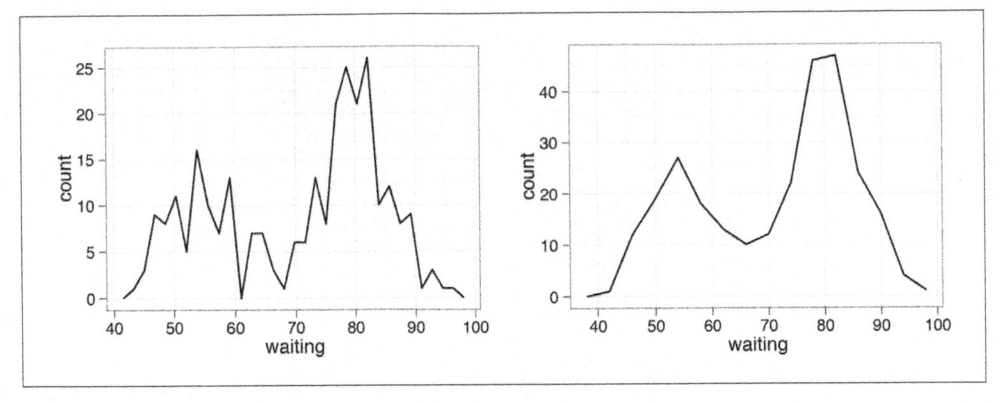

Figure 6-14. Left: a frequency polygon; right: with wider bins

Or, instead of setting the width of each bin directly, you can divide the *x* range into a particular number of bins:

```
# Use 15 bins
binsize <- diff(range(faithful$waiting))/15
ggplot(faithful, aes(x=waiting)) + geom_freqpoly(binwidth=binsize)
```

See Also

Histograms display the same information, but with bars instead of lines. See Recipe 6.1.

6.6. Making a Basic Box Plot

Problem

You want to make a box (or box-and-whiskers) plot.

Solution

Use `geom_boxplot()`, mapping a continuous variable to `y` and a discrete variable to `x` (Figure 6-15):

```
library(MASS) # For the data set

ggplot(birthwt, aes(x=factor(race), y=bwt)) + geom_boxplot()
# Use factor() to convert numeric variable to discrete
```

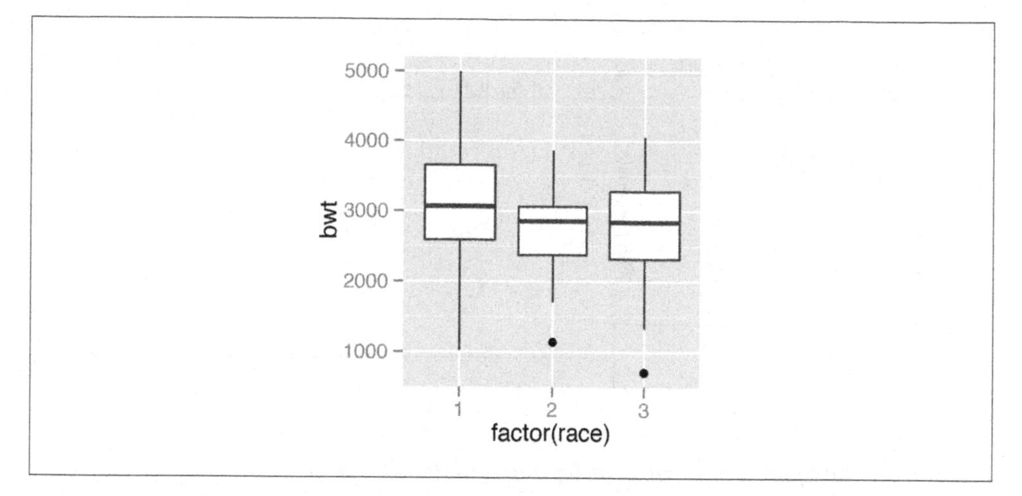

Figure 6-15. A box plot

Discussion

For this example, we used the `birthwt` data set from the `MASS` library. It contains data about birth weights and a number of risk factors for low birth weight:

```
birthwt

low age lwt race smoke ptl ht ui ftv  bwt
  0  19 182    2     0   0  0  1   0 2523
  0  33 155    3     0   0  0  0   3 2551
  0  20 105    1     1   0  0  0   1 2557
...
```

In Figure 6-15, the data is divided into groups by `race`, and we visualize the distributions of `bwt` for each group. The value of `race` is 1, 2, or 3, but since it's stored as a numeric vector, `ggplot()` doesn't know how to use it as a grouping variable. To make this work, we can modify the data frame by converting `race` to a factor, or tell `ggplot()` to treat it as a factor by using `factor(race)` inside of the `aes()` statement. In the preceding example, we used `factor(race)`.

A box plot consists of a box and "whiskers." The box goes from the 25th percentile to the 75th percentile of the data, also known as the *inter-quartile range* (IQR). There's a line indicating the median, or 50th percentile of the data. The whiskers start from the edge of the box and extend to the furthest data point that is within 1.5 times the IQR. If there are any data points that are past the ends of the whiskers, they are considered outliers and displayed with dots. Figure 6-16 shows the relationship between a histogram, a density curve, and a box plot, using a skewed data set.

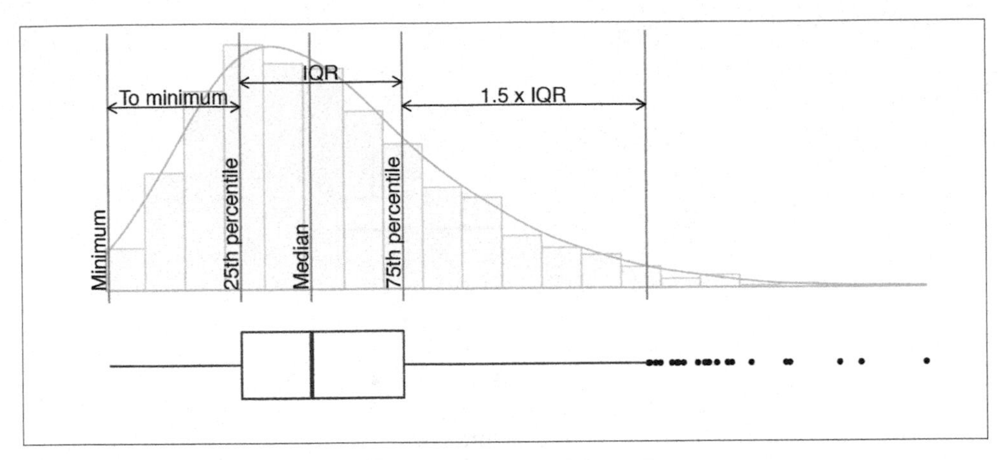

Figure 6-16. Box plot compared to histogram and density curve

To change the width of the boxes, you can set `width` (Figure 6-17, left):

```
ggplot(birthwt, aes(x=factor(race), y=bwt)) + geom_boxplot(width=.5)
```

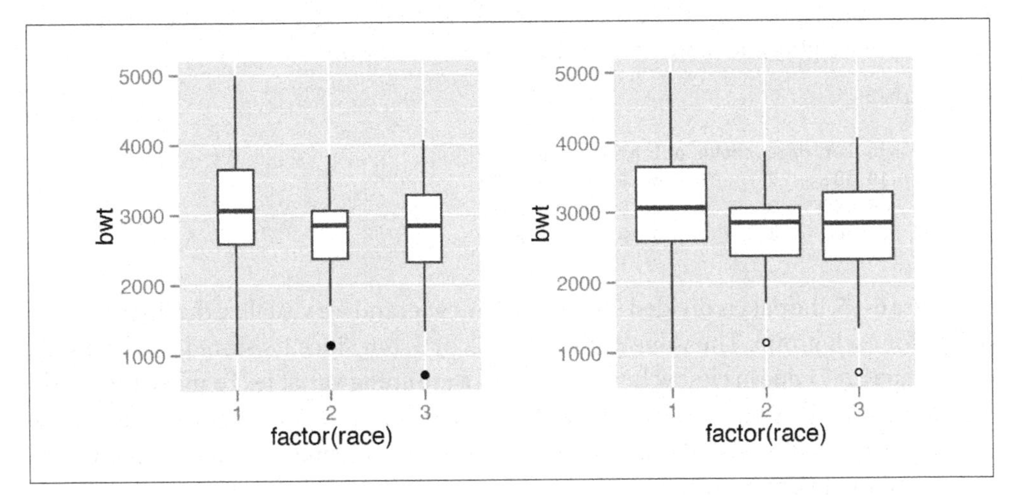

Figure 6-17. Left: box plot with narrower boxes; right: with smaller, hollow outlier points

If there are many outliers and there is overplotting, you can change the size and shape of the outlier points with `outlier.size` and `outlier.shape`. The default size is 2 and the default shape is 16. This will use smaller points, and hollow circles (Figure 6-17, right):

```
ggplot(birthwt, aes(x=factor(race), y=bwt)) +
    geom_boxplot(outlier.size=1.5, outlier.shape=21)
```

To make a box plot of just a single group, we have to provide some arbitrary value for x; otherwise, `ggplot()` won't know what *x* coordinate to use for the box plot. In this case, we'll set it to 1 and remove the x-axis tick markers and label (Figure 6-18):

```
ggplot(birthwt, aes(x=1, y=bwt)) + geom_boxplot() +
    scale_x_continuous(breaks=NULL) +
    theme(axis.title.x = element_blank())
```

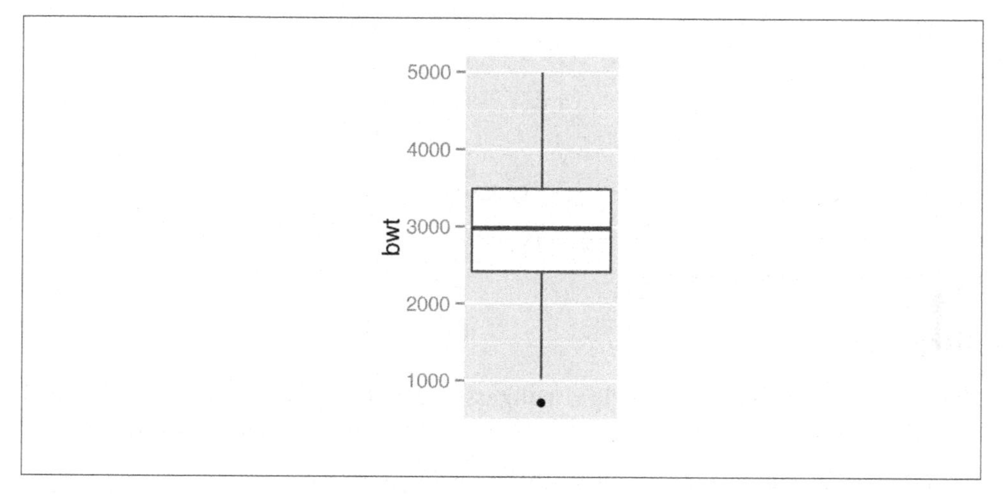

Figure 6-18. Box plot of a single group

The calculation of quantiles works slightly differently from the `boxplot()` function in base R. This can sometimes be noticeable for small sample sizes. See `?geom_boxplot` for detailed information about how the calculations differ.

6.7. Adding Notches to a Box Plot

Problem

You want to add notches to a box plot to assess whether the medians are different.

Solution

Use `geom_boxplot()` and set `notch=TRUE` (Figure 6-19):

```
library(MASS) # For the data set

ggplot(birthwt, aes(x=factor(race), y=bwt)) + geom_boxplot(notch=TRUE)
```

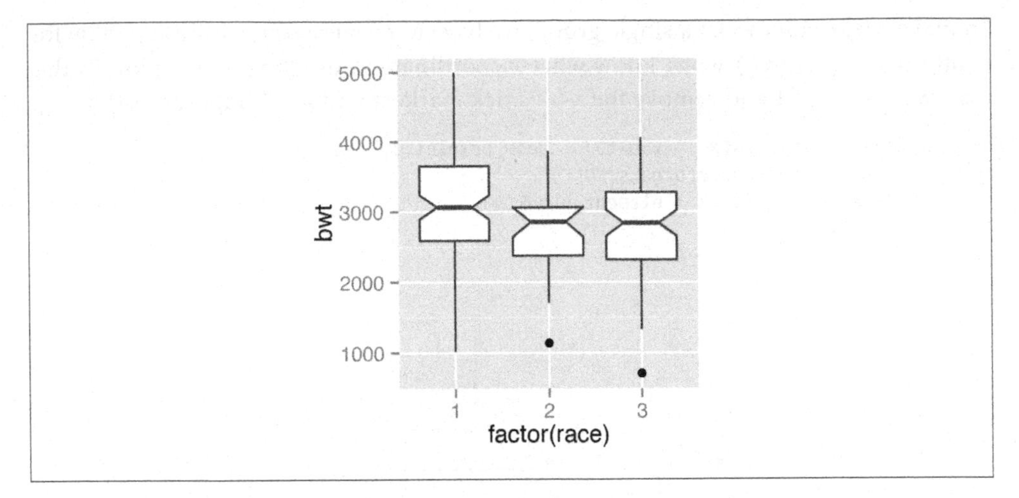

Figure 6-19. A notched box plot

Discussion

Notches are used in box plots to help visually assess whether the medians of distributions differ. If the notches do not overlap, this is evidence that the medians are different.

With this particular data set, you'll see the following message:

```
Notch went outside hinges. Try setting notch=FALSE.
```

This means that the confidence region (the notch) went past the bounds (or hinges) of one of the boxes. In this case, the upper part of the notch in the middle box goes just barely outside the box body, but it's by such a small amount that you can't see it in the final output. There's nothing inherently wrong with a notch going outside the hinges, but it can look strange in more extreme cases.

6.8. Adding Means to a Box Plot

Problem

You want to add markers for the mean to a box plot.

Solution

Use `stat_summary()`. The mean is often shown with a diamond, so we'll use shape 23 with a white fill. We'll also make the diamond slightly larger by setting `size=3` (Figure 6-20):

```
library(MASS) # For the data set
```

```
ggplot(birthwt, aes(x=factor(race), y=bwt)) + geom_boxplot() +
    stat_summary(fun.y="mean", geom="point", shape=23, size=3, fill="white")
```

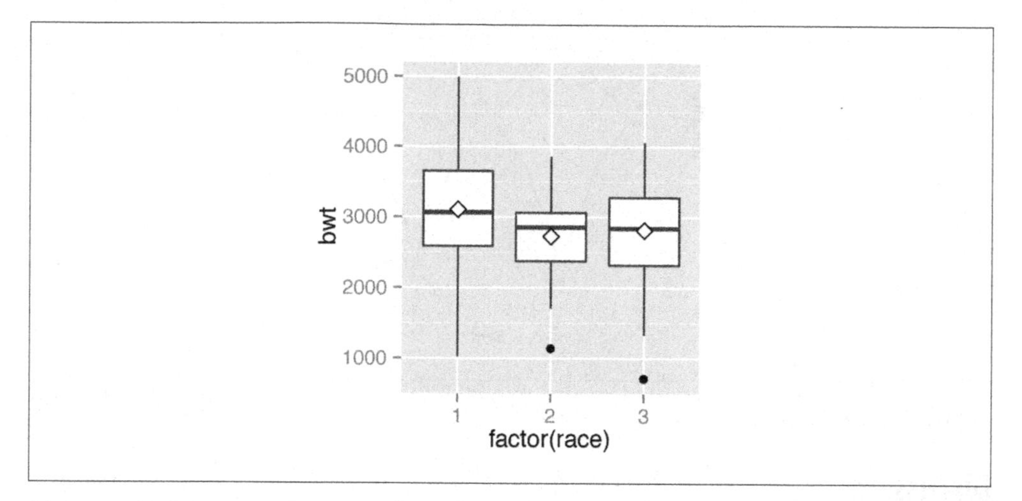

Figure 6-20. Mean markers on a box plot

Discussion

The horizontal line in the middle of a box plot displays the median, not the mean. For data that is normally distributed, the median and mean will be about the same, but for skewed data these values will differ.

6.9. Making a Violin Plot

Problem

You want to make a violin plot to compare density estimates of different groups.

Solution

Use `geom_violin()` (Figure 6-21):

```
library(gcookbook) # For the data set

# Base plot
p <- ggplot(heightweight, aes(x=sex, y=heightIn))

p + geom_violin()
```

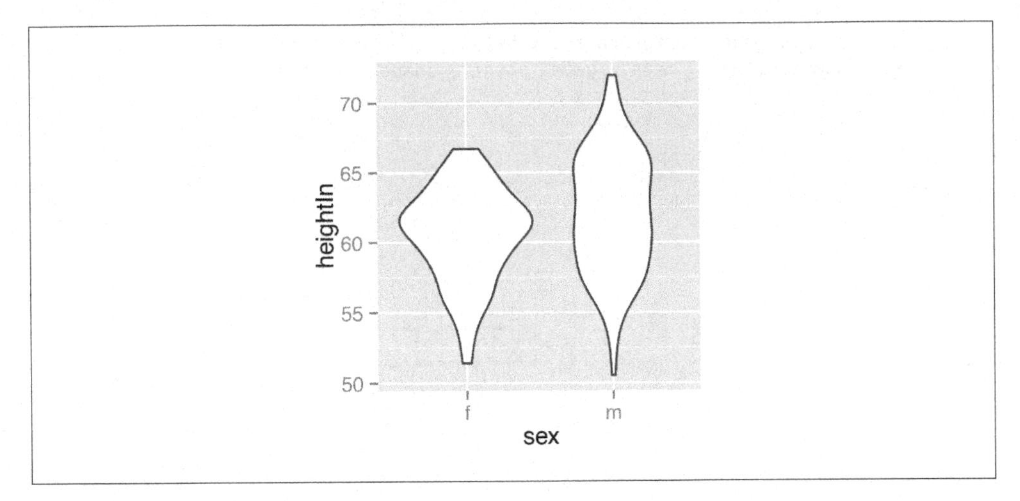

Figure 6-21. A violin plot

Discussion

Violin plots are a way of comparing multiple data distributions. With ordinary density curves, it is difficult to compare more than just a few distributions because the lines visually interfere with each other. With a violin plot, it's easier to compare several distributions since they're placed side by side.

A violin plot is a kernel density estimate, mirrored so that it forms a symmetrical shape. Traditionally, they also have narrow box plots overlaid, with a white dot at the median, as shown in Figure 6-22. Additionally, the box plot outliers are not displayed, which we do by setting `outlier.colour=NA`.:

```
p + geom_violin() + geom_boxplot(width=.1, fill="black", outlier.colour=NA) +
    stat_summary(fun.y=median, geom="point", fill="white", shape=21, size=2.5)
```

In this example we layered the objects from the bottom up, starting with the violin, then the box plot, then the white dot at the median, which is calculated using `stat_summary()`.

The default range goes from the minimum to maximum data values; the flat ends of the violins are at the extremes of the data. It's possible to keep the tails, by setting `trim=FALSE` (Figure 6-23):

```
p + geom_violin(trim=FALSE)
```

By default, the violins are scaled so that the total area of each one is the same (if `trim=TRUE`, then it scales what the area *would be* including the tails). Instead of equal areas, you can use `scale="count"` to scale the areas proportionally to the number of observations in each group (Figure 6-24). In this example, there are slightly fewer females than males, so the `f` violin is slightly narrower:

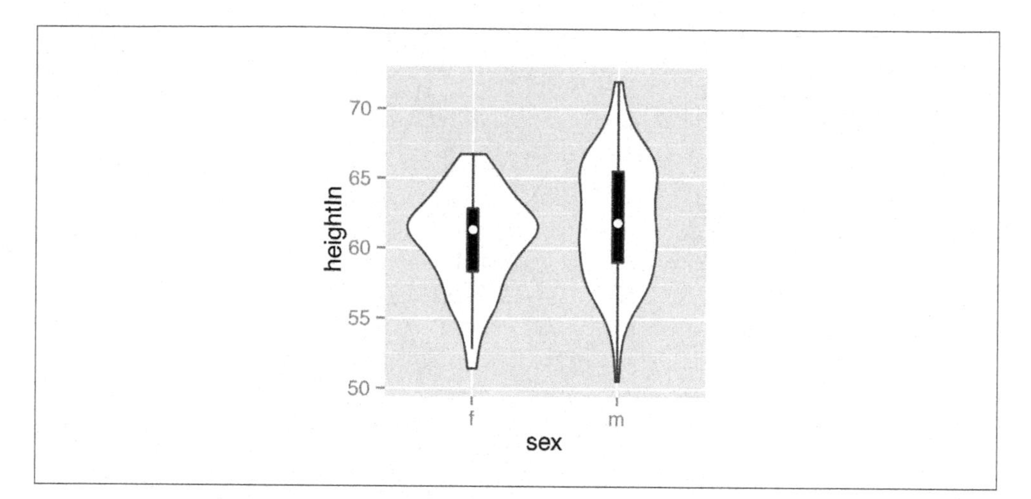

Figure 6-22. A violin plot with box plot overlaid on it

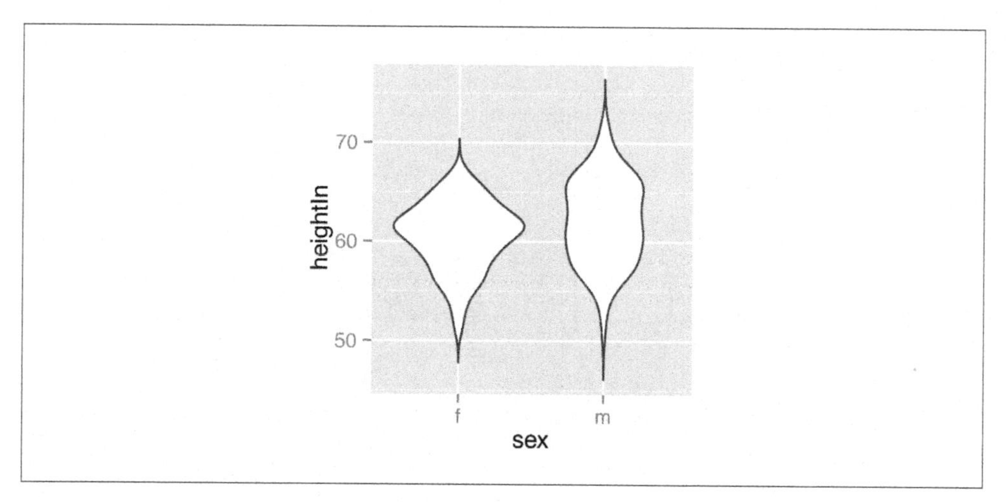

Figure 6-23. A violin plot with tails

```
# Scaled area proportional to number of observations
p + geom_violin(scale="count")
```

To change the amount of smoothing, use the `adjust` parameter, as described in Recipe 6.3. The default value is 1; use larger values for more smoothing and smaller values for less smoothing (Figure 6-25):

```
# More smoothing
p + geom_violin(adjust=2)
```

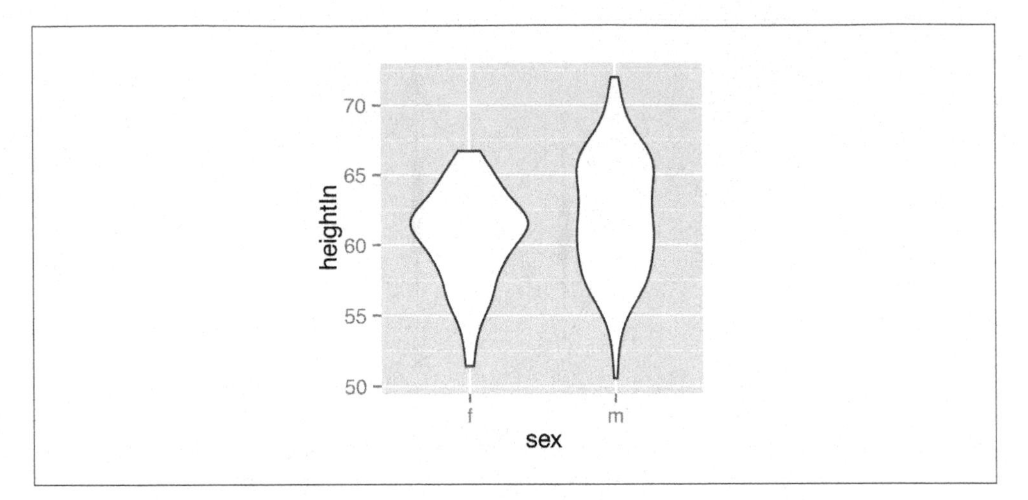

Figure 6-24. Violin plot with area proportional to number of observations

```
# Less smoothing
p + geom_violin(adjust=.5)
```

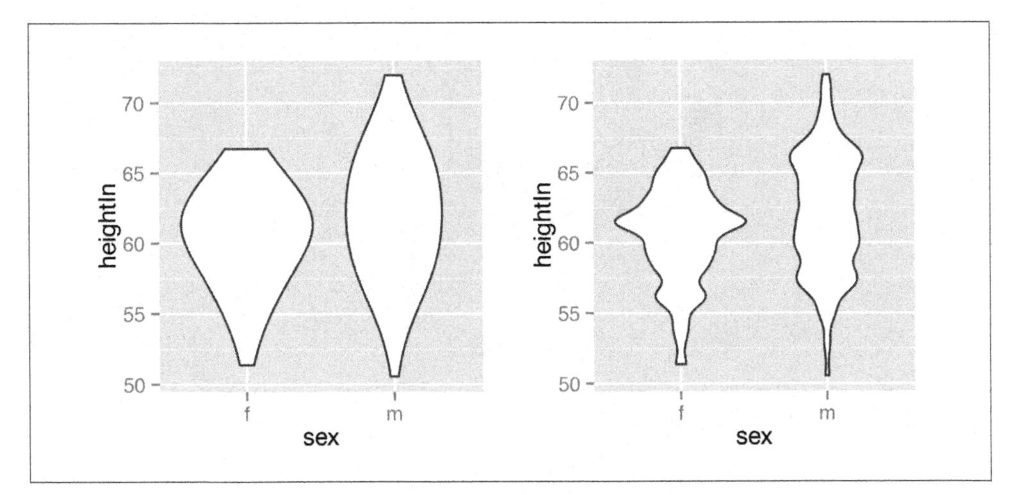

Figure 6-25. Left: violin plot with more smoothing; right: with less smoothing

See Also

To create a traditional density curve, see Recipe 6.3.

To use different point shapes, see Recipe 4.5.

6.10. Making a Dot Plot

Problem

You want to make a Wilkinson dot plot, which shows each data point.

Solution

Use `geom_dotplot()`. For this example (Figure 6-26), we'll use a subset of the `countries` data set:

```
library(gcookbook) # For the data set
countries2009 <- subset(countries, Year==2009 & healthexp>2000)

p <- ggplot(countries2009, aes(x=infmortality))

p + geom_dotplot()
```

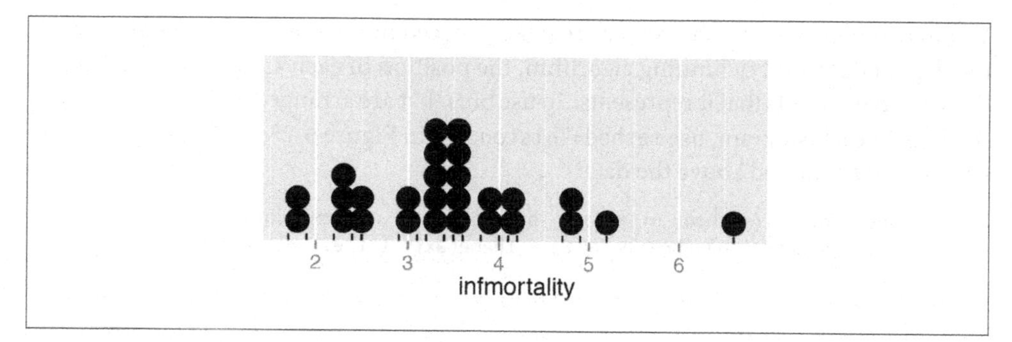

Figure 6-26. A dot plot

Discussion

This kind of dot plot is sometimes called a *Wilkinson* dot plot. It's different from the Cleveland dot plots shown in Recipe 3.10. In these dot plots, the placement of the bins depends on the data, and the width of each dot corresponds to the maximum width of each bin. The maximum bin size defaults to 1/30 of the range of the data, but it can be changed with `binwidth`.

By default, `geom_dotplot()` bins the data along the x-axis and stacks on the y-axis. The dots are stacked visually, and for reasons related to technical limitations of ggplot2, the resulting graph has y-axis tick marks that aren't meaningful. The y-axis labels can be removed by using `scale_y_continuous()`. In this example, we'll also use `geom_rug()` to show exactly where each data point is (Figure 6-27):

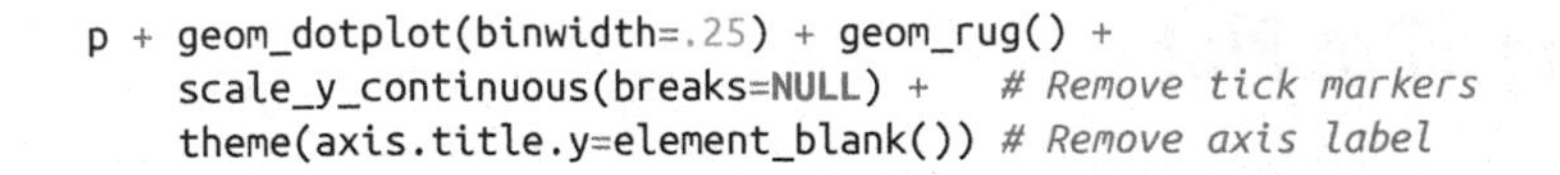

```
p + geom_dotplot(binwidth=.25) + geom_rug() +
    scale_y_continuous(breaks=NULL) +    # Remove tick markers
    theme(axis.title.y=element_blank()) # Remove axis label
```

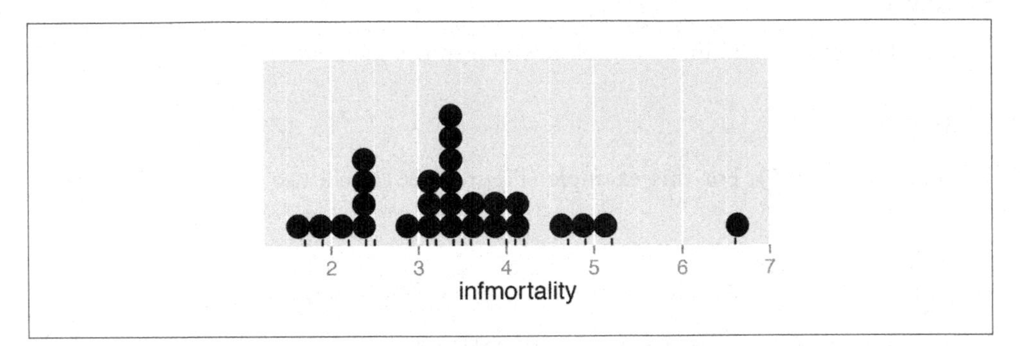

Figure 6-27. Dot plot with no y labels, max bin size of .25, and a rug showing each data point

You may notice that the stacks aren't regularly spaced in the horizontal direction. With the default `dotdensity` binning algorithm, the position of each stack is centered above the set of data points that it represents. To use bins that are arranged with a fixed, regular spacing, like a histogram, use `method="histodot"`. In Figure 6-28, you'll notice that the stacks *aren't* centered above the data:

```
p + geom_dotplot(method="histodot", binwidth=.25) + geom_rug() +
    scale_y_continuous(breaks=NULL) + theme(axis.title.y=element_blank())
```

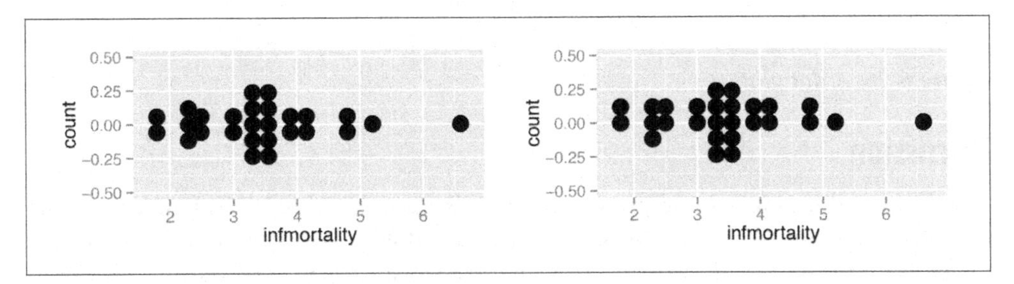

Figure 6-28. Dot plot with histodot (fixed-width) binning

The dots can also be stacked centered, or centered in such a way that stacks with even and odd quantities stay aligned. This can by done by setting `stackdir="center"` or `stackdir="centerwhole"`, as illustrated in Figure 6-29:

```
p + geom_dotplot(binwidth=.25, stackdir="center")
    scale_y_continuous(breaks=NULL) + theme(axis.title.y=element_blank())

p + geom_dotplot(binwidth=.25, stackdir="centerwhole")
    scale_y_continuous(breaks=NULL) + theme(axis.title.y=element_blank())
```

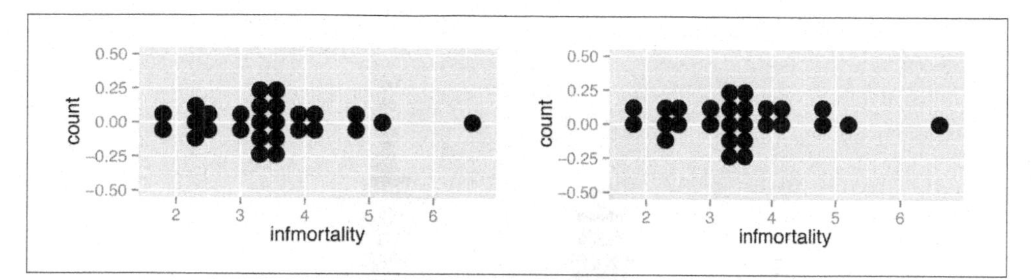

Figure 6-29. Left: dot plot with stackdir="center"; right: with stackdir="centerwhole"

See Also

Leland Wilkinson, "Dot Plots," *The American Statistician* 53 (1999): 276–281, *http://www.cs.uic.edu/~wilkinson/Publications/dots.pdf.*

6.11. Making Multiple Dot Plots for Grouped Data

Problem

You want to make multiple dot plots from grouped data.

Solution

To compare multiple groups, it's possible to stack the dots along the y-axis, and group them along the x-axis, by setting `binaxis="y"`. For this example, we'll use the `heightweight` data set (Figure 6-30):

```
library(gcookbook) # For the data set

ggplot(heightweight, aes(x=sex, y=heightIn)) +
    geom_dotplot(binaxis="y", binwidth=.5, stackdir="center")
```

Discussion

Dot plots are sometimes overlaid on box plots. In these cases, it may be helpful to make the dots hollow and have the box plots *not* show outliers, since the outlier points will be shown as part of the dot plot (Figure 6-31):

```
ggplot(heightweight, aes(x=sex, y=heightIn)) +
    geom_boxplot(outlier.colour=NA, width=.4) +
    geom_dotplot(binaxis="y", binwidth=.5, stackdir="center", fill=NA)
```

It's also possible to show the dot plots next to the box plots, as shown in Figure 6-32. This requires using a bit of a hack, by treating the *x* variable as a numeric variable and subtracting or adding a small quantity to shift the box plots and dot plots left and right.

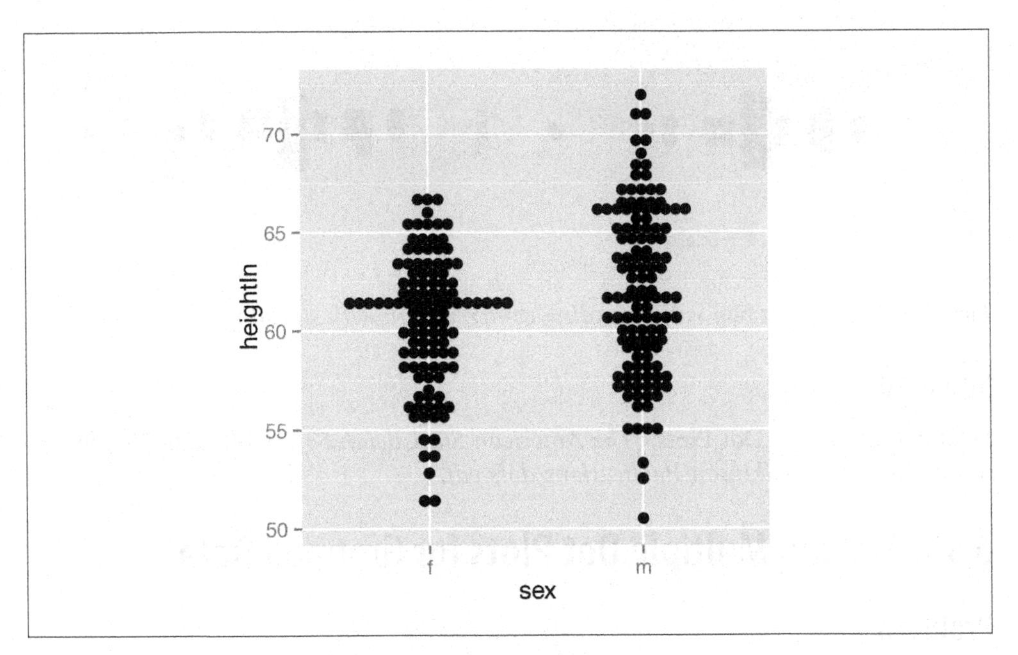

Figure 6-30. Dot plot of multiple groups, binning along the y-axis

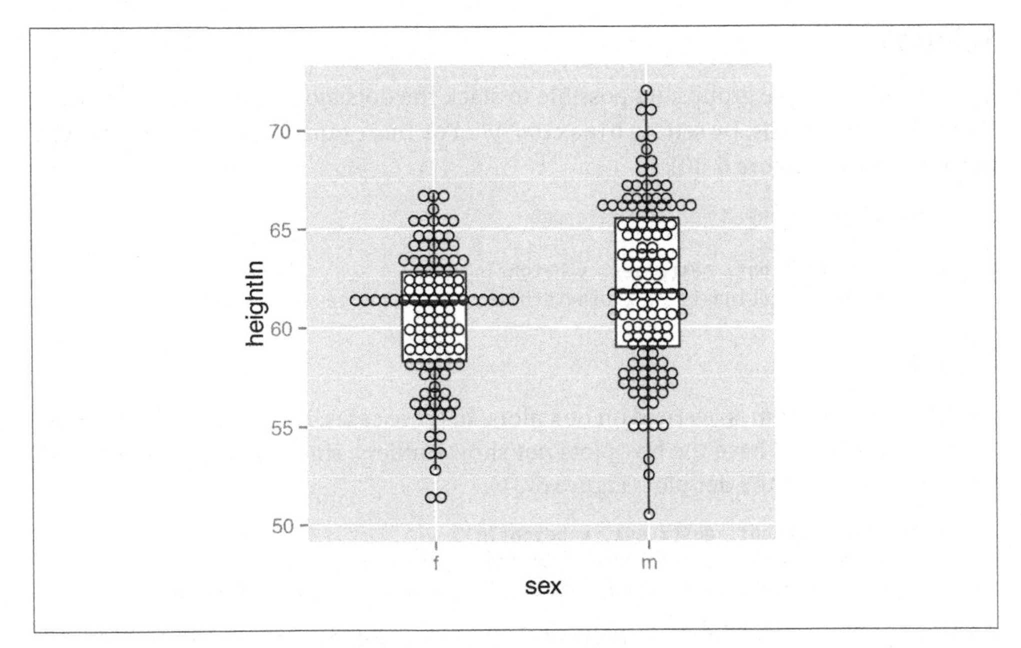

Figure 6-31. Dot plot overlaid on box plot

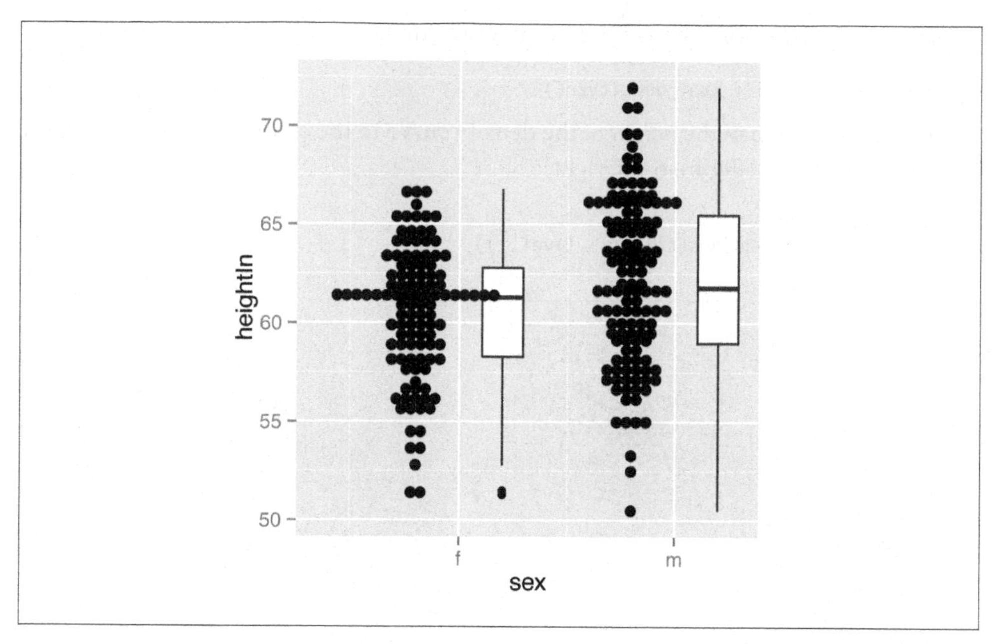

Figure 6-32. Dot plot next to box plot

When the *x* variable is treated as numeric you must also specify the `group`, or else the data will be treated as a single group, with just one box plot and dot plot. Finally, since the x-axis is treated as numeric, it will by default show numbers for the x-axis tick labels; they must be modified with `scale_x_continuous()` to show *x* tick labels as text corresponding to the factor levels:

```
ggplot(heightweight, aes(x=sex, y=heightIn)) +
    geom_boxplot(aes(x=as.numeric(sex) + .2, group=sex), width=.25) +
    geom_dotplot(aes(x=as.numeric(sex) - .2, group=sex), binaxis="y",
                 binwidth=.5, stackdir="center") +
    scale_x_continuous(breaks=1:nlevels(heightweight$sex),
                       labels=levels(heightweight$sex))
```

6.12. Making a Density Plot of Two-Dimensional Data

Problem

You want to plot the density of two-dimensional (2D) data.

Solution

Use `stat_density2d()`. This makes a 2D kernel density estimate from the data. First we'll plot the density contour along with the data points (Figure 6-33, left):

```
# The base plot
p <- ggplot(faithful, aes(x=eruptions, y=waiting))

p + geom_point() + stat_density2d()
```

It's also possible to map the *height* of the density curve to the color of the contour lines, by using `..level..` (Figure 6-33, right):

```
# Contour lines, with "height" mapped to color
p + stat_density2d(aes(colour=..level..))
```

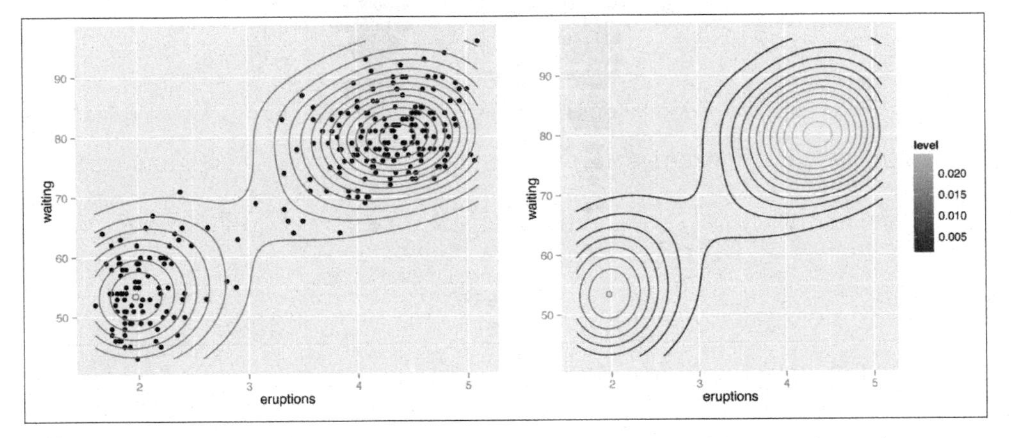

Figure 6-33. Left: points and density contour; right: with ..level.. mapped to color

Discussion

The two-dimensional kernel density estimate is analogous to the one-dimensional density estimate generated by `stat_density()`, but of course, it needs to be viewed in a different way. The default is to use contour lines, but it's also possible to use tiles and map the density estimate to the fill color, or to the transparency of the tiles, as shown in Figure 6-34:

```
# Map density estimate to fill color
p + stat_density2d(aes(fill=..density..), geom="raster", contour=FALSE)

# With points, and map density estimate to alpha
p + geom_point() +
    stat_density2d(aes(alpha=..density..), geom="tile", contour=FALSE)
```

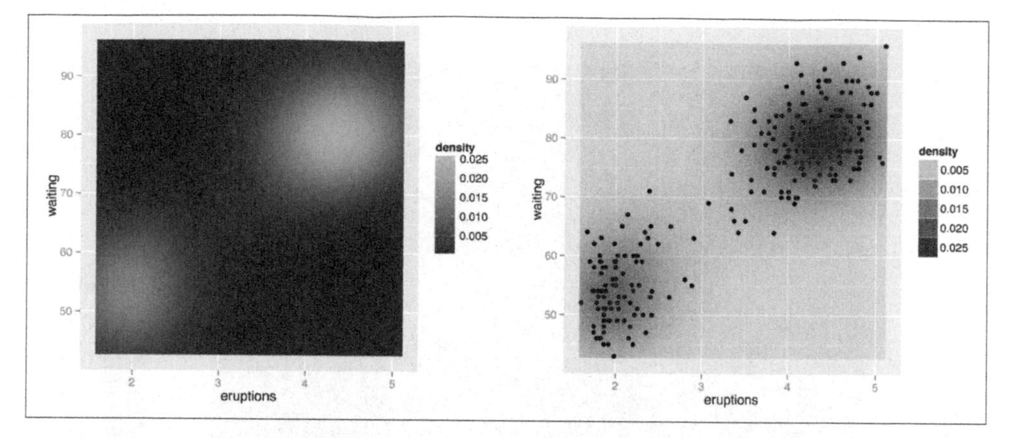

Figure 6-34. Left: with ..density.. mapped to fill; right: with points, and ..density.. mapped to alpha

We used `geom="raster"` in the first of the preceding examples and `geom="tile"` in the second. The main difference is that the raster geom renders more efficiently than the tile geom. In theory they *should* appear the same, but in practice they often do not. If you are writing to a PDF file, the appearance depends on the PDF viewer. On some viewers, when `tile` is used there may be faint lines between the tiles, and when `raster` is used the edges of the tiles may appear blurry (although it doesn't matter in this particular case).

As with the one-dimensional density estimate, you can control the bandwidth of the estimate. To do this, pass a vector for the *x* and *y* bandwidths to `h`. This argument gets passed on to the function that actually generates the density estimate, `kde2d()`. In this example (Figure 6-35), we'll use a smaller bandwidth in the *x* and *y* directions, so that the density estimate is more closely fitted (perhaps overfitted) to the data:

```
p + stat_density2d(aes(fill=..density..), geom="raster",
                   contour=FALSE, h=c(.5,5))
```

See Also

The relationship between `stat_density2d()` and `stat_bin2d()` is the same as the relationship between their one-dimensional counterparts, the density curve and the histogram. The density curve is an *estimate* of the distribution under certain assumptions, while the binned visualization represents the observed data directly. See Recipe 5.5 for more about binning data.

If you want to use a different color palette, see Recipe 12.6.

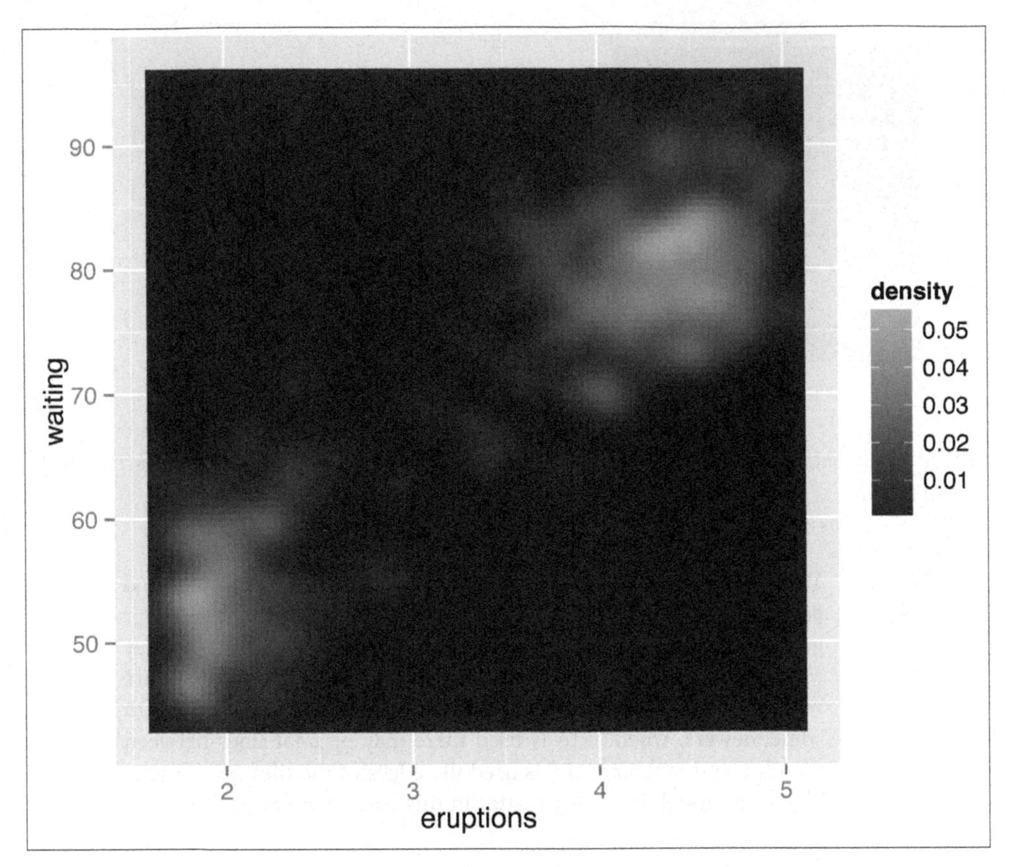

Figure 6-35. Density plot with a smaller bandwidth in the x and y directions

`stat_density2d()` passes options to `kde2d()`; see `?kde2d` for information on the available options.

CHAPTER 7
Annotations

Displaying just your data usually isn't enough—there's all sorts of other information that can help the viewer interpret the data. In addition to the standard repertoire of axis labels, tick marks, and legends, you can also add individual graphical or text elements to your plot. These can be used to add extra contextual information, highlight an area of the plot, or add some descriptive text about the data.

7.1. Adding Text Annotations

Problem

You want to add a text annotation to a plot.

Solution

Use `annotate()` and a text geom (Figure 7-1):

```
p <- ggplot(faithful, aes(x=eruptions, y=waiting)) + geom_point()

p + annotate("text", x=3, y=48, label="Group 1") +
    annotate("text", x=4.5, y=66, label="Group 2")
```

Discussion

The `annotate()` function can be used to add any type of geometric object. In this case, we used `geom="text"`.

Other text properties can be specified, as shown in Figure 7-2:

```
p + annotate("text", x=3, y=48, label="Group 1", family="serif",
             fontface="italic", colour="darkred", size=3) +
    annotate("text", x=4.5, y=66, label="Group 2", family="serif",
             fontface="italic", colour="darkred", size=3)
```

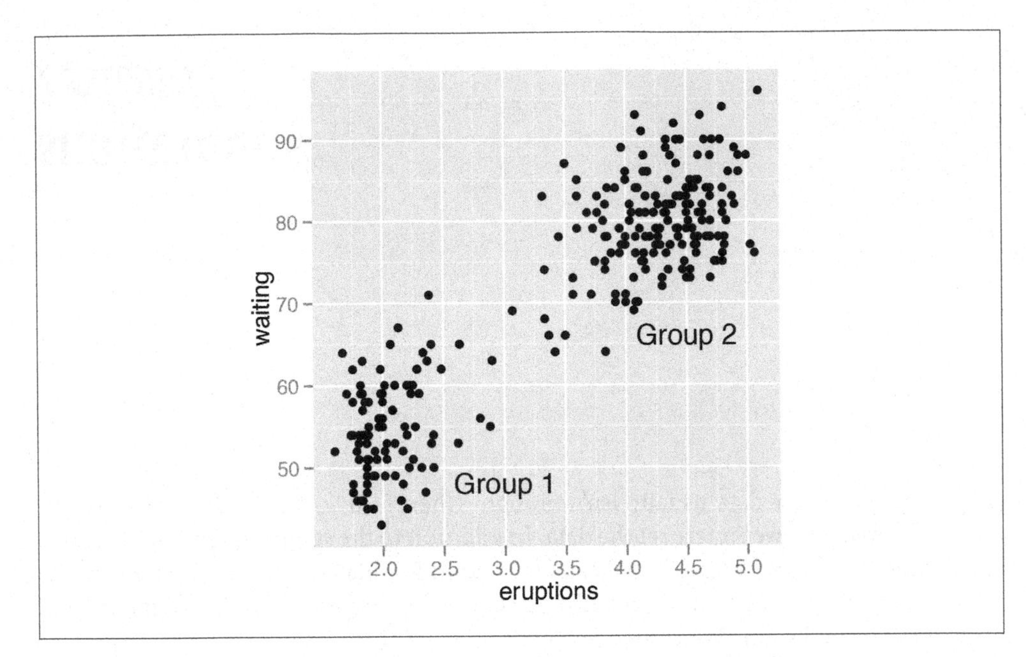

Figure 7-1. Text annotations

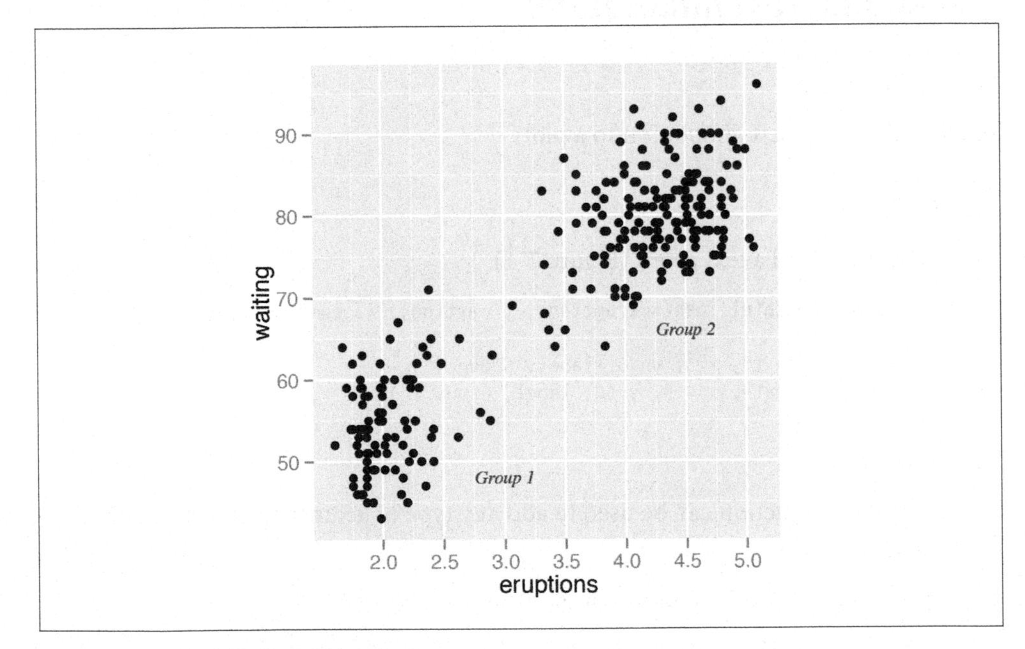

Figure 7-2. Modified text properties

Be careful not to use `geom_text()` when you want to add individual text objects. While `annotate(geom="text")` will add a single text object to the plot, `geom_text()` will create many text objects based on the data, as discussed in Recipe 5.11.

If you use `geom_text()`, the text will be heavily overplotted on the same location, with one copy per data point:

```
p + annotate("text", x=3, y=48, label="Group 1", alpha=.1) +     # Normal
    geom_text(x=4.5, y=66, label="Group 2", alpha=.1)            # Overplotted
```

In Figure 7-3, each text label is 90% transparent, making it clear which one is overplotted. The overplotting can lead to output with aliased (jagged) edges when outputting to a bitmap.

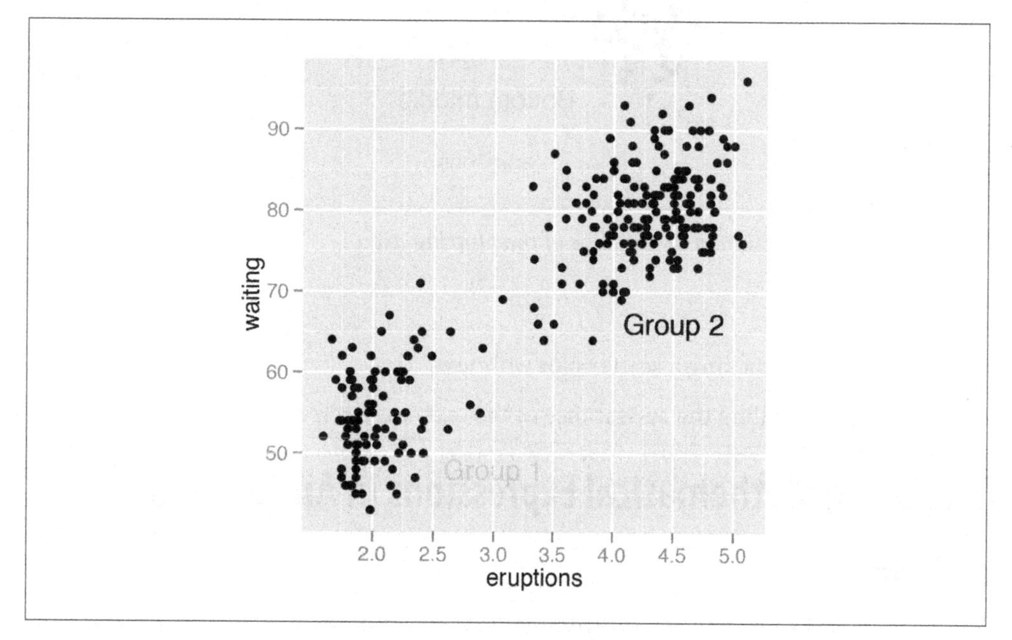

Figure 7-3. Overplotting one of the labels—both should be 90% transparent

If the axes are continuous, you can use the special values `Inf` and `-Inf` to place text annotations at the edge of the plotting area, as shown in Figure 7-4. You will also need to adjust the position of the text relative to the corner using `hjust` and `vjust`—if you leave them at their default values, the text will be centered on the edge. It may take a little experimentation with these values to get the text positioned to your liking:

```
p + annotate("text", x=-Inf, y=Inf, label="Upper left", hjust=-.2, vjust=2) +
    annotate("text", x=mean(range(faithful$eruptions)), y=-Inf, vjust=-0.4,
             label="Bottom middle")
```

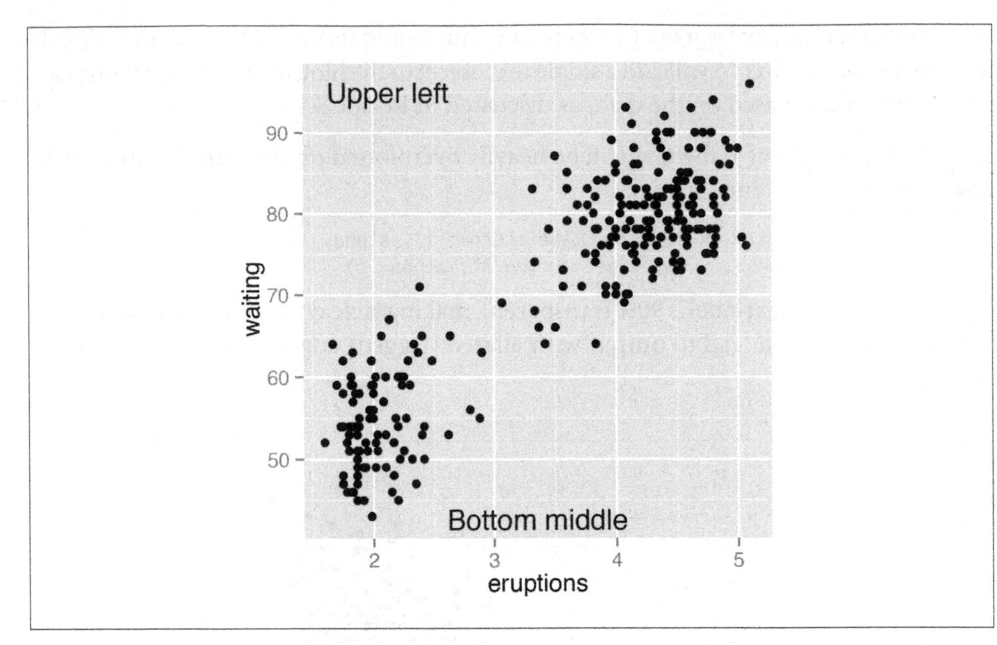

Figure 7-4. Text positioned at the edge of the plotting area

See Also

See Recipe 5.11 for making a scatter plot with text.

For more on controlling the appearance of the text, see Recipe 9.2.

7.2. Using Mathematical Expressions in Annotations

Problem

You want to add a text annotation with mathematical notation.

Solution

Use `annotate(geom="text")` and set `parse=TRUE` (Figure 7-5):

```
# A normal curve
p <- ggplot(data.frame(x=c(-3,3)), aes(x=x)) + stat_function(fun = dnorm)

p + annotate("text", x=2, y=0.3, parse=TRUE,
             label="frac(1, sqrt(2 * pi)) * e ^ {-x^2 / 2}")
```

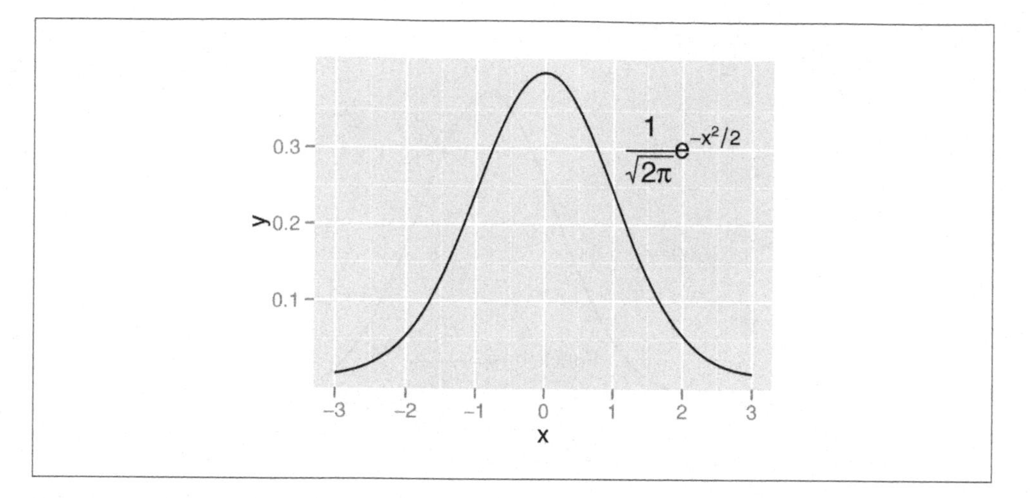

Figure 7-5. Annotation with mathematical expressions

Discussion

Mathematical expressions made with text geoms using `parse=TRUE` in ggplot2 have a format similar to those made with `plotmath` and `expression` in base R, except that they are stored as strings, rather than as expression objects.

To mix regular text with expressions, use single quotes within double quotes (or vice versa) to mark the plain-text parts. Each block of text enclosed by the inner quotes is treated as a variable in a mathematical expression. Bear in mind that, in R's syntax for mathematical expressions, you can't simply put a variable right next to another without something else in between. To display two variables next to each other, as in Figure 7-6, put a `*` operator between them; when displayed in a graphic, this is treated as an invisible multiplication sign (for a visible multiplication sign, use `%*%`):

```
p + annotate("text", x=0, y=0.05, parse=TRUE, size=4,
             label="'Function:  ' * y==frac(1, sqrt(2*pi)) * e^{-x^2/2}")
```

See Also

See `?plotmath` for many examples of mathematical expressions, and `?demo(plotmath)` for graphical examples of mathematical expressions.

See Recipe 5.9 for adding regression coefficients to a graph.

For using other fonts in mathematical expressions, see Recipe 14.6.

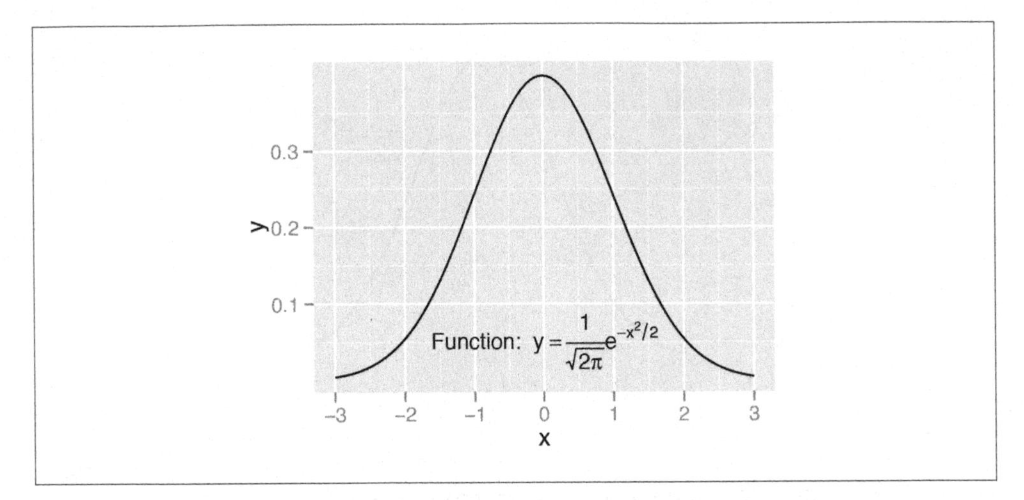

Figure 7-6. Mathematical expression with regular text

7.3. Adding Lines

Problem

You want to add lines to a plot.

Solution

For horizontal and vertical lines, use `geom_hline()` and `geom_vline()`, and for angled lines, use `geom_abline()` (Figure 7-7). For this example, we'll use the `heightweight` data set:

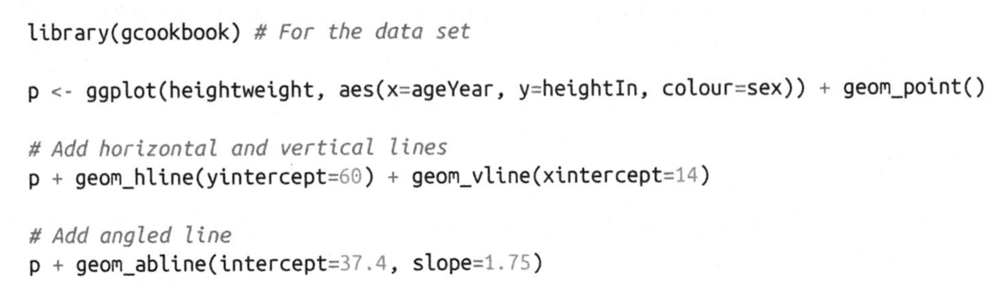

```
library(gcookbook) # For the data set

p <- ggplot(heightweight, aes(x=ageYear, y=heightIn, colour=sex)) + geom_point()

# Add horizontal and vertical lines
p + geom_hline(yintercept=60) + geom_vline(xintercept=14)

# Add angled line
p + geom_abline(intercept=37.4, slope=1.75)
```

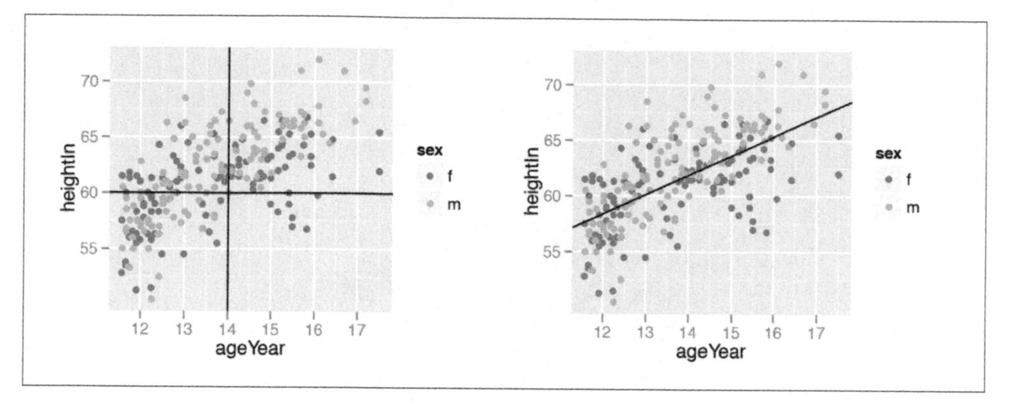

Figure 7-7. Left: horizontal and vertical lines; right: angled line

Discussion

The previous examples demonstrate setting the positions of the lines manually, resulting in one line drawn for each geom added. It is also possible to *map* values from the data to `xintercept`, `yintercept`, and so on, and even draw them from another data frame.

Here we'll take the average height for males and females and store it in a data frame, `hw_means`. Then we'll draw a horizontal line for each, and set the `linetype` and `size` (Figure 7-8):

```
library(plyr) # For the ddply() function
hw_means <- ddply(heightweight, "sex", summarise, heightIn=mean(heightIn))
hw_means

 sex heightIn
   f 60.52613
   m 62.06000

p + geom_hline(aes(yintercept=heightIn, colour=sex), data=hw_means,
               linetype="dashed", size=1)
```

If one of the axes is discrete rather than continuous, you can't specify the intercepts as just a character string—they must still be specified as numbers. If the axis represents a factor, the first level has a numeric value of 1, the second level has a value of 2, and so on. You can specify the numerical intercept manually, or calculate the numerical value using `which(levels(...))` (Figure 7-9):

```
pg <- ggplot(PlantGrowth, aes(x=group, y=weight)) + geom_point()

pg + geom_vline(xintercept = 2)

pg + geom_vline(xintercept = which(levels(PlantGrowth$group)=="ctrl"))
```

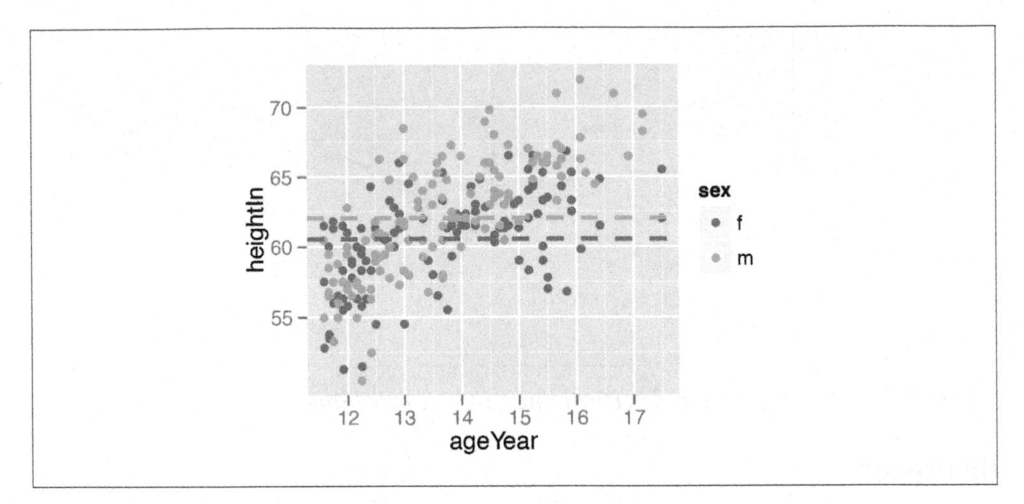

Figure 7-8. Multiple lines, drawn at the mean of each group

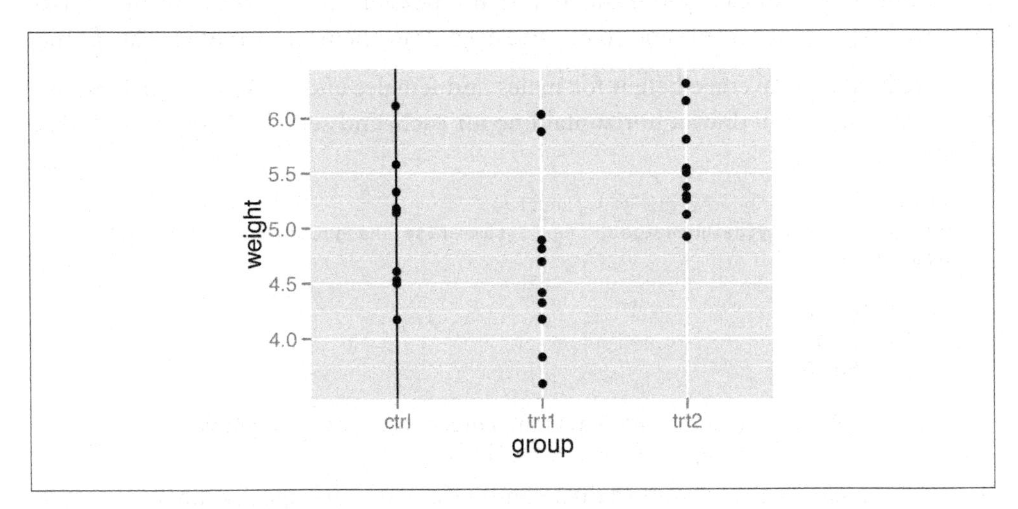

Figure 7-9. Lines with a discrete axis

You may have noticed that adding lines differs from adding other annotations. Instead of using the `annotate()` function, we've used `geom_hline()` and friends. This is because previous versions of ggplot2 didn't have the `annotate()` function. The line geoms had code to handle the special cases where they were used to add a single line, and changing it would break backward compatibility. In a future version of ggplot2, this will change, and `annotate()` will work with line geoms.

See Also

For adding regression lines, see Recipes Recipe 5.6 and 5.7.

Lines are often used to indicate summarized information about data. See Recipe 15.17 for more on how to summarize data by groups.

7.4. Adding Line Segments and Arrows

Problem

You want to add line segments or arrows to a plot.

Solution

Use `annotate("segment")`. In this example, we'll use the `climate` data set and use a subset of data from the Berkeley source (Figure 7-10):

```
library(gcookbook) # For the data set

p <- ggplot(subset(climate, Source=="Berkeley"), aes(x=Year, y=Anomaly10y)) +
     geom_line()

p + annotate("segment", x=1950, xend=1980, y=-.25, yend=-.25)
```

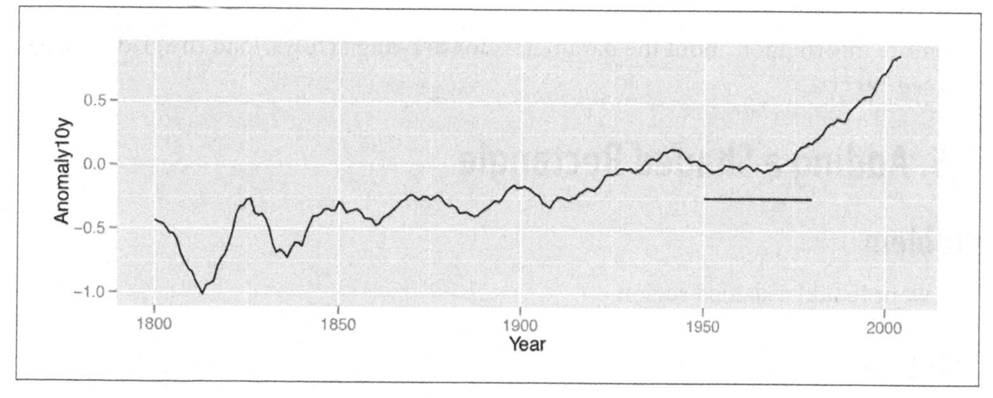

Figure 7-10. Line segment annotation

Discussion

It's possible to add arrowheads or flat ends to the line segments, using `arrow()` from the grid package. In this example, we'll do both (Figure 7-11):

```
library(grid)
p + annotate("segment", x=1850, xend=1820, y=-.8, yend=-.95, colour="blue",
```

```
             size=2, arrow=arrow()) +
  annotate("segment", x=1950, xend=1980, y=-.25, yend=-.25,
           arrow=arrow(ends="both", angle=90, length=unit(.2,"cm")))
```

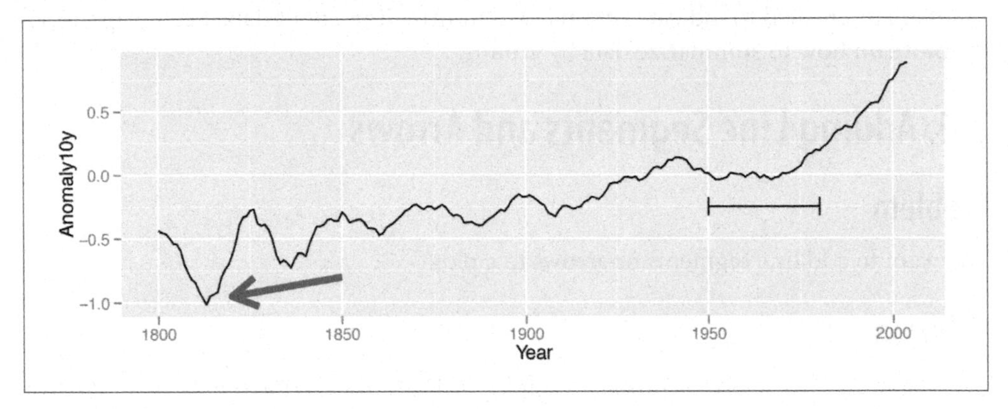

Figure 7-11. Line segments with arrow heads

The default `angle` is 30, and the default `length` of the arrowhead lines is 0.2 inches.

If one or both axes are discrete, the *x* and *y* positions are such that the categorical items have coordinate values 1, 2, 3, and so on.

See Also

For more information about the parameters for drawing arrows, load the grid package and see `?arrow`.

7.5. Adding a Shaded Rectangle

Problem

You want to add a shaded region.

Solution

Use `annotate("rect")` (Figure 7-12):

```
library(gcookbook) # For the data set

p <- ggplot(subset(climate, Source=="Berkeley"), aes(x=Year, y=Anomaly10y)) +
  geom_line()

p + annotate("rect", xmin=1950, xmax=1980, ymin=-1, ymax=1, alpha=.1,
             fill="blue")
```

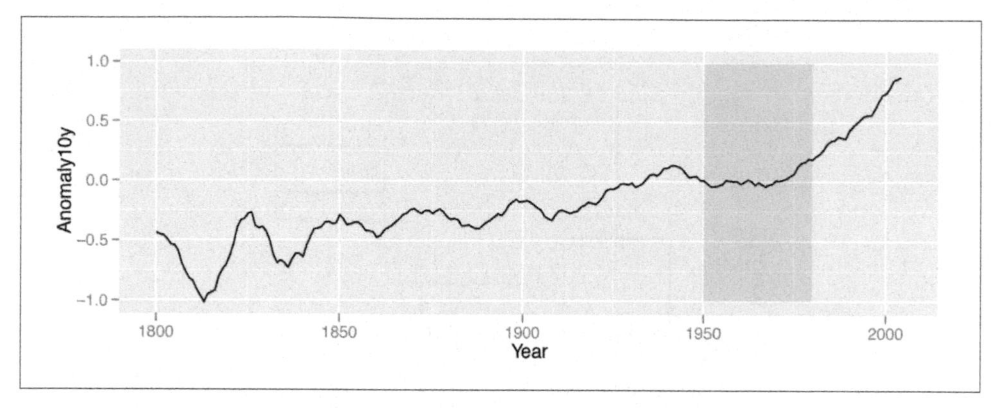

Figure 7-12. A shaded rectangle

Discussion

Each layer is drawn in the order that it's added to the `ggplot` object, so in the preceding example, the rectangle is drawn on top of the line. It's not a problem in that case, but if you'd like to have the line above the rectangle, add the rectangle first, and then the line.

Any geom can be used with `annotate()`, as long as you pass in the proper parameters. In this case, `geom_rect()` requires min and max values for `x` and `y`.

7.6. Highlighting an Item

Problem

You want to change the color of an item to make it stand out.

Solution

To highlight one or more items, create a new column in the data and map it to the color. In this example, we'll create a new column, `hl`, and set its value based on the value of `group`:

```
pg <- PlantGrowth                      # Make a copy of the PlantGrowth data
pg$hl <- "no"                          # Set all to "no"
pg$hl[pg$group=="trt2"] <- "yes"  # If group is "trt2", set to "yes"
```

Then we'll plot it with manually specified colors and with no legend (Figure 7-13):

```
ggplot(pg, aes(x=group, y=weight, fill=hl)) + geom_boxplot() +
    scale_fill_manual(values=c("grey85", "#FFDDCC"), guide=FALSE)
```

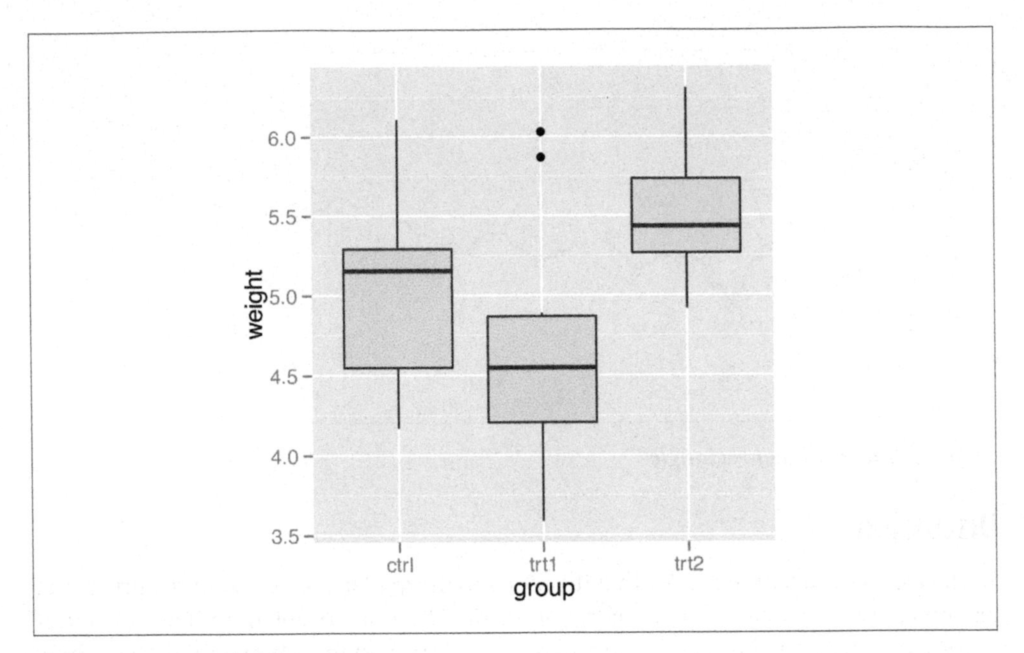

Figure 7-13. Highlighting one item

Discussion

If you have a small number of items, as in this example, instead of creating a new column you could use the original one and specify the colors for every level of that variable. For example, the following code will use the `group` column from `PlantGrowth` and manually set the colors for each of the three levels. The result will appear the same as with the preceding code:

```
ggplot(PlantGrowth, aes(x=group, y=weight, fill=group)) + geom_boxplot() +
    scale_fill_manual(values=c("grey85", "grey85", "#FFDDCC"), guide=FALSE)
```

See Also

See Chapter 12 for more information about specifying colors.

For more information about removing the legend, see Recipe 10.1.

7.7. Adding Error Bars

Problem

You want to add error bars to a graph.

Solution

Use `geom_errorbar` and map variables to the values for `ymin` and `ymax`. Adding the error bars is done the same way for bar graphs and line graphs, as shown in Figure 7-14 (notice that default *y* range is different for bars and lines, though):

```
library(gcookbook) # For the data set
# Take a subset of the cabbage_exp data for this example
ce <- subset(cabbage_exp, Cultivar == "c39")

# With a bar graph
ggplot(ce, aes(x=Date, y=Weight)) +
    geom_bar(fill="white", colour="black") +
    geom_errorbar(aes(ymin=Weight-se, ymax=Weight+se), width=.2)

# With a line graph
ggplot(ce, aes(x=Date, y=Weight)) +
    geom_line(aes(group=1)) +
    geom_point(size=4) +
    geom_errorbar(aes(ymin=Weight-se, ymax=Weight+se), width=.2)
```

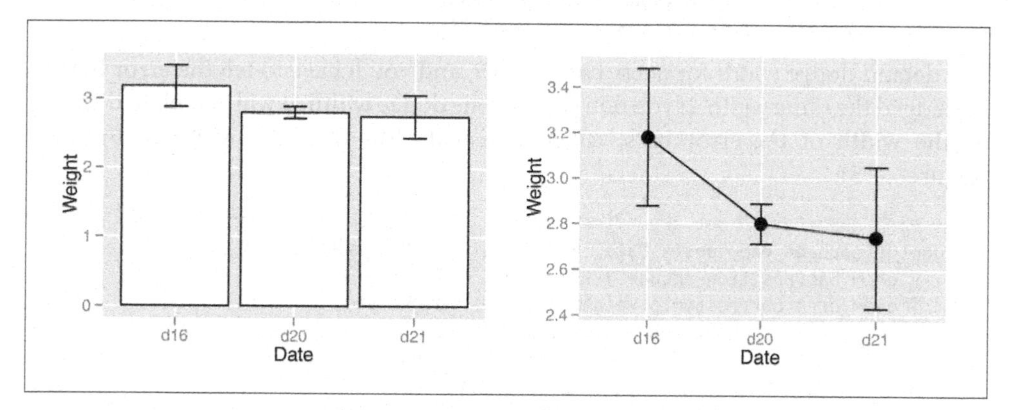

Figure 7-14. Left: error bars on a bar graph; right: on a line graph

Discussion

In this example, the data already has values for the standard error of the mean (`se`), which we'll use for the error bars (it also has values for the standard deviation, `sd`, but we're not using that here):

```
ce

Cultivar Date Weight        sd  n         se
     c39  d16   3.18 0.9566144 10 0.30250803
     c39  d20   2.80 0.2788867 10 0.08819171
     c39  d21   2.74 0.9834181 10 0.31098410
```

To get the values for `ymax` and `ymin`, we took the y variable, `Weight`, and added/subtracted `se`.

We also specified the width of the ends of the error bars, with `width=.2`. It's best to play around with this to find a value that looks good. If you don't set the width, the error bars will be very wide, spanning all the space between items on the x-axis.

For a bar graph with groups of bars, the error bars must also be *dodged*; otherwise, they'll have the exact same *x* coordinate and won't line up with the bars. (See Recipe 3.2 for more information about grouped bars and dodging.)

We'll work with the full `cabbage_exp` data set this time:

```
cabbage_exp

 Cultivar Date Weight        sd  n         se
      c39  d16   3.18 0.9566144 10 0.30250803
      c39  d20   2.80 0.2788867 10 0.08819171
      c39  d21   2.74 0.9834181 10 0.31098410
      c52  d16   2.26 0.4452215 10 0.14079141
      c52  d20   3.11 0.7908505 10 0.25008887
      c52  d21   1.47 0.2110819 10 0.06674995
```

The default dodge width for `geom_bar()` is 0.9, and you'll have to tell the error bars to be dodged the same width. If you don't specify the dodge width, it will default to dodging by the width of the error bars, which is usually less than the width of the bars (Figure 7-15):

```
# Bad: dodge width not specified
ggplot(cabbage_exp, aes(x=Date, y=Weight, fill=Cultivar)) +
    geom_bar(position="dodge") +
    geom_errorbar(aes(ymin=Weight-se, ymax=Weight+se),
                  position="dodge", width=.2)

# Good: dodge width set to same as bar width (0.9)
ggplot(cabbage_exp, aes(x=Date, y=Weight, fill=Cultivar)) +
    geom_bar(position="dodge") +
    geom_errorbar(aes(ymin=Weight-se, ymax=Weight+se),
                  position=position_dodge(0.9), width=.2)
```

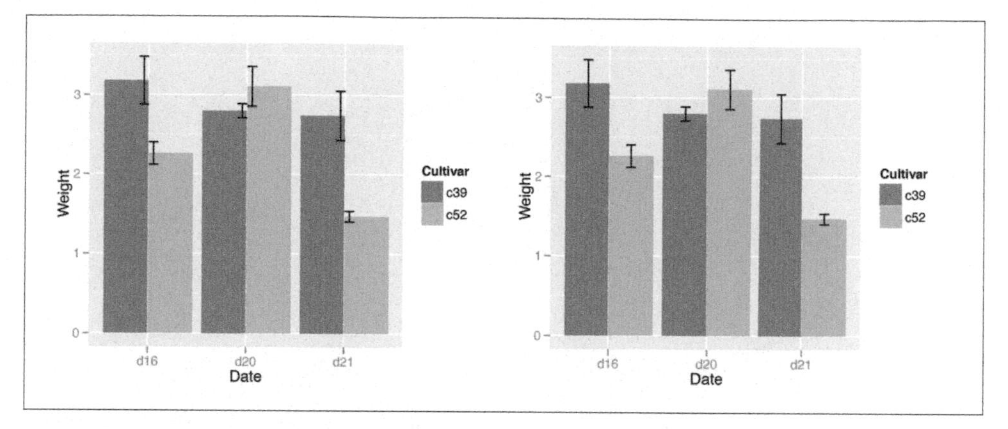

Figure 7-15. Left: error bars on a grouped bar graph without dodging width specified; right: with dodging width specified

Notice that we used `position="dodge"`, which is shorthand for `position=position_dodge()`, in the first version. But to pass a specific value, we have to spell it out, as in `position_dodge(0.9)`.

For line graphs, if the error bars are a different color than the lines and points, you should draw the error bars first, so that they are underneath the points and lines. Otherwise the error bars will be drawn on top of the points and lines, which won't look right.

Additionally, you should dodge all the geometric elements so that they will align with the error bars, as shown in Figure 7-16:

```
pd <- position_dodge(.3)  # Save the dodge spec because we use it repeatedly

ggplot(cabbage_exp, aes(x=Date, y=Weight, colour=Cultivar, group=Cultivar)) +
    geom_errorbar(aes(ymin=Weight-se, ymax=Weight+se),
                  width=.2, size=0.25, colour="black", position=pd) +
    geom_line(position=pd) +
    geom_point(position=pd, size=2.5)

# Thinner error bar lines with size=0.25, and larger points with size=2.5
```

Notice that we set `colour="black"` to make the error bars black; otherwise, they would inherit `colour`. We also made sure the `Cultivar` was used as a grouping variable by mapping it to `group`.

When a discrete variable is *mapped* to an aesthetic like `colour` or `fill` (as in the case of the bars), that variable is used for grouping the data. But by *setting* the `colour` of the error bars, we made it so that the variable for `colour` was not used for grouping, and

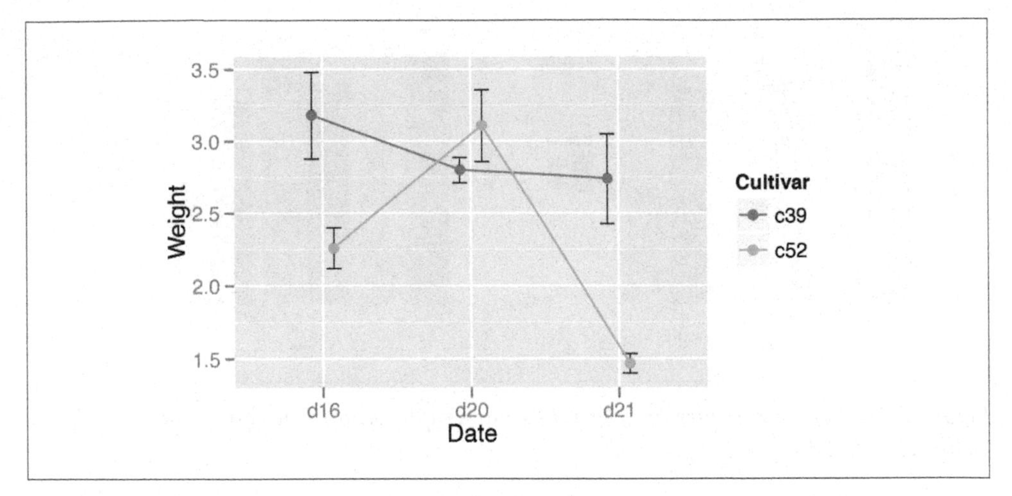

Figure 7-16. Error bars on a line graph, dodged so they don't overlap

we needed some other way to inform `ggplot()` that the two data entries at each *x* were in different groups so that they would be dodged.

See Also

See Recipe 3.2 for more about creating grouped bar graphs, and Recipe 4.3 for more about creating line graphs with multiple lines.

See Recipe 15.18 for calculating summaries with means, standard deviations, standard errors, and confidence intervals.

See Recipe 4.9 for adding a confidence region when the data has a higher density along the x-axis.

7.8. Adding Annotations to Individual Facets

Problem

You want to add annotations to each facet in a plot.

Solution

Create a new data frame with the faceting variable(s), and a value to use in each facet. Then use `geom_text()` with the new data frame (Figure 7-17):

```
# The base plot
p <- ggplot(mpg, aes(x=displ, y=hwy)) + geom_point() + facet_grid(. ~ drv)
```

```
# A data frame with labels for each facet
f_labels <- data.frame(drv = c("4", "f", "r"), label = c("4wd", "Front", "Rear"))

p + geom_text(x=6, y=40, aes(label=label), data=f_labels)

# If you use annotate(), the label will appear in all facets
p + annotate("text", x=6, y=42, label="label text")
```

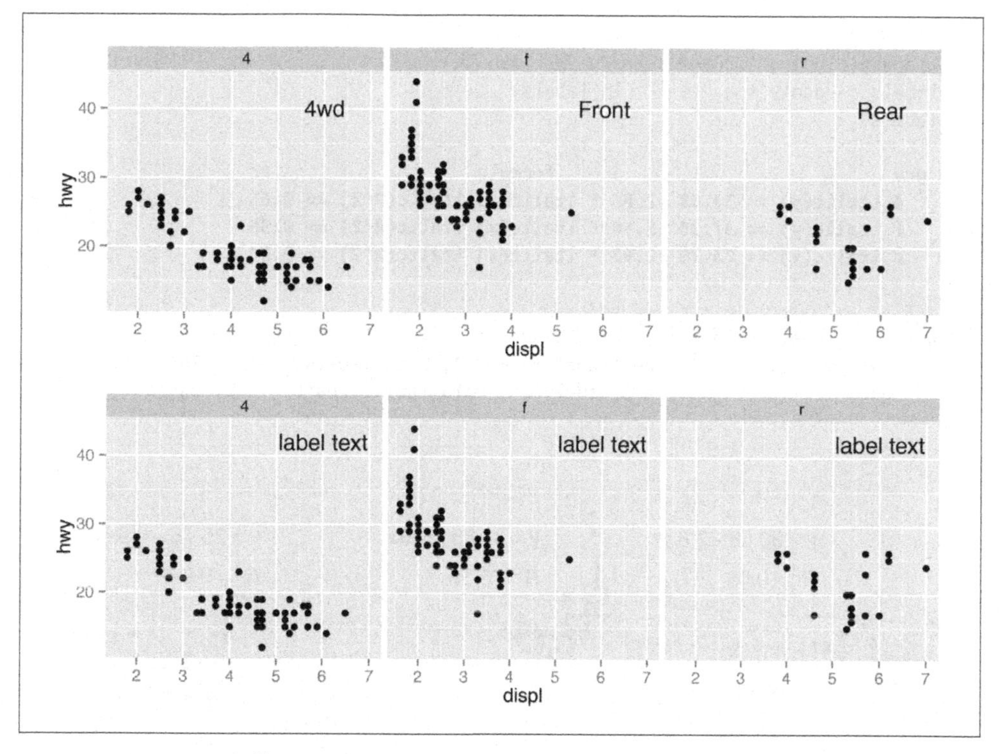

Figure 7-17. Top: different annotations in each facet; bottom: the same annotation in each facet

Discussion

This method can be used to display information about the data in each facet, as shown in Figure 7-18. For example, in each facet we can show linear regression lines, the formula for each line, and the r^2 value (). To do this, we'll write a function that takes a data frame and returns another data frame containing a string for a regression equation, and a string for the r^2 value. Then we'll use `ddply()` to apply that function to each group of the data:

```
# This function returns a data frame with strings representing the regression
# equation, and the r^2 value
# These strings will be treated as R math expressions
```

```
lm_labels <- function(dat) {
  mod <- lm(hwy ~ displ, data=dat)
  formula <- sprintf("italic(y) == %.2f %+.2f * italic(x)",
                     round(coef(mod)[1], 2), round(coef(mod)[2], 2))

  r <- cor(dat$displ, dat$hwy)
  r2 <- sprintf("italic(R^2) == %.2f", r^2)
  data.frame(formula=formula, r2=r2, stringsAsFactors=FALSE)
}

library(plyr) # For the ddply() function
labels <- ddply(mpg, "drv", lm_labels)
labels
```

```
 drv                              formula                 r2
   4 italic(y) == 30.68 -2.88 * italic(x) italic(R^2) == 0.65
   f italic(y) == 37.38 -3.60 * italic(x) italic(R^2) == 0.36
   r italic(y) == 25.78 -0.92 * italic(x) italic(R^2) == 0.04
```

```
# Plot with formula and R^2 values
p + geom_smooth(method=lm, se=FALSE) +
    geom_text(x=3, y=40, aes(label=formula), data=labels, parse=TRUE, hjust=0) +
    geom_text(x=3, y=35, aes(label=r2), data=labels, parse=TRUE, hjust=0)
```

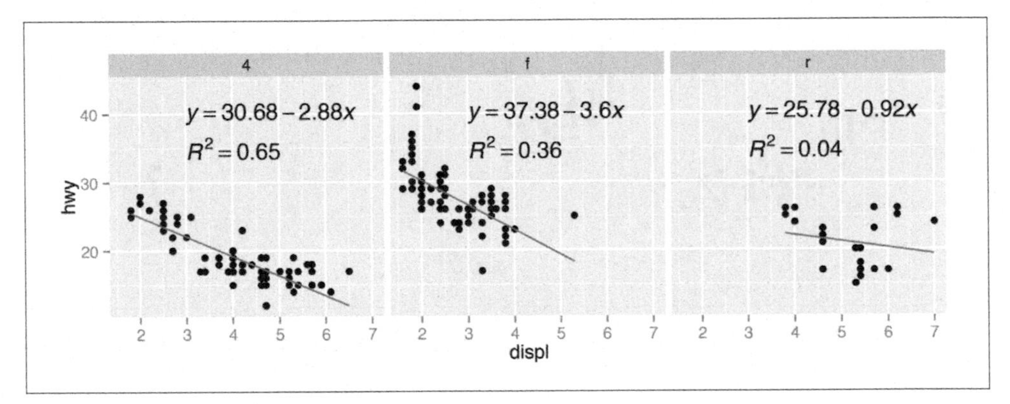

Figure 7-18. Annotations in each facet with information about the data

We needed to write our own function here because generating the linear model and extracting the coefficients requires operating on each subset data frame directly. If you just want to display the r^2 values, it's possible to do something simpler, by using `ddply()` with the `summarise()` function and then passing additional arguments for `summarise()`:

```
# Find r^2 values for each group
labels <- ddply(mpg, "drv", summarise, r2 = cor(displ, hwy)^2)
labels$r2 <- sprintf("italic(R^2) == %.2f", labels$r2)
```

Text geoms aren't the only kind that can be added individually for each facet. Any geom can be used, as long as the input data is structured correctly.

See Also

See Recipe 7.2 for more about using math expressions in plots.

If you want to make prediction lines from your own model objects, instead of having ggplot2 do it for you with `stat_smooth()`, see Recipe 5.8.

CHAPTER 8

Axes

The x- and y-axes provide context for interpreting the displayed data. Ggplot2 will display the axes with defaults that look good in most cases, but you might want to control, for example, the axis labels, the number and placement of tick marks, the tick mark labels, and so on. In this chapter, I'll cover how to fine-tune the appearance of the axes.

8.1. Swapping X- and Y-Axes

Problem

You want to swap the x- and y-axes on a graph.

Solution

Use `coord_flip()` to flip the axes (Figure 8-1):

```
ggplot(PlantGrowth, aes(x=group, y=weight)) + geom_boxplot()

ggplot(PlantGrowth, aes(x=group, y=weight)) + geom_boxplot() + coord_flip()
```

Discussion

For a scatter plot, it is trivial to change what goes on the vertical axis and what goes on the horizontal axis: just exchange the variables mapped to `x` and `y`. But not all the geoms in ggplot2 treat the x- and y-axes equally. For example, box plots summarize the data along the y-axis, the lines in line graphs move in only one direction along the x-axis, error bars have a single *x* value and a range of *y* values, and so on. If you're using these geoms and want them to behave as though the axes are swapped, `coord_flip()` is what you need.

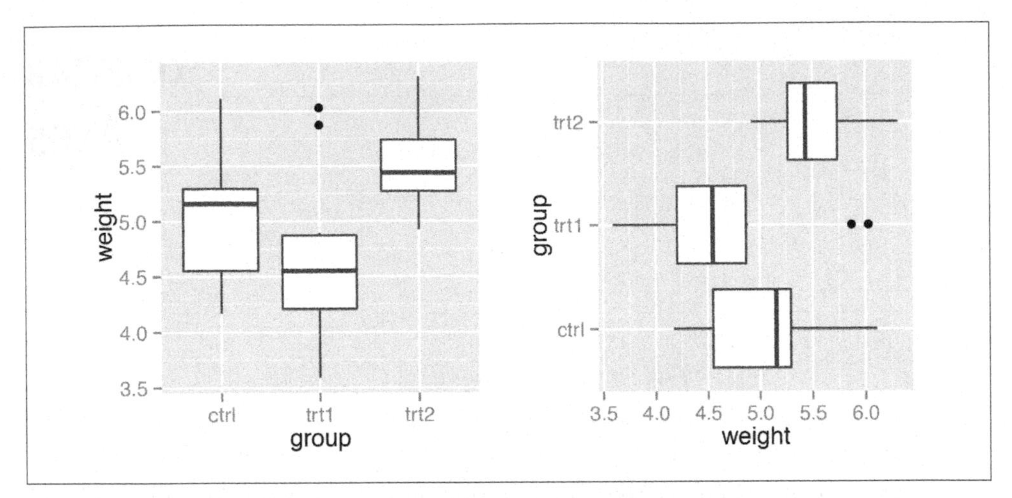

Figure 8-1. Left: a box plot with regular axes; right: with swapped axes

Sometimes when the axes are swapped, the order of items will be the reverse of what you want. On a graph with standard x- and y-axes, the *x* items start at the left and go to the right, which corresponds to the normal way of reading, from left to right. When you swap the axes, the items still go from the origin outward, which in this case will be from bottom to top—but this conflicts with the normal way of reading, from top to bottom. Sometimes this is a problem, and sometimes it isn't. If the *x* variable is a factor, the order can be reversed by using `scale_x_discrete()` with `limits=rev(levels(...))`, as in Figure 8-2:

```
ggplot(PlantGrowth, aes(x=group, y=weight)) + geom_boxplot() + coord_flip() +
    scale_x_discrete(limits=rev(levels(PlantGrowth$group)))
```

See Also

If the variable is continuous, see Recipe 8.3 to reverse the direction.

8.2. Setting the Range of a Continuous Axis

Problem

You want to set the range (or limits) of an axis.

Solution

You can use `xlim()` or `ylim()` to set the minimum and maximum values of a continuous axis. Figure 8-3 shows one graph with the default *y* limits, and one with manually set *y* limits:

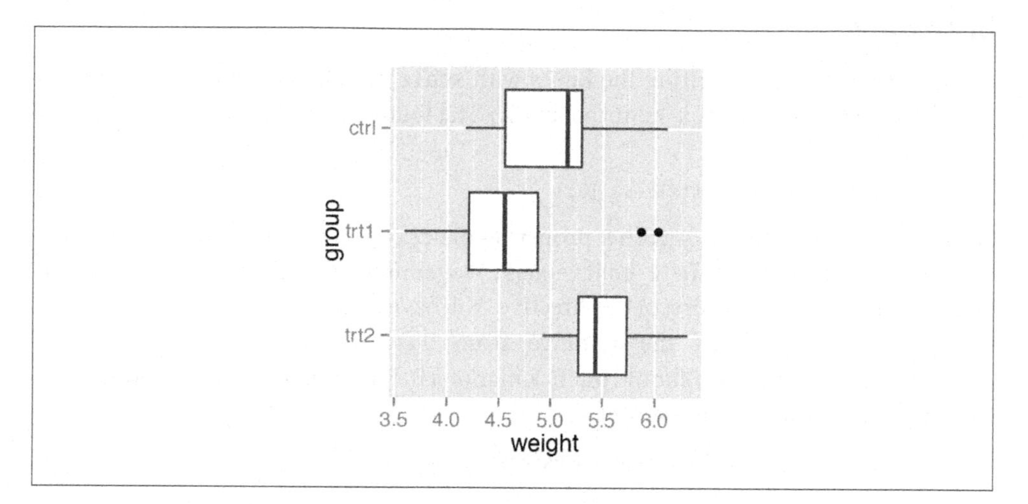

Figure 8-2. A box plot with swapped axes and x-axis order reversed

```
p <- ggplot(PlantGrowth, aes(x=group, y=weight)) + geom_boxplot()
# Display the basic graph
p

p + ylim(0, max(PlantGrowth$weight))
```

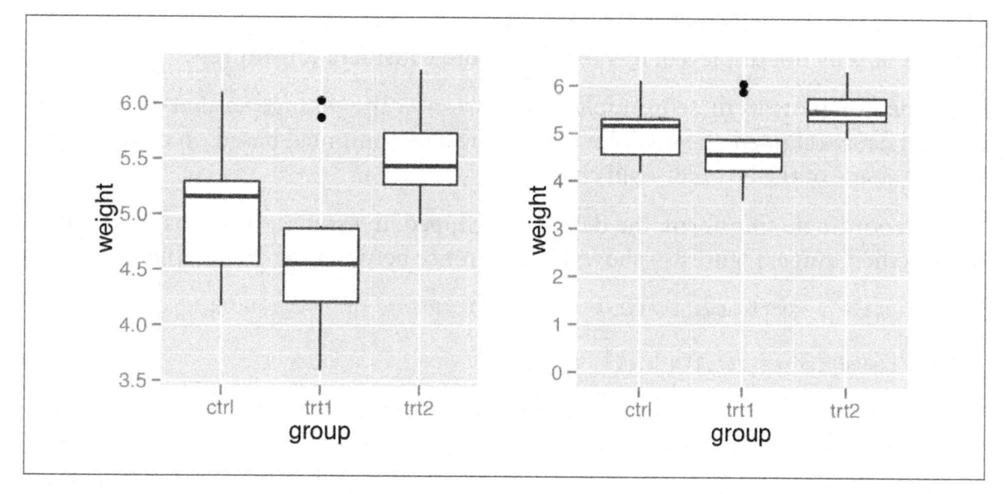

Figure 8-3. Left: box plot with default range; right: with manually set range

The latter example sets the *y* range from 0 to the maximum value of the `weight` column, though a constant value (like 10) could instead be used as the maximum.

Discussion

`ylim()` is shorthand for setting the limits with `scale_y_continuous()`. (The same is true for `xlim()` and `scale_x_continuous()`.) The following are equivalent:

```
ylim(0, 10)
scale_y_continuous(limits=c(0, 10))
```

Sometimes you will need to set other properties of `scale_y_continuous()`, and in these cases using `xlim()` and `scale_y_continuous()` together may result in some unexpected behavior, because only the first of the directives will have an effect. In these two examples, `ylim(0, 10)` should set the *y* range from 0 to 10, and `scale_y_continuous(breaks=c(0, 5, 10))` should put tick marks at 0, 5, and 10. However, in both cases, only the second directive has any effect:

```
p + ylim(0, 10) + scale_y_continuous(breaks=NULL)

p + scale_y_continuous(breaks=NULL) + ylim(0, 10)
```

To make both changes work, get rid of `ylim()` and set both `limits` and `breaks` in `scale_y_continuous()`:

```
p + scale_y_continuous(limits=c(0, 10), breaks=NULL)
```

In ggplot2, there are two ways of setting the range of the axes. The first way is to modify the *scale*, and the second is to apply a *coordinate transform*. When you modify the limits of the *x* or *y* scale, any data outside of the limits is removed—that is, the out-of-range data is not only not displayed, it is removed from consideration entirely.

With the box plots in these examples, if you restrict the *y* range so that some of the original data is clipped, the box plot statistics will be computed based on clipped data, and the shape of the box plots will change.

With a coordinate transform, the data is not clipped; in essence, it zooms in or out to the specified range. Figure 8-4 shows the difference between the two methods:

```
p + scale_y_continuous(limits = c(5, 6.5))  # Same as using ylim()

p + coord_cartesian(ylim = c(5, 6.5))
```

Finally, it's also possible to *expand* the range in one direction, using `expand_limits()` (Figure 8-5). You can't use this to shrink the range, however:

```
p + expand_limits(y=0)
```

8.3. Reversing a Continuous Axis

Problem

You want to reverse the direction of a continuous axis.

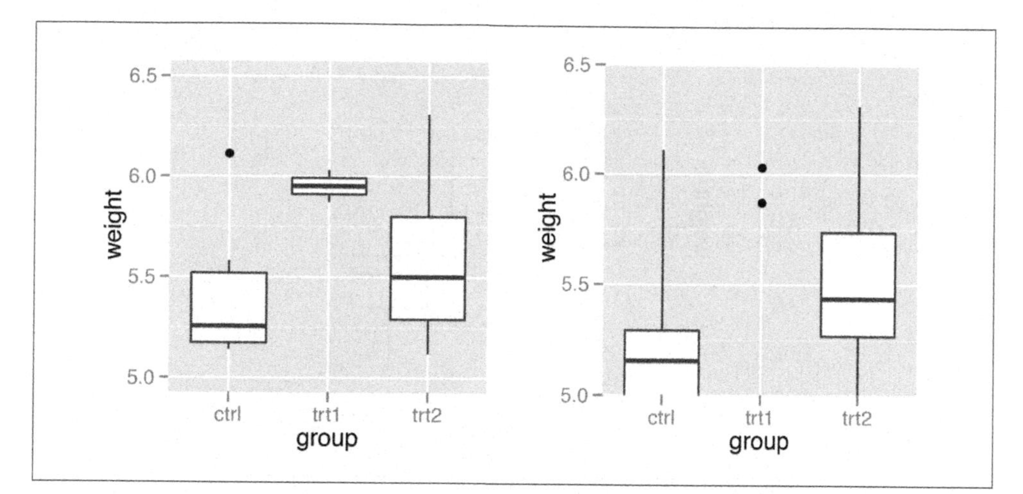

Figure 8-4. Left: smaller y range using a scale (data has been dropped, so the box plots have changed shape); right: "zooming in" using a coordinate transform

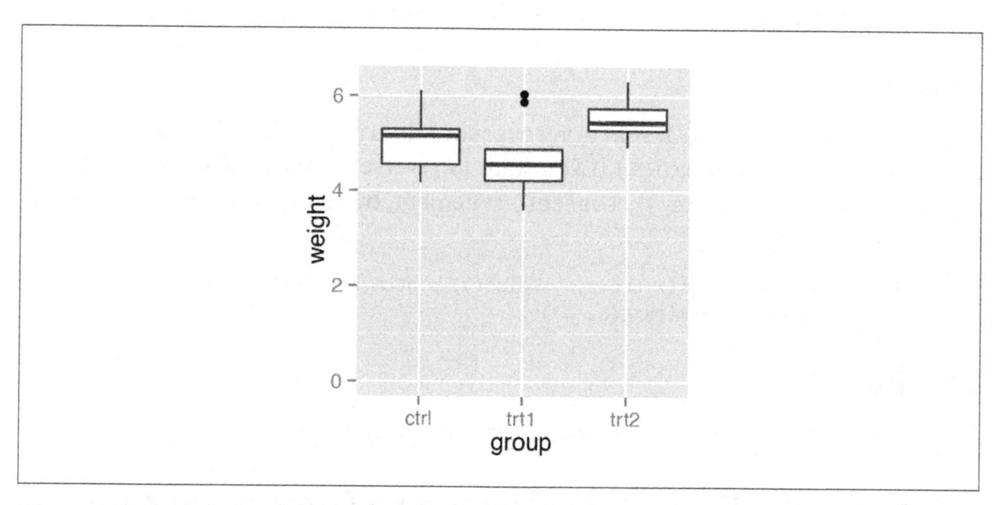

Figure 8-5. Box plot on which y range has been expanded to include 0

Solution

Use `scale_y_reverse` or `scale_x_reverse` (Figure 8-6). The direction of an axis can also be reversed by specifying the limits in reversed order, with the maximum first, then the minimum:

```
ggplot(PlantGrowth, aes(x=group, y=weight)) + geom_boxplot() + scale_y_reverse()
```

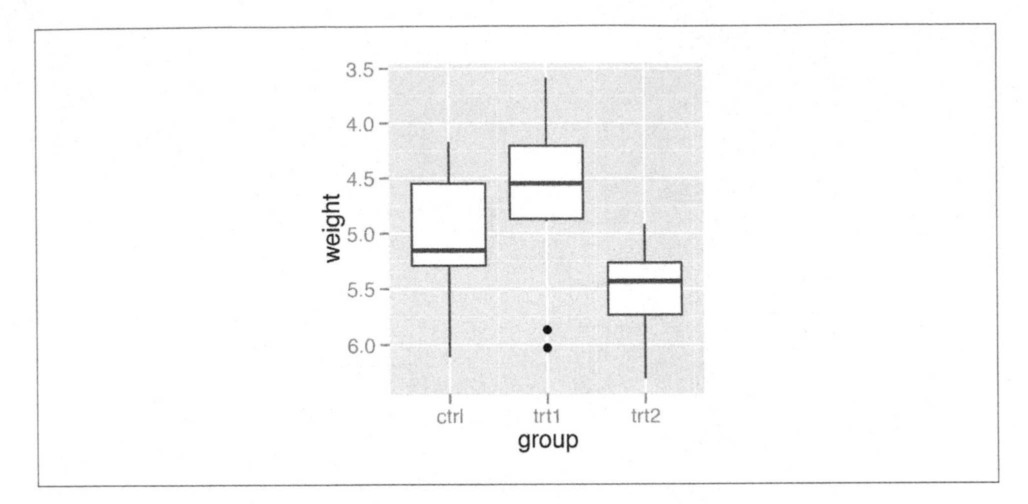

Figure 8-6. Box plot with reversed y-axis

```
# Similar effect by specifying limits in reversed order
ggplot(PlantGrowth, aes(x=group, y=weight)) + geom_boxplot() + ylim(6.5, 3.5)
```

Discussion

Like `scale_y_continuous()`, `scale_y_reverse()` does not work with `ylim`. (The same is true for the x-axis properties.) If you want to reverse an axis *and* set its range, you must do it within the `scale_y_reverse()` statement, by setting the limits in reversed order (Figure 8-7):

```
ggplot(PlantGrowth, aes(x=group, y=weight)) + geom_boxplot() +
    scale_y_reverse(limits=c(8, 0))
```

See Also

To reverse the order of items on a *discrete* axis, see Recipe 8.4.

8.4. Changing the Order of Items on a Categorical Axis

Problem

You want to change the order of items on a categorical axis.

Solution

For a categorical (or discrete) axis—one with a factor mapped to it—the order of items can be changed by setting `limits` in `scale_x_discrete()` or `scale_y_discrete()`.

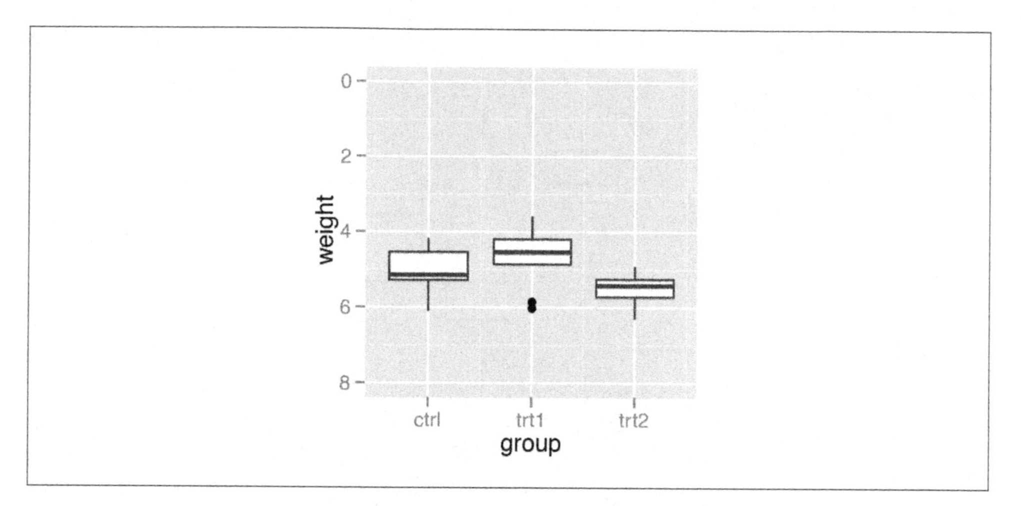

Figure 8-7. Box plot with reversed y-axis with manually set limits

To manually set the order of items on the axis, specify `limits` with a vector of the levels in the desired order. You can also omit items with this vector, as shown in Figure 8-8:

```
p <- ggplot(PlantGrowth, aes(x=group, y=weight)) + geom_boxplot()

p + scale_x_discrete(limits=c("trt1","ctrl","trt2"))
```

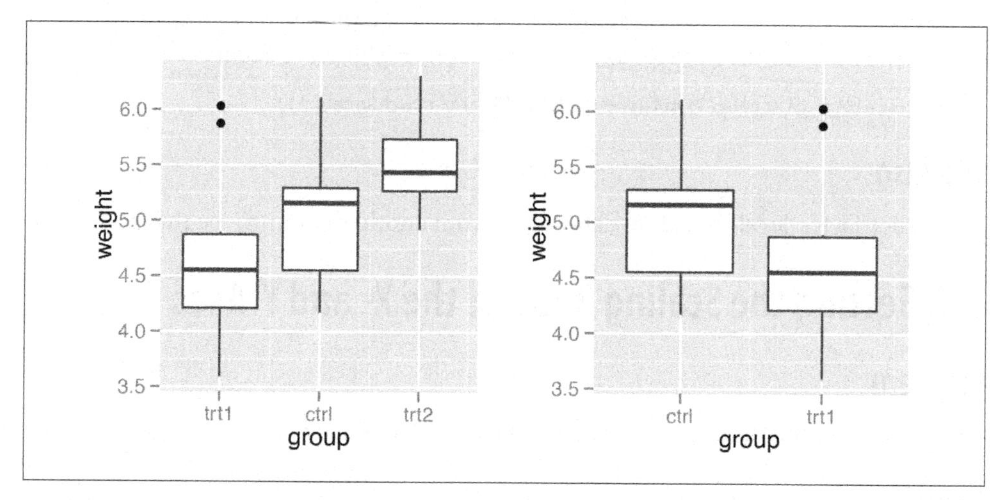

Figure 8-8. Left: box plot with manually specified items on the x-axis; right: with only two items

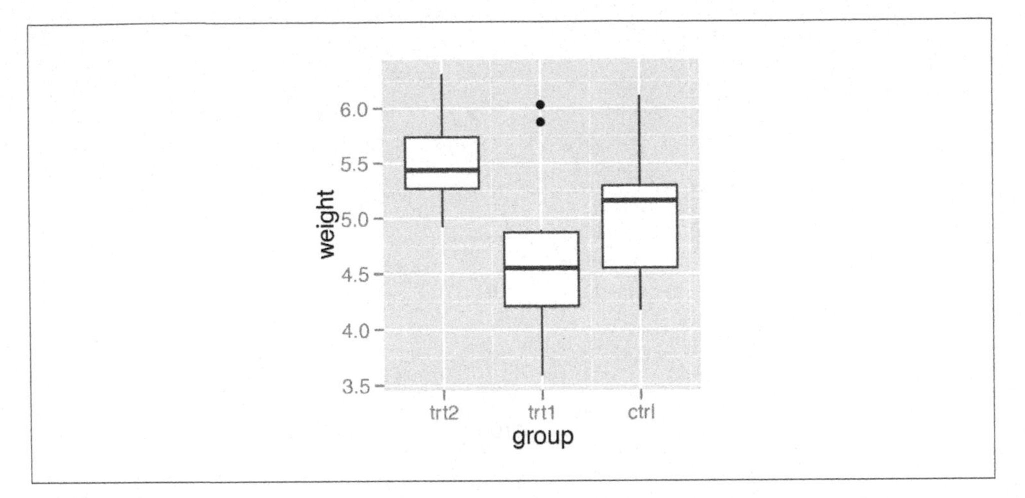

Figure 8-9. Box plot with order reversed on the x-axis

Discussion

You can also use this method to display a subset of the items on the axis. This will show only `ctrl` and `trt1` (Figure 8-8, right):

```
p + scale_x_discrete(limits=c("ctrl","trt1"))
```

To reverse the order, set `limits=rev(levels(...))`, and put the factor inside. This will reverse the order of the `PlantGrowth$group` factor, as shown in Figure 8-9:

```
p + scale_x_discrete(limits=rev(levels(PlantGrowth$group)))
```

See Also

To reorder factor levels based on data values from another column, see Recipe 15.9.

8.5. Setting the Scaling Ratio of the X- and Y-Axes

Problem

You want to set the ratio at which the x- and y-axes are scaled.

Solution

Use `coord_fixed()`. This will result in a 1:1 scaling between the x- and y-axes, as shown in Figure 8-10:

```
library(gcookbook) # For the data set
```

```
sp <- ggplot(marathon, aes(x=Half,y=Full)) + geom_point()

sp + coord_fixed()
```

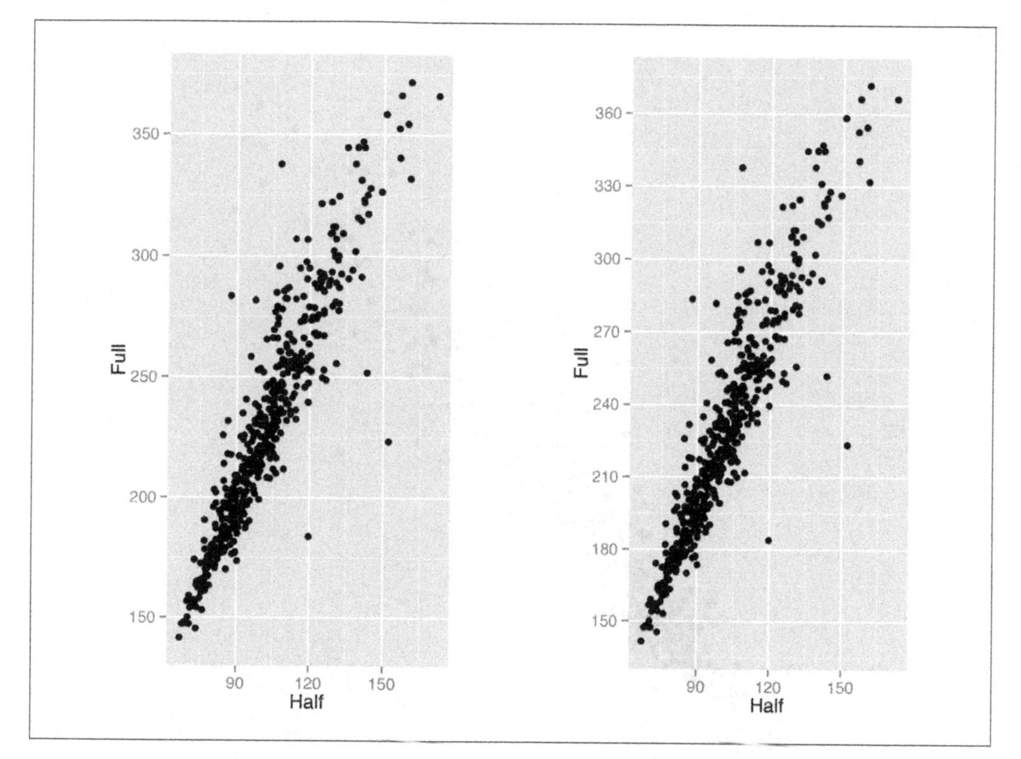

Figure 8-10. Left: scatter plot with equal scaling of axes; right: with tick marks at specified positions

Discussion

The `marathon` data set contains runners' marathon and half-marathon times. In this case it might be useful to force the x- and y-axes to have the same scaling.

It's also helpful to set the tick spacing to be the same, by setting `breaks` in `scale_y_continuous()` and `scale_x_continuous()` (also in Figure 8-10):

```
sp + coord_fixed() +
    scale_y_continuous(breaks=seq(0, 420, 30)) +
    scale_x_continuous(breaks=seq(0, 420, 30))
```

If, instead of an equal ratio, you want some other fixed ratio between the axes, set the `ratio` parameter. With the `marathon` data, we might want the axis with half-marathon times stretched out to twice that of the axis with the marathon times (Figure 8-11). We'll also add tick marks twice as often on the x-axis:

```
sp + coord_fixed(ratio=1/2) +
    scale_y_continuous(breaks=seq(0, 420, 30)) +
    scale_x_continuous(breaks=seq(0, 420, 15))
```

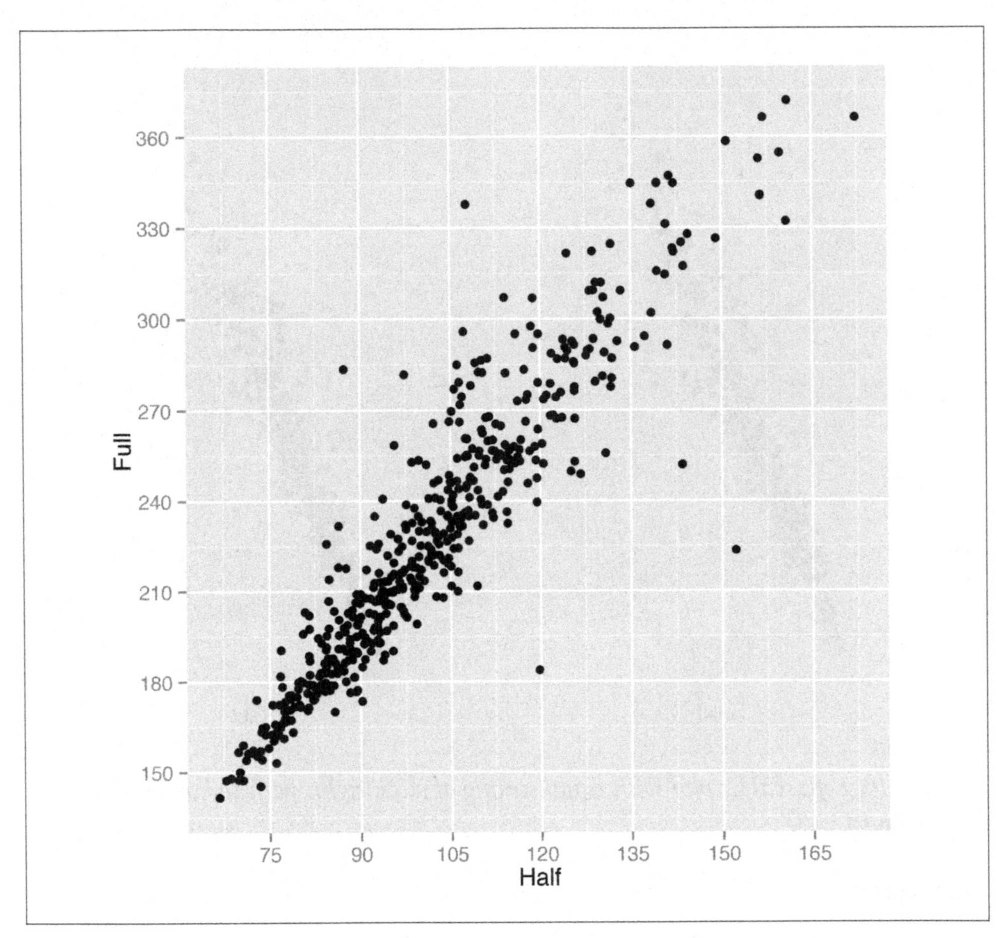

Figure 8-11. Scatter plot with a 1/2 scaling ratio for the axes

8.6. Setting the Positions of Tick Marks

Problem

You want to set where the tick marks appear on the axis.

Solution

Usually `ggplot()` does a good job of deciding where to put the tick marks, but if you want to change them, set `breaks` in the scale (Figure 8-12):

```
ggplot(PlantGrowth, aes(x=group, y=weight)) + geom_boxplot()

ggplot(PlantGrowth, aes(x=group, y=weight)) + geom_boxplot() +
    scale_y_continuous(breaks=c(4, 4.25, 4.5, 5, 6, 8))
```

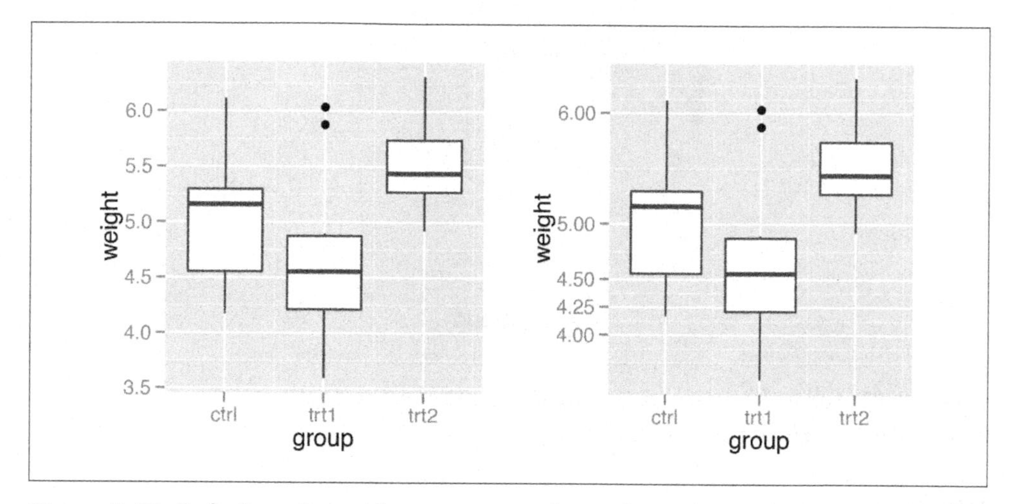

Figure 8-12. Left: box plot with automatic tick marks; right: with manually set tick marks

Discussion

The location of the tick marks defines where *major* grid lines are drawn. If the axis represents a continuous variable, *minor* grid lines, which are fainter and unlabeled, will by default be drawn halfway between each major grid line.

You can also use the `seq()` function or the `:` operator to generate vectors for tick marks:

```
seq(4, 7, by=.5)

4.0 4.5 5.0 5.5 6.0 6.5 7.0

5:10

5  6  7  8  9 10
```

If the axis is discrete instead of continuous, then there is by default a tick mark for each item. For discrete axes, you can change the order of items or remove them by specifying the limits (see Recipe 8.4). Setting `breaks` will change which of the levels are labeled, but will not remove them or change their order. Figure 8-13 shows what happens when you set `limits` and `breaks`:

```
# Set both breaks and labels for a discrete axis
ggplot(PlantGrowth, aes(x=group, y=weight)) + geom_boxplot() +
    scale_x_discrete(limits=c("trt2", "ctrl"), breaks="ctrl")
```

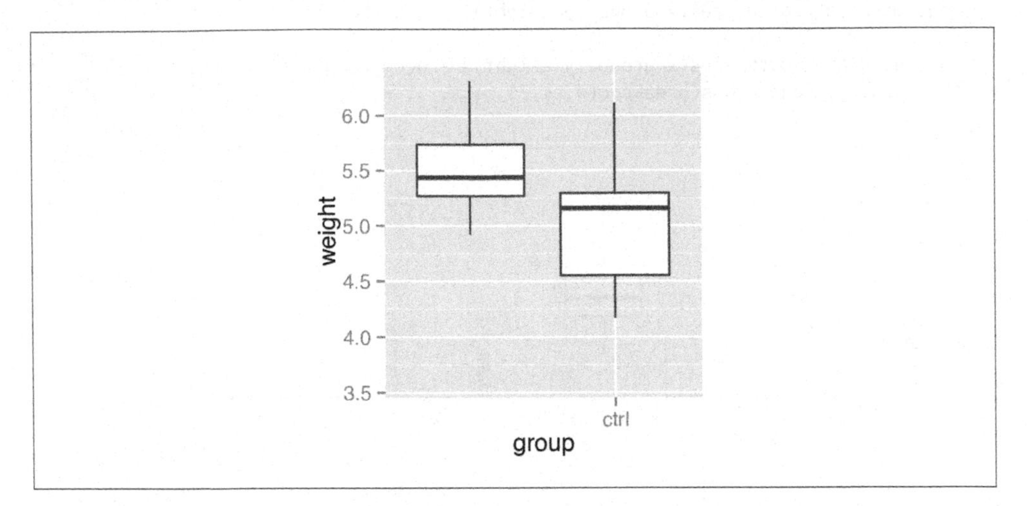

Figure 8-13. For a discrete axis, setting limits reorders and removes items, and setting breaks controls which items have labels

See Also

To remove the tick marks and labels (but not the data) from the graph, see Recipe 8.7.

8.7. Removing Tick Marks and Labels

Problem

You want to remove tick marks and labels.

Solution

To remove just the tick labels, as in Figure 8-14 (left), use `theme(axis.text.y = element_blank())` (or do the same for `axis.text.x`). This will work for both continuous and categorical axes:

```
p <- ggplot(PlantGrowth, aes(x=group, y=weight)) + geom_boxplot()

p + theme(axis.text.y = element_blank())
```

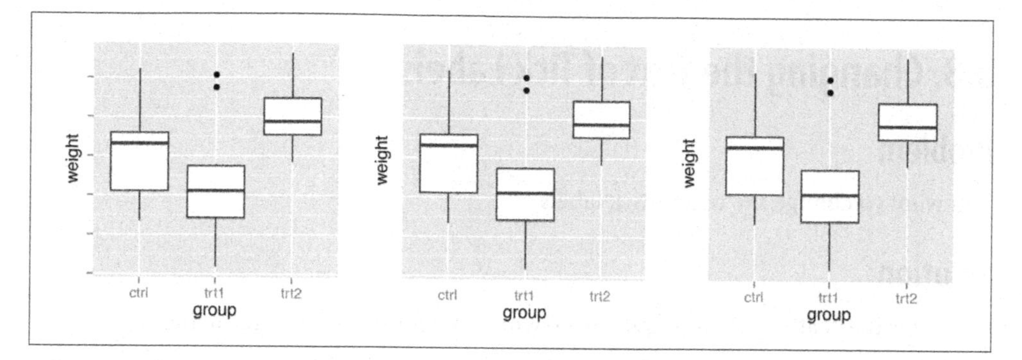

Figure 8-14. Left: no tick labels on y-axis; middle: no tick marks and no tick labels on y-axis; right: with breaks=NULL

To remove the tick marks, use `theme(axis.ticks=element_blank())`. This will remove the tick marks on both axes. (It's not possible to hide the tick marks on just one axis.) In this example, we'll hide all tick marks as well as the *y* tick labels (Figure 8-14, center):

```
p + theme(axis.ticks = element_blank(), axis.text.y = element_blank())
```

To remove the tick marks, the labels, and the grid lines, set `breaks` to `NULL` (Figure 8-14, right):

```
p + scale_y_continuous(breaks=NULL)
```

This will work for continuous axes only; if you remove items from a categorical axis using `limits`, as in Recipe 8.4, the data with that value won't be shown at all.

Discussion

There are actually three related items that can be controlled: tick labels, tick marks, and the grid lines. For continuous axes, `ggplot()` normally places a tick label, tick mark, and major grid line at each value of `breaks`. For categorical axes, these things go at each value of `limits`.

The tick labels on each axis can be controlled independently. However, the tick marks and grid lines must be controlled all together.

8.8. Changing the Text of Tick Labels

Problem

You want to change the text of tick labels.

Solution

Consider the scatter plot in Figure 8-15, where height is reported in inches:

```
library(gcookbook) # For the data set

hwp <- ggplot(heightweight, aes(x=ageYear, y=heightIn)) +
         geom_point()

hwp
```

To set arbitrary labels, as in Figure 8-15 (right), pass values to `breaks` and `labels` in the scale. One of the labels has a newline (`\n`) character, which tells `ggplot()` to put a line break there:

```
hwp + scale_y_continuous(breaks=c(50, 56, 60, 66, 72),
                         labels=c("Tiny", "Really\nshort", "Short",
                                  "Medium", "Tallish"))
```

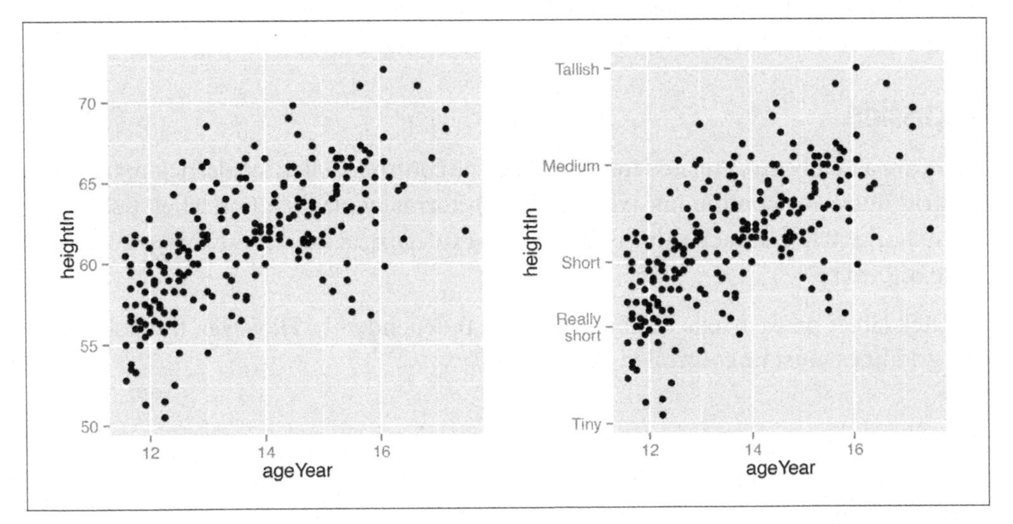

Figure 8-15. Left: scatter plot with automatic tick labels; right: with manually specified labels on the y-axis

Discussion

Instead of setting completely arbitrary labels, it is more common to have your data stored in one format, while wanting the labels to be displayed in another. We might, for example, want heights to be displayed in feet and inches (like 5'6") instead of just inches. To do this, we can define a *formatter* function, which takes in a value and returns the corresponding string. For example, this function will convert inches to feet and inches:

```
footinch_formatter <- function(x) {
    foot <- floor(x/12)
    inch <- x %% 12
    return(paste(foot, "'", inch, "\"", sep=""))
}
```

Here's what it returns for values 56–64 (the backslashes are there as escape characters, to distinguish the quotes *in* a string from the quotes that *delimit* a string):

```
footinch_formatter(56:64)

 "4'8\""  "4'9\""  "4'10\"" "4'11\"" "5'0\""  "5'1\""  "5'2\""  "5'3\""  "5'4\""
```

Now we can pass our function to the scale, using the `labels` parameter (Figure 8-16):

```
hwp + scale_y_continuous(labels=footinch_formatter)
```

Here, the automatic tick marks were placed every five inches, but that looks a little off for this data. We can instead have `ggplot()` set tick marks every four inches, by specifying `breaks` (Figure 8-16, right):

```
hwp + scale_y_continuous(breaks=seq(48, 72, 4), labels=footinch_formatter)
```

Another common task is to convert time measurements to HH:MM:SS format, or something similar. This function will take numeric minutes and convert them to this format, rounding to the nearest second (it can be customized for your particular needs):

```
timeHMS_formatter <- function(x) {
    h <- floor(x/60)
    m <- floor(x %% 60)
    s <- round(60*(x %% 1))                   # Round to nearest second
    lab <- sprintf("%02d:%02d:%02d", h, m, s) # Format the strings as HH:MM:SS
    lab <- gsub("^00:", "", lab)              # Remove leading 00: if present
    lab <- gsub("^0", "", lab)                # Remove leading 0 if present
    return(lab)
}
```

Running it on some sample numbers yields:

```
timeHMS_formatter(c(.33, 50, 51.25, 59.32, 60, 60.1, 130.23))

 "0:20"    "50:00"    "51:15"    "59:19"    "1:00:00" "1:00:06" "2:10:14"
```

The scales package, which is installed with ggplot2, comes with some built-in formatting functions:

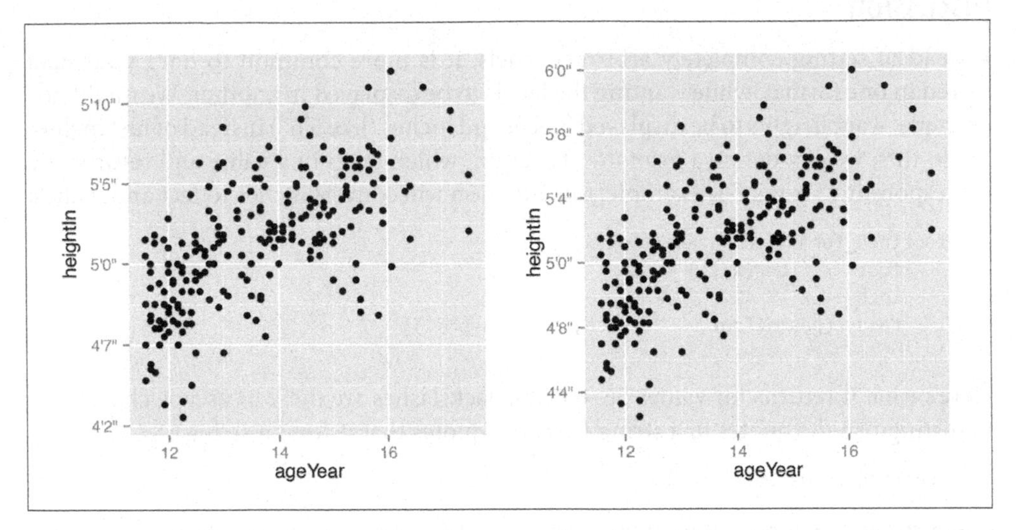

Figure 8-16. Left: scatter plot with a formatter function; right: with manually specified breaks on the y-axis

- `comma()` adds commas to numbers, in the thousand, million, billion, etc. places.
- `dollar()` adds a dollar sign and rounds to the nearest cent.
- `percent()` multiplies by 100, rounds to the nearest integer, and adds a percent sign.
- `scientific()` gives numbers in scientific notation, like `3.30e+05`, for large and small numbers.

If you want to use these functions, you must first load the scales package, with `library(scales)`.

8.9. Changing the Appearance of Tick Labels

Problem

You want to change the appearance of tick labels.

Solution

In Figure 8-17 (left), we've manually set the labels to be long—long enough that they overlap:

```
bp <- ggplot(PlantGrowth, aes(x=group, y=weight)) + geom_boxplot() +
      scale_x_discrete(breaks=c("ctrl", "trt1", "trt2"),
                       labels=c("Control", "Treatment 1", "Treatment 2"))
bp
```

To rotate the text 90 degrees counterclockwise (Figure 8-17, middle), use:

```
bp + theme(axis.text.x = element_text(angle=90, hjust=1, vjust=.5))
```

Rotating the text 30 degrees (Figure 8-17, right) uses less vertical space and makes the labels easier to read without tilting your head:

```
bp + theme(axis.text.x = element_text(angle=30, hjust=1, vjust=1))
```

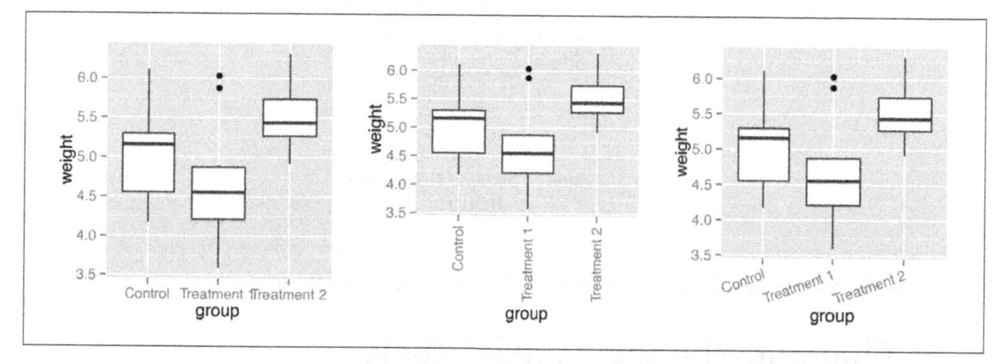

Figure 8-17. X-axis tick labels rotated 0 (left), 90 (middle), and 30 degrees (right)

The `hjust` and `vjust` settings specify the horizontal alignment (left/center/right) and vertical alignment (top/middle/bottom).

Discussion

Besides rotation, other text properties, such as size, style (bold/italic/normal), and the font family (such as Times or Helvetica) can be set with `element_text()`, as shown in Figure 8-18:

```
bp + theme(axis.text.x = element_text(family="Times", face="italic",
                                      colour="darkred", size=rel(0.9)))
```

In this example, the `size` is set to `rel(0.9)`, which means that it is 0.9 times the size of the base font size for the theme.

These commands control the appearance of only the tick labels, on only one axis. They don't affect the other axis, the axis label, the overall title, or the legend. To control all of these at once, you can use the theming system, as discussed in Recipe 9.3.

See Also

See Recipe 9.2 for more about controlling the appearance of the text.

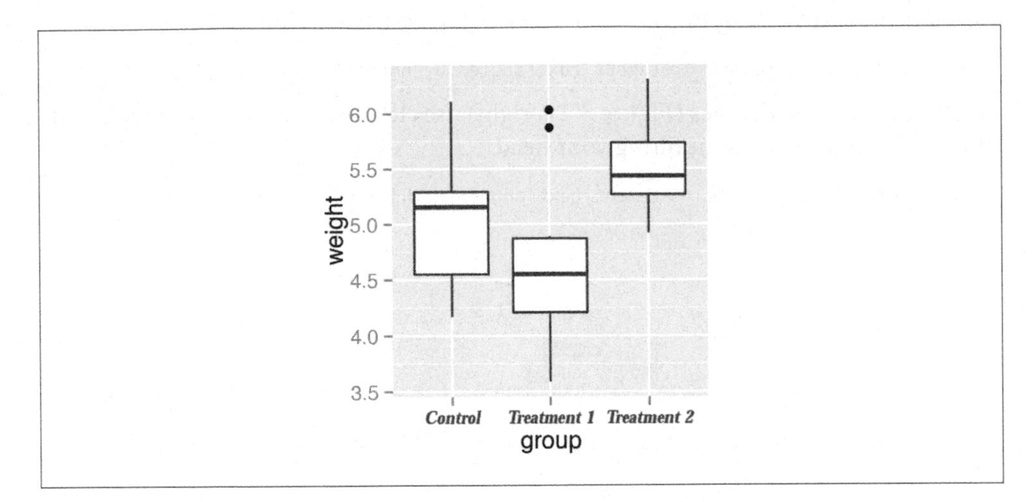

Figure 8-18. X-axis tick labels with manually specified appearance

8.10. Changing the Text of Axis Labels

Problem

You want to change the text of axis labels.

Solution

Use `xlab()` or `ylab()` to change the text of the axis labels (Figure 8-19):

```
library(gcookbook) # For the data set

hwp <- ggplot(heightweight, aes(x=ageYear, y=heightIn, colour=sex)) +
       geom_point()
# With default axis labels
hwp

# Set the axis labels
hwp + xlab("Age in years") + ylab("Height in inches")
```

Discussion

By default the graphs will just use the column names from the data frame as axis labels. This might be fine for exploring data, but for presenting it, you may want more descriptive axis labels.

Instead of `xlab()` and `ylab()`, you can use `labs()`:

```
hwp + labs(x = "Age in years", y = "Height in inches")
```

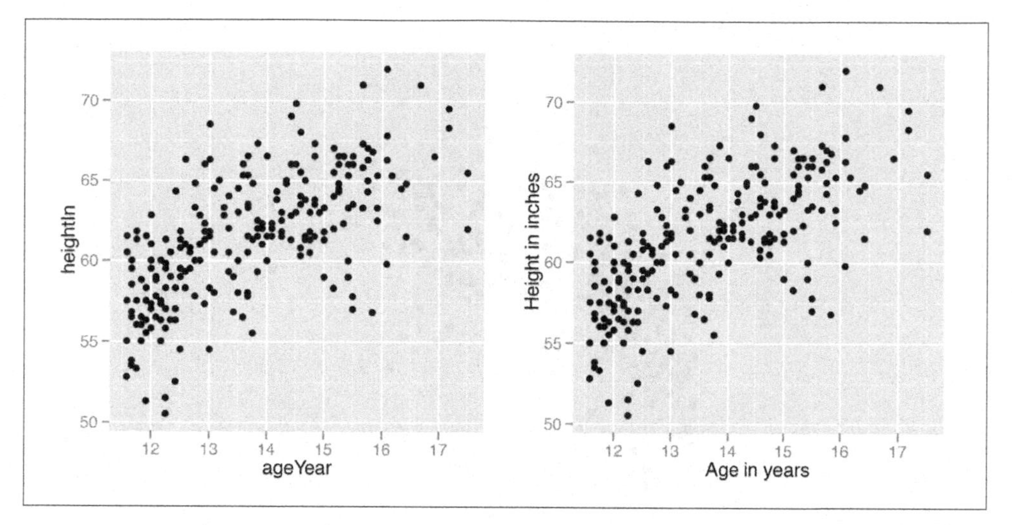

Figure 8-19. Left: scatter plot with the default axis labels; right: manually specified labels for the x- and y-axes

Another way of setting the axis labels is in the scale specification, like this:

```
hwp + scale_x_continuous(name="Age in years")
```

This may look a bit awkward, but it can be useful if you're also setting other properties of the scale, such as the tick mark placement, range, and so on.

This also applies, of course, to other axis scales, such as `scale_y_continuous()`, `scale_x_discrete()`, and so on.

You can also add line breaks with `\n`, as shown in Figure 8-20:

```
hwp + scale_x_continuous(name="Age\n(years)")
```

8.11. Removing Axis Labels

Problem

You want to remove the label on an axis.

Solution

For the x-axis label, use `theme(axis.title.x=element_blank())`. For the y-axis label, do the same with `axis.title.y`.

We'll hide the x-axis in this example (Figure 8-21):

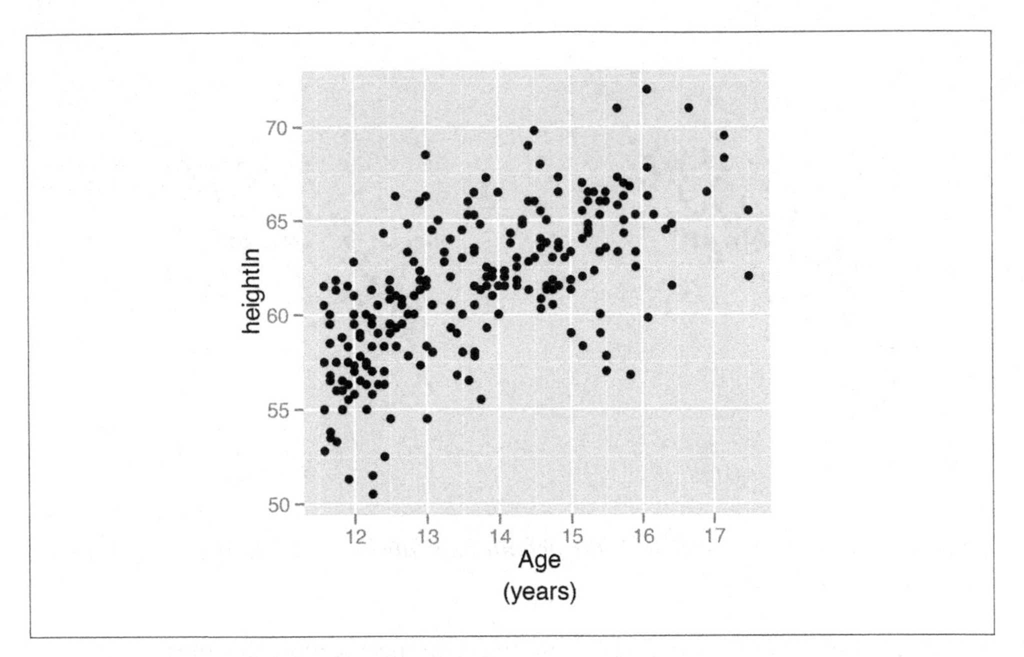

Figure 8-20. X-axis label with a line break

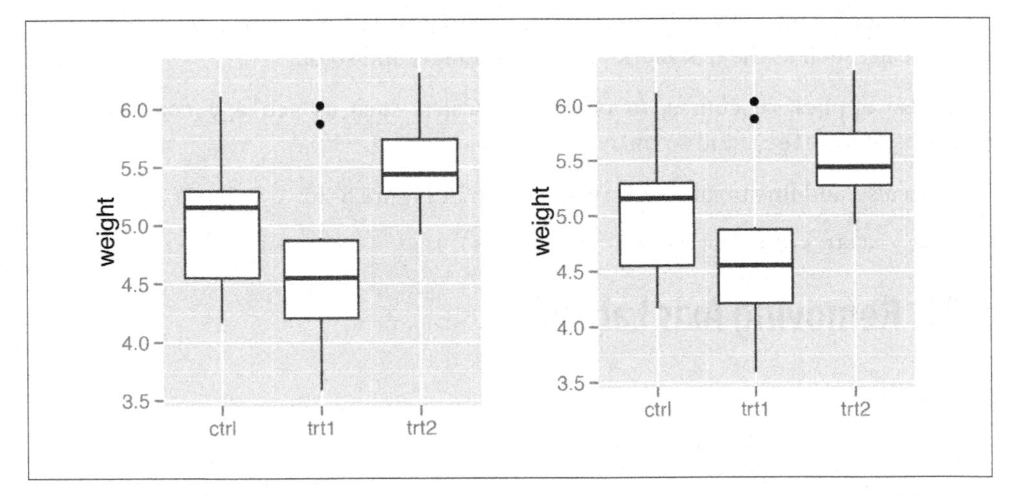

Figure 8-21. Left: x-axis label with element_blank(); right: with the label set to ""

```
p <- ggplot(PlantGrowth, aes(x=group, y=weight)) + geom_boxplot()

p + theme(axis.title.x=element_blank())
```

Discussion

Sometimes axis labels are redundant or obvious from the context, and don't need to be displayed. In the example here, the x-axis represents `group`, but this should be obvious from the context. Similarly, if the *y* tick labels had *kg* or some other unit in each label, the axis label "weight" would be unnecessary.

Another way to remove the axis label is to set it to an empty string. However, if you do it this way, the resulting graph will still have space reserved for the text, as shown in the graph on the right in Figure 8-21:

```
p + xlab("")
```

When you use `theme()` to set `axis.title.x=element_blank()`, the name of the *x* or *y* scale is unchanged, but the text is not displayed and no space is reserved for it. When you set the label to `""`, the name of the scale is changed and the (empty) text does display.

8.12. Changing the Appearance of Axis Labels

Problem

You want to change the appearance of axis labels.

Solution

To change the appearance of the x-axis label (Figure 8-22), use `axis.title.x`:

```
library(gcookbook) # For the data set

hwp <- ggplot(heightweight, aes(x=ageYear, y=heightIn)) + geom_point()

hwp + theme(axis.title.x=element_text(face="italic", colour="darkred", size=14))
```

Discussion

For the y-axis label, it might also be useful to display the text unrotated, as shown in Figure 8-23 (left). The `\n` in the label represents a newline character:

```
hwp + ylab("Height\n(inches)") +
    theme(axis.title.y=element_text(angle=0, face="italic", size=14))
```

When you call `element_text()`, the default angle is 0, so if you set `axis.title.y` but don't specify the angle, it will show in this orientation, with the top of the text pointing up. If you change any other properties of `axis.title.y` and want it to be displayed in its usual orientation, rotated 90 degrees, you must manually specify the angle (Figure 8-23, right):

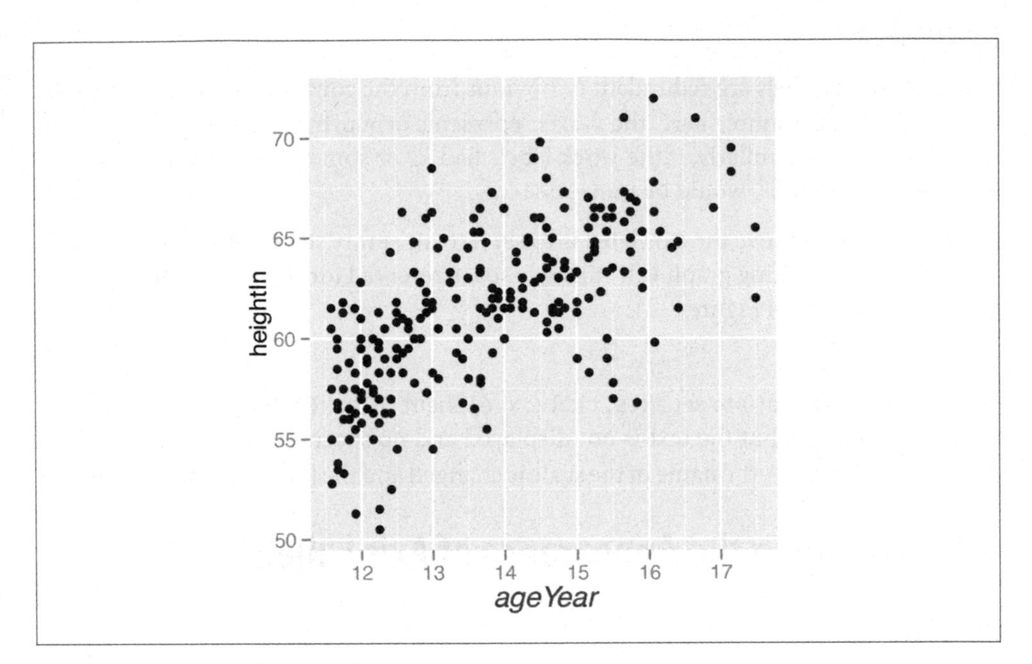

Figure 8-22. X-axis label with customized appearance

```
hwp + ylab("Height\n(inches)") +
    theme(axis.title.y=element_text(angle=90, face="italic", colour="darkred",
                                     size=14))
```

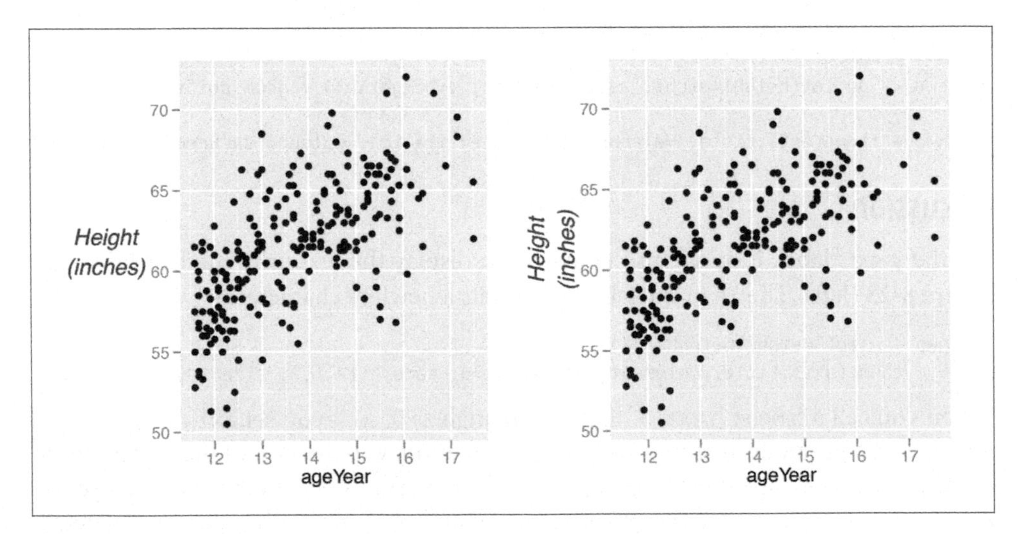

Figure 8-23. Left: y-axis label with angle=0; right: with angle=90

See Also

See Recipe 9.2 for more about controlling the appearance of the text.

8.13. Showing Lines Along the Axes

Problem

You want to display lines along the x- and y-axes, but not on the other sides of the graph.

Solution

Using themes, use `axis.line` (Figure 8-24):

```
library(gcookbook) # For the data set

p <- ggplot(heightweight, aes(x=ageYear, y=heightIn)) + geom_point()

p + theme(axis.line = element_line(colour="black"))
```

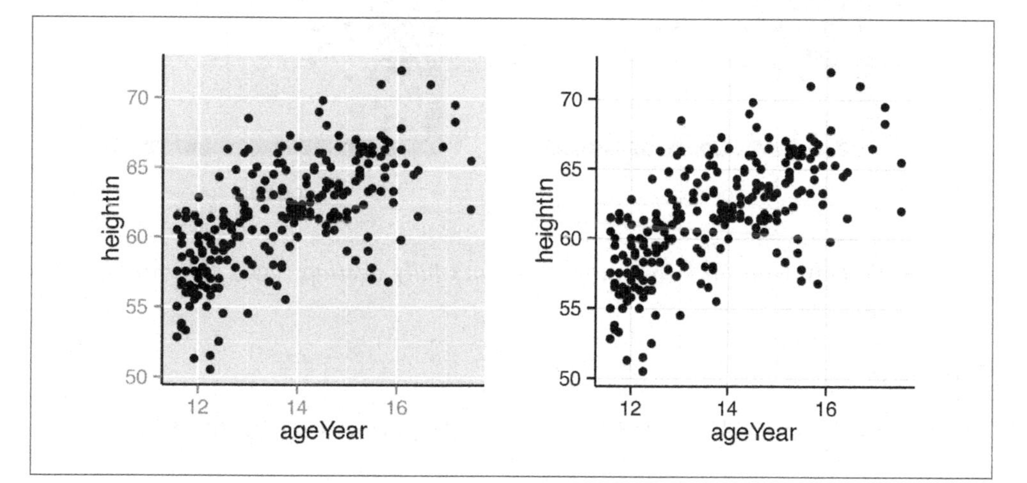

Figure 8-24. Left: scatter plot with axis lines; right: with theme_bw(), panel.border must also be made blank

Discussion

If you are starting with a theme that has a border around the plotting area, like `theme_bw()`, you will also need to unset `panel.border` (Figure 8-24, right):

```
p + theme_bw() +
    theme(panel.border = element_blank(),
          axis.line = element_line(colour="black"))
```

If the lines are thick, the ends will only partially overlap (Figure 8-25, left). To make them fully overlap (Figure 8-25, right), set `lineend="square"`:

```
# With thick lines, only half overlaps
p + theme_bw() +
    theme(panel.border = element_blank(),
          axis.line = element_line(colour="black", size=4))

# Full overlap
p + theme_bw() +
    theme(panel.border = element_blank(),
          axis.line = element_line(colour="black", size=4, lineend="square"))
```

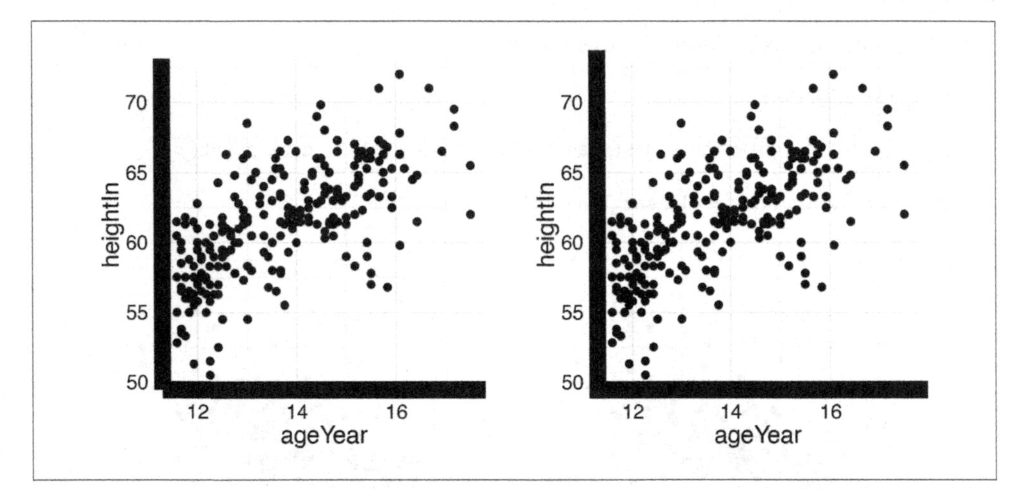

Figure 8-25. Left: with thick lines, the ends don't fully overlap; right: full overlap with lineend="square"

See Also

For more information about how the theming system works, see Recipe 9.3.

8.14. Using a Logarithmic Axis

Problem

You want to use a logarithmic axis for a graph.

Solution

Use `scale_x_log10()` and/or `scale_y_log10()` (Figure 8-26):

```
library(MASS) # For the data set

# The base plot
p <- ggplot(Animals, aes(x=body, y=brain, label=rownames(Animals))) +
    geom_text(size=3)
p

# With logarithmic x and y scales
p + scale_x_log10() + scale_y_log10()
```

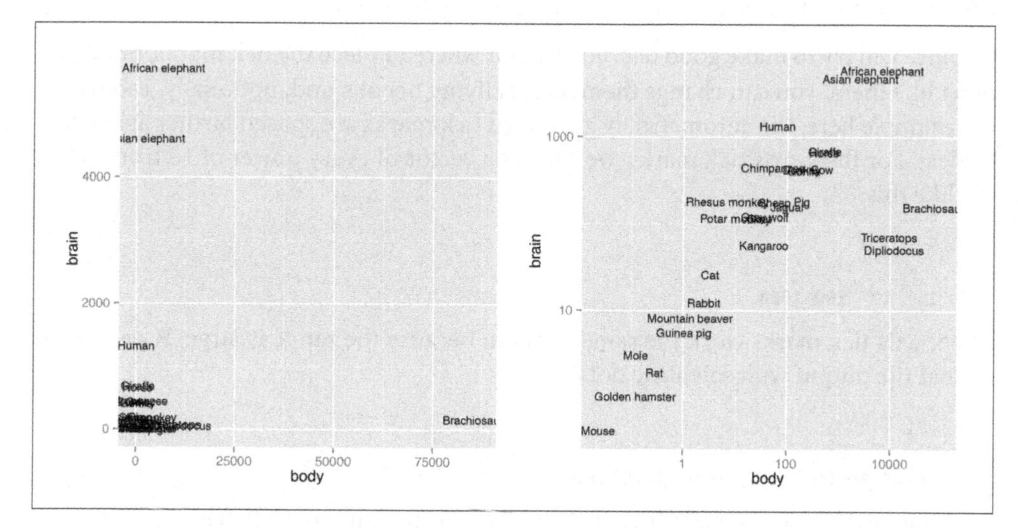

Figure 8-26. Left: exponentially distributed data with linear-scaled axes; right: with logarithmic axes

Discussion

With a log axis, a given visual distance represents a constant *proportional* change; for example, each centimeter on the y-axis might represent a multiplication of the quantity by 10. In contrast, with a linear axis, a given visual distance represents a constant quantity change; each centimeter might represent adding 10 to the quantity.

Some data sets are exponentially distributed on the x-axis, and others on the y-axis (or both). For example, the `Animals` data set from the `MASS` library contains data on the average brain mass (in g) and body mass (in kg) of various mammals, with a few dinosaurs thrown in for comparison:

```
Animals

                    body  brain
Mountain beaver    1.350    8.1
Cow              465.000  423.0
Grey wolf         36.330  119.5
```

```
...
Brachiosaurus    87000.000  154.5
Mole                 0.122    3.0
Pig                192.000  180.0
```

As shown in Figure 8-26, we can make a scatter plot to visualize the relationship between brain and body mass. With the default linearly scaled axes, it's hard to make much sense of this graph. Because of a few very large animals, the rest of the animals get squished into the lower-left corner—a mouse barely looks different from a triceratops! This is a case where the data is distributed exponentially on both axes.

Ggplot2 will try to make good decisions about where to place the tick marks, but if you don't like them, you can change them by specifying `breaks` and, optionally, `labels`. In the example here, the automatically generated tick marks are spaced farther apart than is ideal. For the y-axis tick marks, we can get a vector of every power of 10 from 10^0 to 10^3 like this:

```
10^(0:3)

   1   10  100 1000
```

The x-axis tick marks work the same way, but because the range is large, R decides to format the output with scientific notation:

```
10^(-1:5)

 1e-01 1e+00 1e+01 1e+02 1e+03 1e+04 1e+05
```

And then we can use those values as the breaks, as in Figure 8-27 (left):

```
p + scale_x_log10(breaks=10^(-1:5)) + scale_y_log10(breaks=10^(0:3))
```

To instead use exponential notation for the break labels (Figure 8-27, right), use the `trans_format()` function, from the scales package:

```
library(scales)
p + scale_x_log10(breaks=10^(-1:5),
                  labels=trans_format("log10", math_format(10^.x))) +
    scale_y_log10(breaks=10^(0:3),
                  labels=trans_format("log10", math_format(10^.x)))
```

Another way to use log axes is to transform the data before mapping it to the *x* and *y* coordinates (Figure 8-28). Technically, the axes are still linear—it's the quantity that is log-transformed:

```
ggplot(Animals, aes(x=log10(body), y=log10(brain), label=rownames(Animals))) +
    geom_text(size=3)
```

The previous examples used a $\log_{10}$ transformation, but it is possible to use other transformations, such as $\log_2$ and natural log, as shown in Figure 8-29. It's a bit more complicated to use these—`scale_x_log10()` is shorthand, but for these other log scales, we need to spell them out:

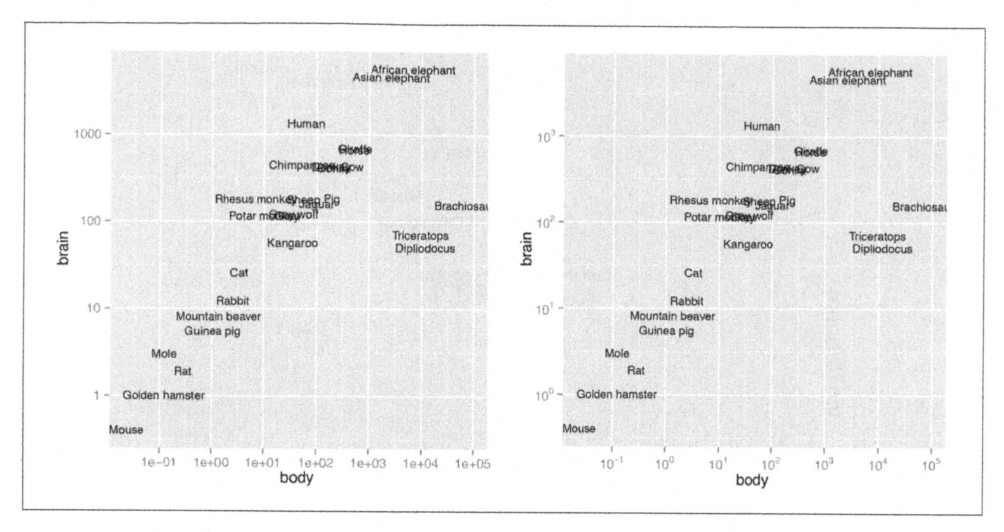

Figure 8-27. Left: scatter plot with log_{10} x- and y-axes, and with manually specified breaks; right: with exponents for the tick labels

```
library(scales)

# Use natural log on x, and log2 on y
p + scale_x_continuous(trans = log_trans(),
                       breaks = trans_breaks("log", function(x) exp(x)),
                       labels = trans_format("log", math_format(e^.x))) +
    scale_y_continuous(trans = log2_trans(),
                       breaks = trans_breaks("log2", function(x) 2^x),
                       labels = trans_format("log2", math_format(2^.x)))
```

It's possible to use a log axis for just one axis. It is often useful to represent financial data this way, because it better represents proportional change. Figure 8-30 shows Apple's stock price with linear and log y-axes. The default tick marks might not be spaced well for your graph; they can be set with the `breaks` in the scale:

```
library(gcookbook) # For the data set

ggplot(aapl, aes(x=date,y=adj_price)) + geom_line()

ggplot(aapl, aes(x=date,y=adj_price)) + geom_line() +
    scale_y_log10(breaks=c(2,10,50,250))
```

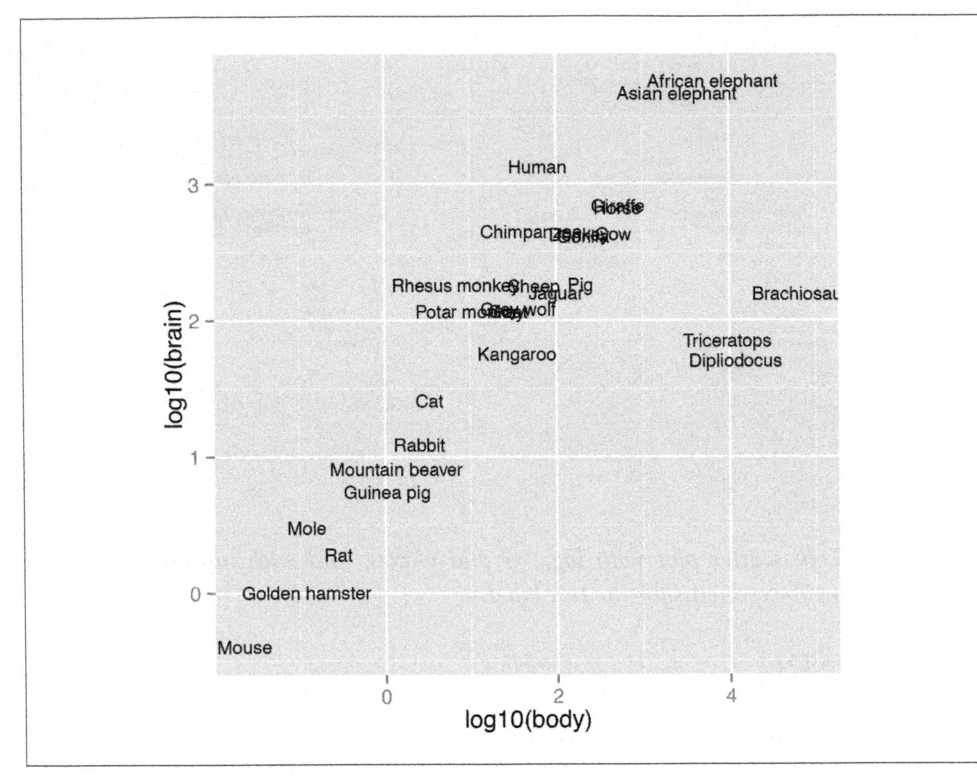

Figure 8-28. Plot with log transform before mapping to x- and y-axes

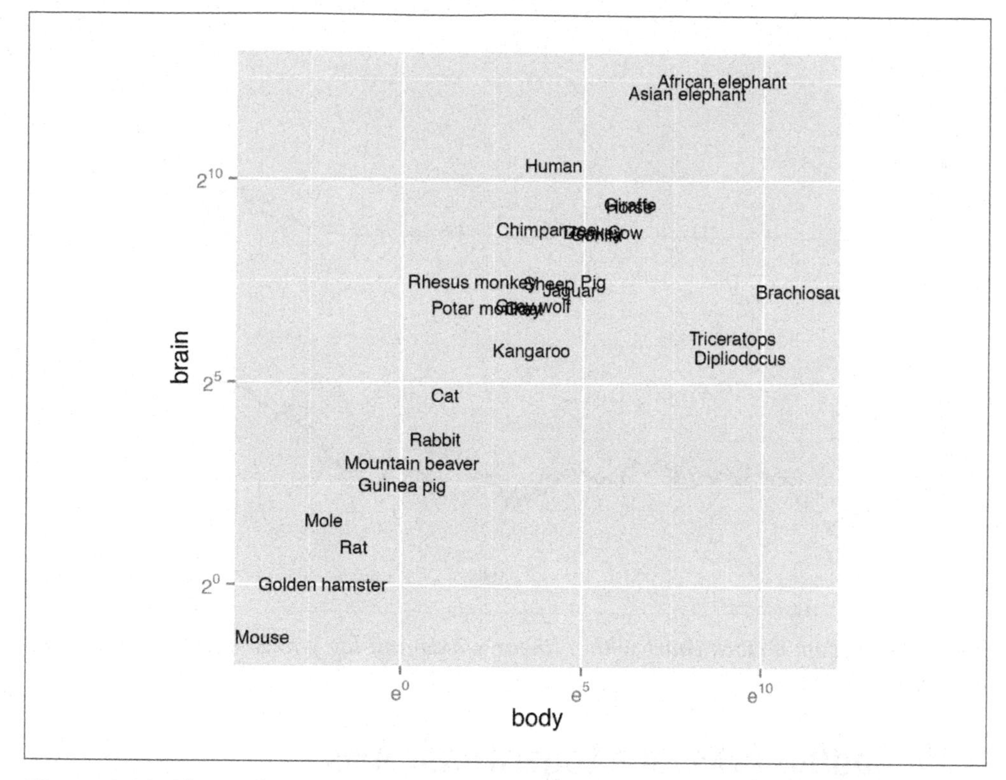

Figure 8-29. Plot with exponents in tick labels. Notice that different bases are used for the x and y axes.

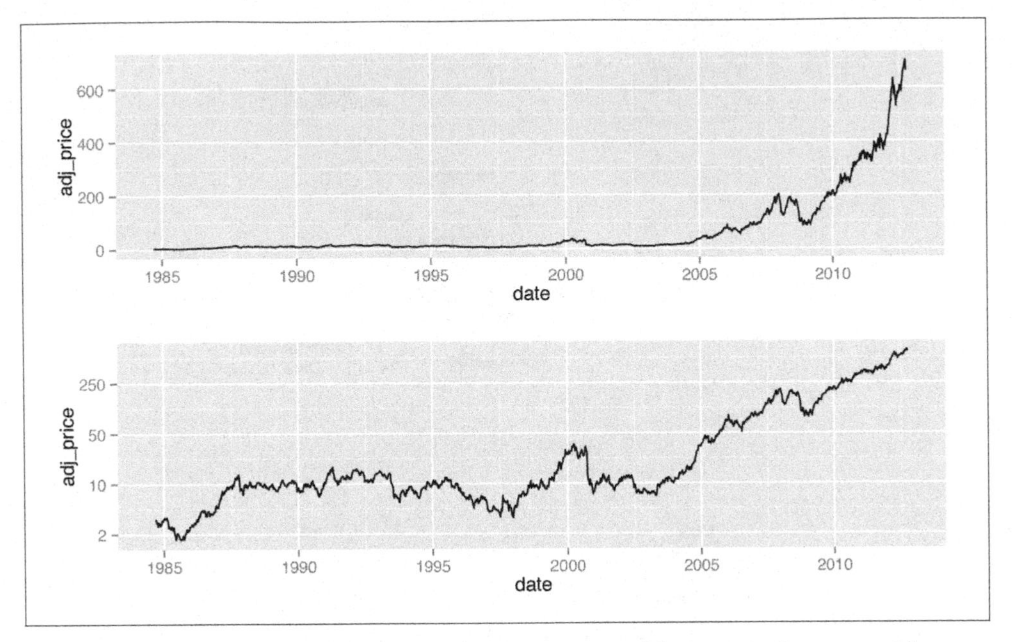

Figure 8-30. Top: a stock chart with a linear x-axis and log y-axis; bottom: with manual breaks

8.15. Adding Ticks for a Logarithmic Axis

Problem

You want to add tick marks with diminishing spacing for a logarithmic axis.

Solution

Use `annotation_logticks()` (Figure 8-31):

```
library(MASS)    # For the data set
library(scales) # For the trans and format functions
ggplot(Animals, aes(x=body, y=brain, label=rownames(Animals))) +
    geom_text(size=3) +
    annotation_logticks() +
    scale_x_log10(breaks = trans_breaks("log10", function(x) 10^x),
                  labels = trans_format("log10", math_format(10^.x))) +
    scale_y_log10(breaks = trans_breaks("log10", function(x) 10^x),
                  labels = trans_format("log10", math_format(10^.x)))
```

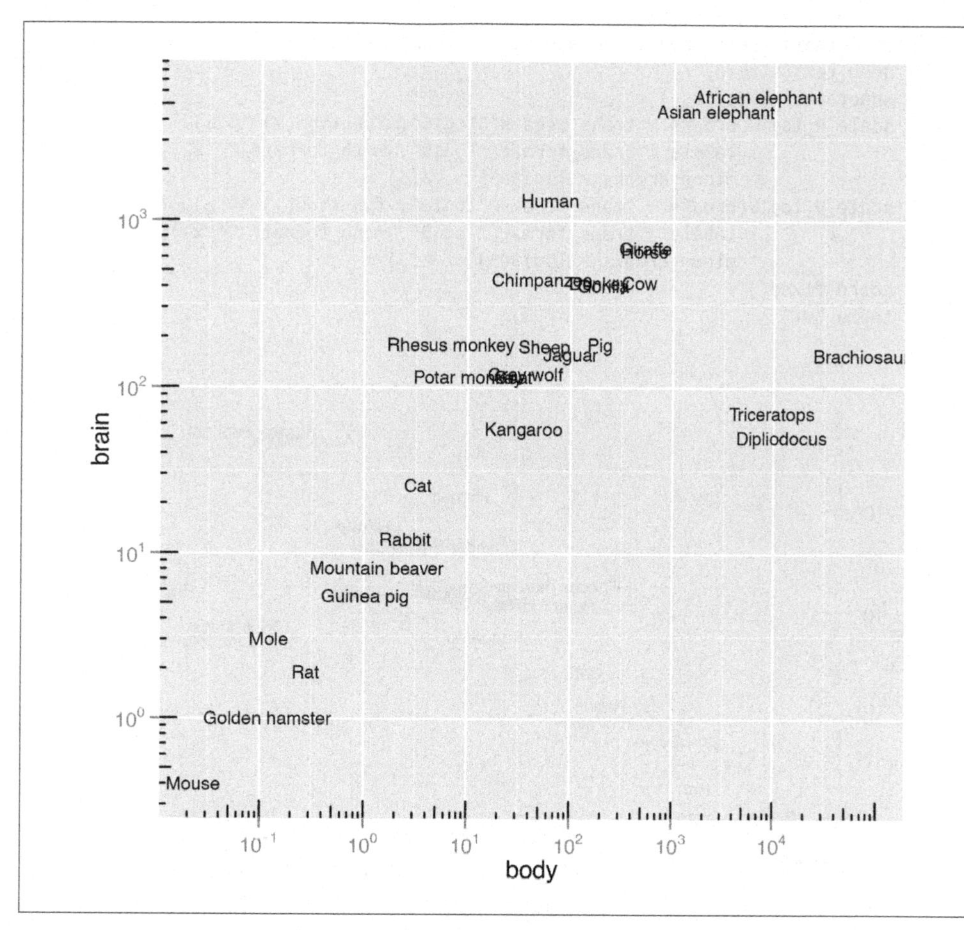

Figure 8-31. Log axes with diminishing tick marks

Discussion

The tick marks created by `annotation_logticks()` are actually geoms inside the plotting area. There is a long tick mark at each power of 10, and a mid-length tick mark at each 5.

To get the colors of the tick marks and the grid lines to match up a bit better, you can use `theme_bw()`.

By default, the minor grid lines appear visually halfway between the major grid lines, but this is not the same place as the "5" tick marks on a logarithmic scale. To get them to be the same, you can manually set the scale's `minor_breaks`. To do this, we need to set them to `log10(5*10^(minpow:maxpow))`, which reduces to `log10(5) + minpow:maxpow` (Figure 8-32):

```
ggplot(Animals, aes(x=body, y=brain, label=rownames(Animals))) +
    geom_text(size=3) +
    annotation_logticks() +
    scale_x_log10(breaks = trans_breaks("log10", function(x) 10^x),
                  labels = trans_format("log10", math_format(10^.x)),
                  minor_breaks = log10(5) + -2:5) +
    scale_y_log10(breaks = trans_breaks("log10", function(x) 10^x),
                  labels = trans_format("log10", math_format(10^.x)),
                  minor_breaks = log10(5) + -1:3) +
    coord_fixed() +
    theme_bw()
```

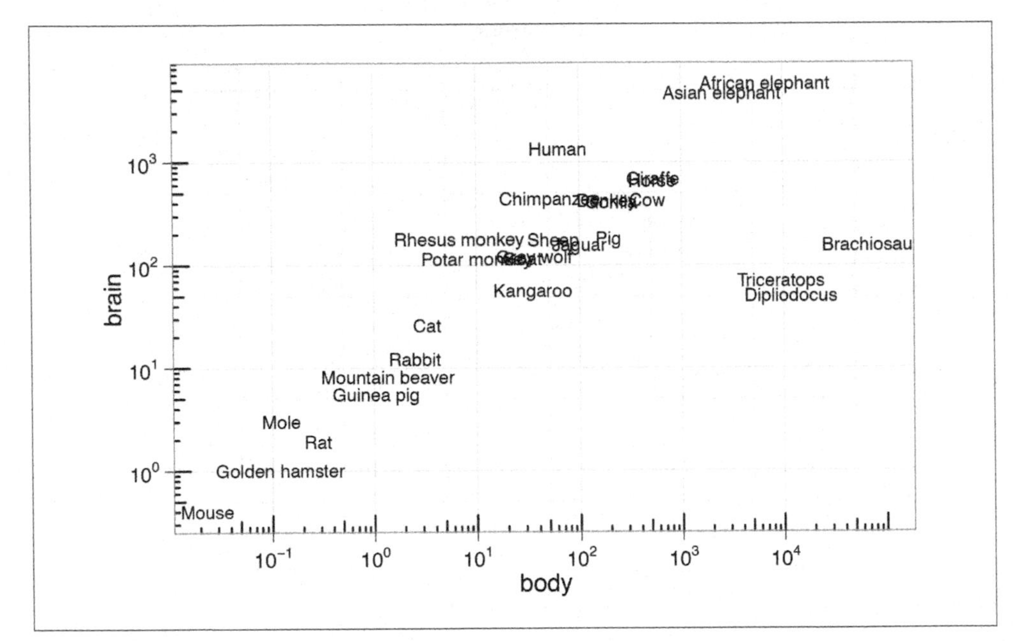

Figure 8-32. Log axes with ticks at each 5, and fixed coordinate ratio

See Also

For more on controlling the scaling ratio of the x- and y-axes, see Recipe 8.5.

8.16. Making a Circular Graph

Problem

You want to make a circular graph.

Solution

Use coord_polar(). For this example we'll use the wind data set from gcookbook. It contains samples of wind speed and direction for every 5 minutes throughout a day. The direction of the wind is categorized into 15-degree bins, and the speed is categorized into 5 m/s increments:

```
library(gcookbook) # For the data set
wind

 TimeUTC  Temp WindAvg WindMax WindDir SpeedCat DirCat
       0  3.54    9.52   10.39      89    10-15     90
       5  3.52    9.10    9.90      92     5-10     90
      10  3.53    8.73    9.51      92     5-10     90
...
    2335  6.74   18.98   23.81     250      >20    255
    2340  6.62   17.68   22.05     252      >20    255
    2345  6.22   18.54   23.91     259      >20    255
```

We'll plot a count of the number of samples at each SpeedCat and DirCat using geom_histogram() (Figure 8-33). We'll set binwidth to 15 and make the origin of the histogram start at –7.5, so that each bin is centered around 0, 15, 30, etc.:

```
ggplot(wind, aes(x=DirCat, fill=SpeedCat)) +
    geom_histogram(binwidth=15, origin=-7.5) +
    coord_polar() +
    scale_x_continuous(limits=c(0,360))
```

Discussion

Be cautious when using polar plots, since they can perceptually distort the data. In the example here, at 210 degrees there are 15 observations with a speed of 15–20 and 13 observations with a speed of >20, but a quick glance at the picture makes it appear that there are more observations at >20. There are also three observations with a speed of 10–15, but they're barely visible.

In this example we can make the plot a little prettier by reversing the legend, using a different palette, adding an outline, and setting the breaks to some more familiar numbers (Figure 8-34):

```
ggplot(wind, aes(x=DirCat, fill=SpeedCat)) +
    geom_histogram(binwidth=15, origin=-7.5, colour="black", size=.25) +
    guides(fill=guide_legend(reverse=TRUE)) +
    coord_polar() +
    scale_x_continuous(limits=c(0,360), breaks=seq(0, 360, by=45),
                       minor_breaks=seq(0, 360, by=15)) +
    scale_fill_brewer()
```

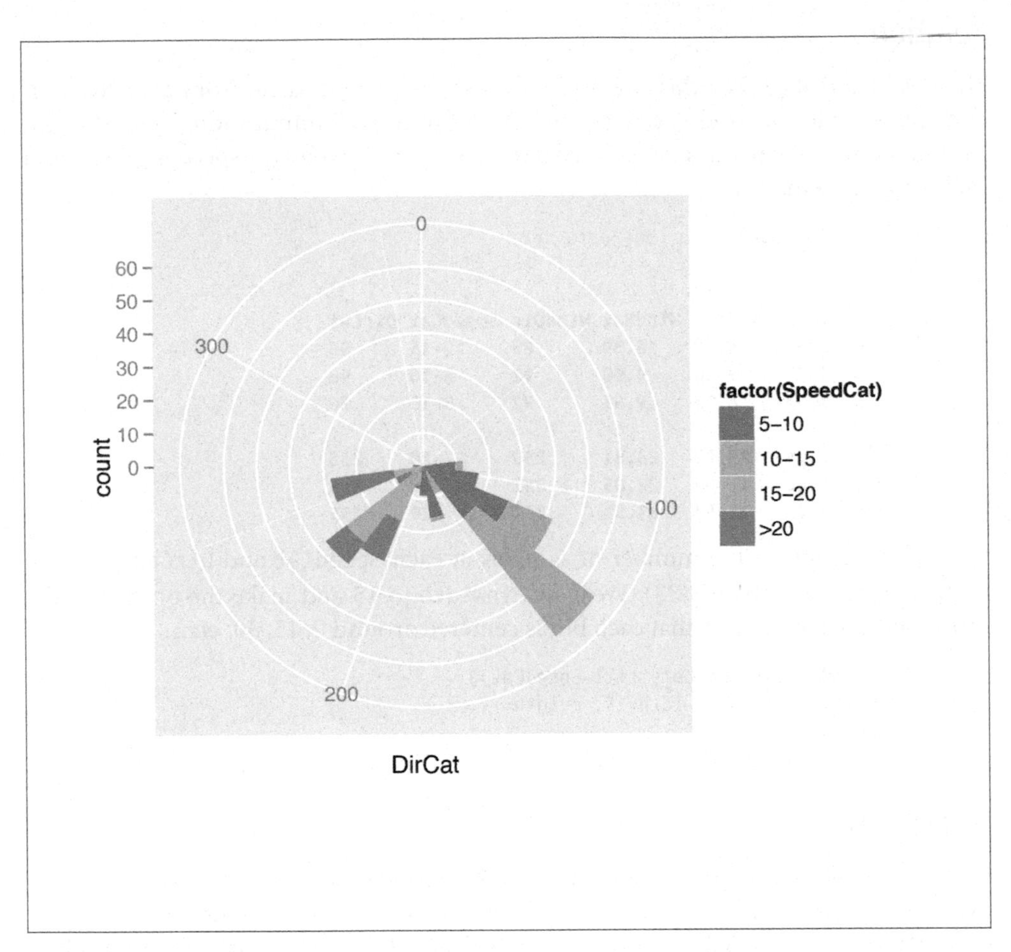

Figure 8-33. Polar plot

It may also be useful to set the starting angle with the `start` argument, especially when using a discrete variable for *theta*. The starting angle is specified in radians, so if you know the adjustment in degrees, you'll have to convert it to radians:

```
coord_polar(start=-45 * pi / 180)
```

Polar coordinates can be used with other geoms, including lines and points. There are a few important things to keep in mind when using these geoms. First, by default, for the variable that is mapped to *y* (or *r*), the smallest actual value gets mapped to the center; in other words, the smallest data value gets mapped to a visual radius value of 0. You may be expecting a data value of 0 to be mapped to a radius of 0, but to make sure this happens, you'll need to set the limits.

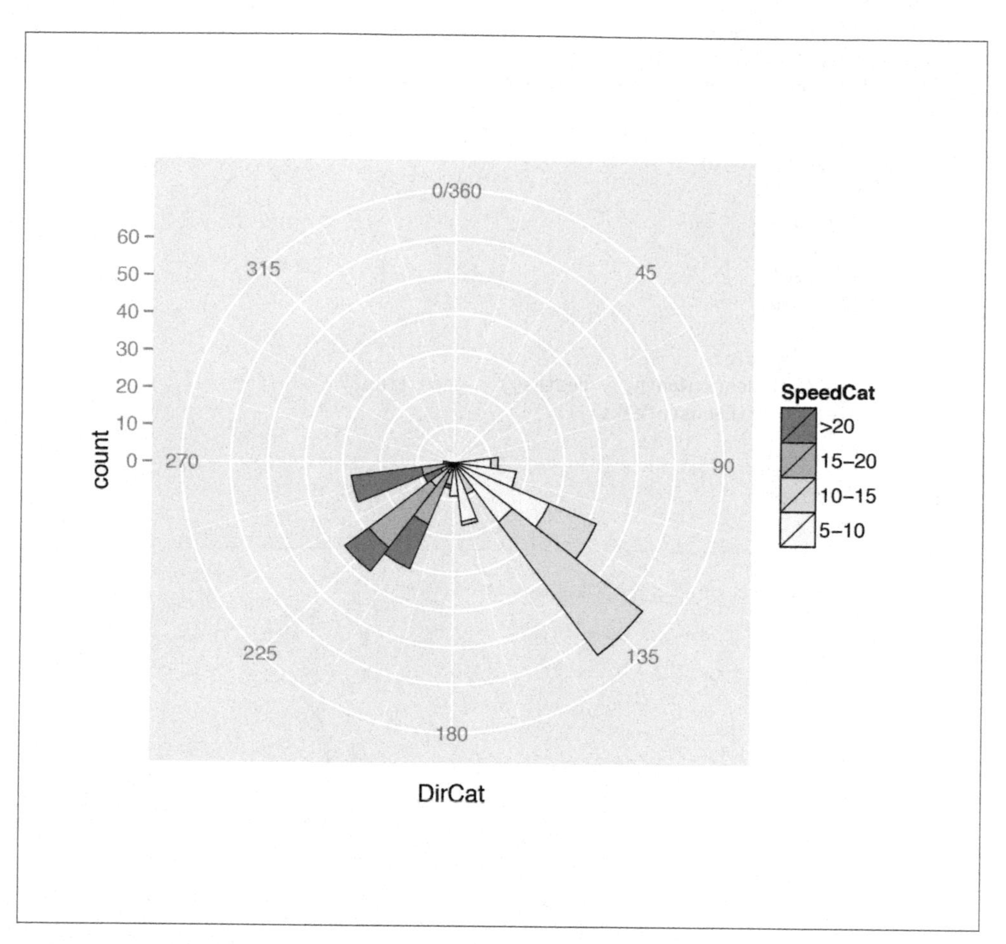

Figure 8-34. Polar plot with different colors and breaks

Next, when using a continuous *x* (or *theta*), the smallest and largest data values are merged. Sometimes this is desirable, sometimes not. To change this behavior, you'll need to set the limits.

Finally, the *theta* values of the polar coordinates do not wrap around—it is presently not possible to have a geom that crosses over the starting angle (usually vertical).

We'll illustrate these issues with an example. The following code creates a data frame from the `mdeaths` time series data set and produces the graph shown on the left in Figure 8-35:

```
# Put mdeaths time series data into a data frame
md <- data.frame(deaths = as.numeric(mdeaths),
                 month  = as.numeric(cycle(mdeaths)))
```

```
# Calculate average number of deaths in each month
library(plyr) # For the ddply() function
md <- ddply(md, "month", summarise, deaths = mean(deaths))
md

 month   deaths
     1 2129.833
     2 2081.333
 ...
    11 1377.667
    12 1796.500

# Make the base plot
p <- ggplot(md, aes(x=month, y=deaths)) + geom_line() +
     scale_x_continuous(breaks=1:12)
```

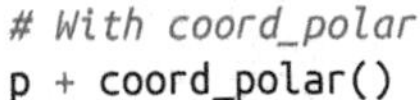

```
# With coord_polar
p + coord_polar()
```

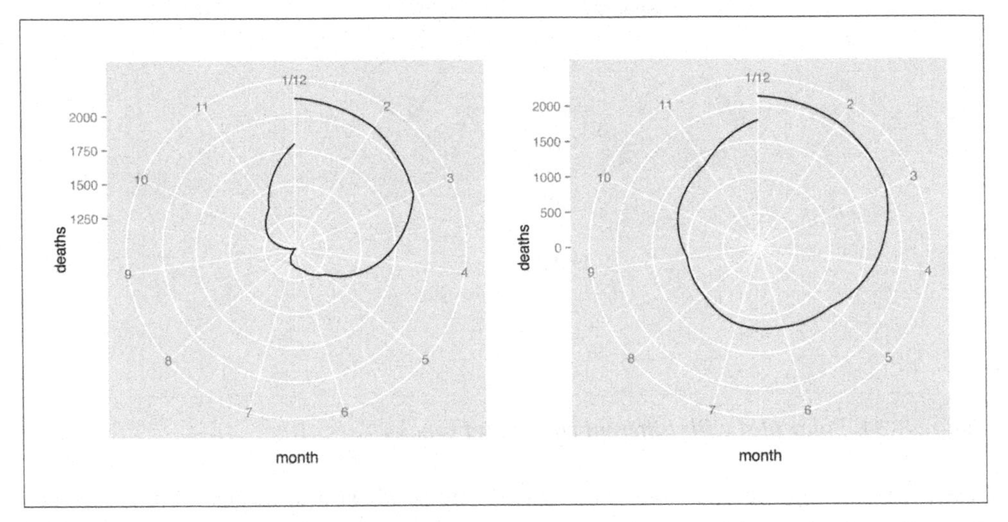

Figure 8-35. Left: polar plot with line (notice the data range of the radius); right: with the radius representing a data range starting from zero

The first problem is that the data values (ranging from about 1000 to 2100) are mapped to the radius such that the smallest data value is at radius 0. We'll fix this by setting the *y* (or *r*) limits from 0 to the maximum data value, as shown in the graph on the right in Figure 8-35:

```
# With coord_polar and y (r) limits going to zero
p + coord_polar() + ylim(0, max(md$deaths))
```

The next problem is that the lowest and highest `month` values, 1 and 12, are shown at the same angle. We'll fix this by setting the *x* limits from 0 to 12, creating the graph on the left in Figure 8-36 (notice that using `xlim()` overrides the `scale_x_continuous()` in `p`, so it no longer displays breaks for each month; see Recipe 8.2 for more information):

```
p + coord_polar() + ylim(0, max(md$deaths)) + xlim(0, 12)
```

There's one last issue, which is that the beginning and end aren't connected. To fix that, we need to modify our data frame by adding one row with a month of 0 that has the same value as the row with month 12. This will make the starting and ending points the same, as in the graph on the right in Figure 8-36 (alternatively, we could add a row with month 13, instead of month 0):

```
# Connect the lines by adding a value for 0 that is the same as 12
mdx <- md[md$month==12, ]
mdx$month <- 0
mdnew <- rbind(mdx, md)

# Make the same plot as before, but with the new data, by using %+%
p %+% mdnew + coord_polar() + ylim(0, max(md$deaths))
```

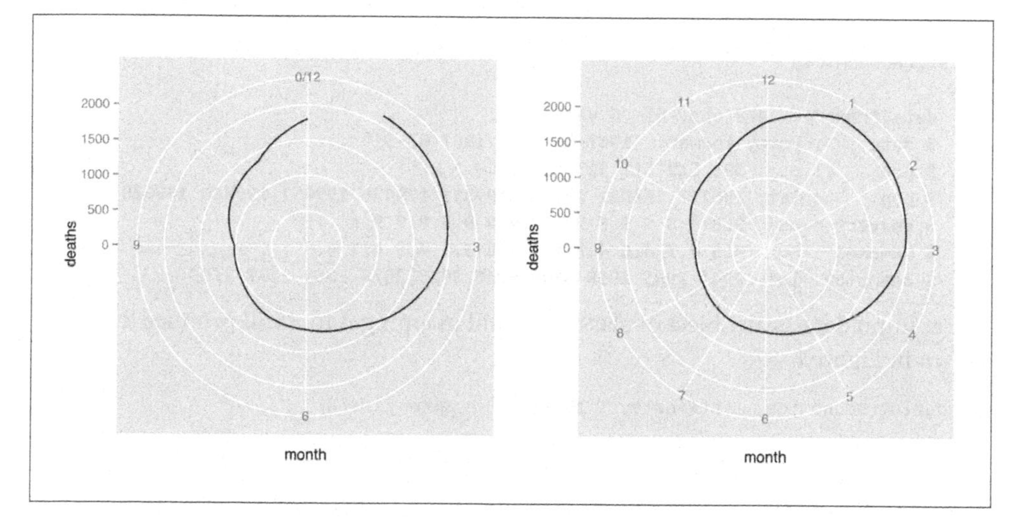

Figure 8-36. Left: polar plot with theta representing x values from 0 to 12; right: the gap is filled in by adding a dummy data point for month 0

Notice the use of the `%+%` operator. When you add a data frame to a `ggplot` object with `%+%`, it replaces the default data frame in the `ggplot` object. In this case, it changed the default data frame for `p` from `md` to `mdnew`.

See Also

See Recipe 10.4 for more about reversing the direction of a legend.

See Recipe 8.6 for more about specifying which values will have tick marks (breaks) and labels.

8.17. Using Dates on an Axis

Problem

You want to use dates on an axis.

Solution

Map a column of class `Date` to the x- or y-axis. We'll use the `economics` data set for this example:

```
# Look at the structure
str(economics)

'data.frame':    478 obs. of  6 variables:
 $ date    : Date, format: "1967-06-30" "1967-07-31" ...
 $ pce     : num  508 511 517 513 518 ...
 $ pop     : int  198712 198911 199113 199311 199498 199657 199808 199920 ...
 $ psavert : num  9.8 9.8 9 9.8 9.7 9.4 9 9.5 8.9 9.6 ...
 $ uempmed : num  4.5 4.7 4.6 4.9 4.7 4.8 5.1 4.5 4.1 4.6 ...
 $ unemploy: int  2944 2945 2958 3143 3066 3018 2878 3001 2877 2709 ...
```

The column `date` is an object of class `Date`, and mapping it to `x` will produce the result shown in Figure 8-37:

```
ggplot(economics, aes(x=date, y=psavert)) + geom_line()
```

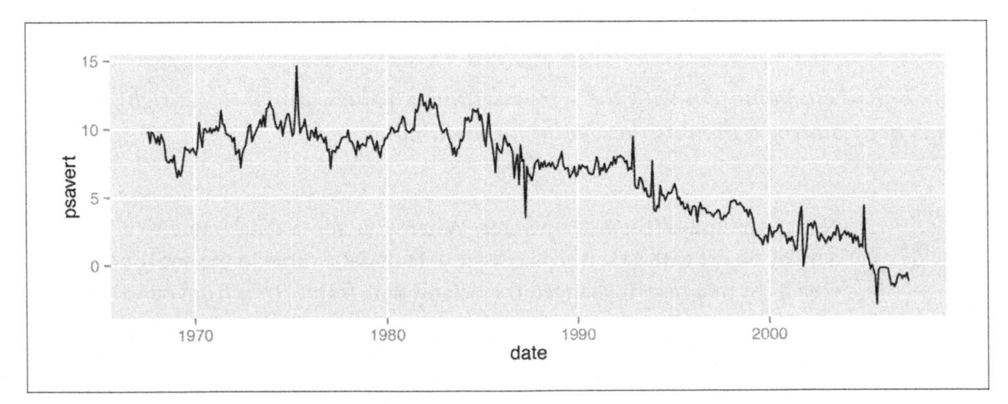

Figure 8-37. Dates on the x-axis

Discussion

ggplot2 handles two kinds of time-related objects: dates (objects of class `Date`) and date-times (objects of class `POSIXt`). The difference between these is that `Date` objects represent dates and have a resolution of one day, while `POSIXt` objects represent moments in time and have a resolution of a fraction of a second.

Specifying the breaks is similar to with a numeric axis—the main difference is in specifying the sequence of dates to use. We'll use a subset of the `economics` data, ranging from mid-1992 to mid-1993. If breaks aren't specified, they will be automatically selected, as shown in Figure 8-38 (top):

```
# Take a subset of economics
econ <- subset(economics, date >= as.Date("1992-05-01") &
                          date <  as.Date("1993-06-01"))

# Base plot - without specifying breaks
p <- ggplot(econ, aes(x=date, y=psavert)) + geom_line()
p
```

The breaks can be created by using the `seq()` function with starting and ending dates, and an interval (Figure 8-38, bottom):

```
# Specify breaks as a Date vector
datebreaks <- seq(as.Date("1992-06-01"), as.Date("1993-06-01"), by="2 month")

# Use breaks, and rotate text labels
p + scale_x_date(breaks=datebreaks) +
    theme(axis.text.x = element_text(angle=30, hjust=1))
```

Notice that the formatting of the breaks changed. You can specify the formatting by using the `date_format()` function from the scales package. Here we'll use `"%Y %b"`, which results in a format like "1992 Jun", as shown in Figure 8-39:

```
library(scales)
p + scale_x_date(breaks=datebreaks, labels=date_format("%Y %b")) +
    theme(axis.text.x = element_text(angle=30, hjust=1))
```

Common date format options are shown in Table 8-1. They are to be put in a string that is passed to `date_format()`, and the format specifiers will be replaced with the appropriate values. For example, if you use `"%B %d, %Y"`, it will result in labels like "June 01, 1992".

Table 8-1. Date format options

Option	Description
%Y	Year with century (2012)
%y	Year without century (12)
%m	Month as a decimal number (08)

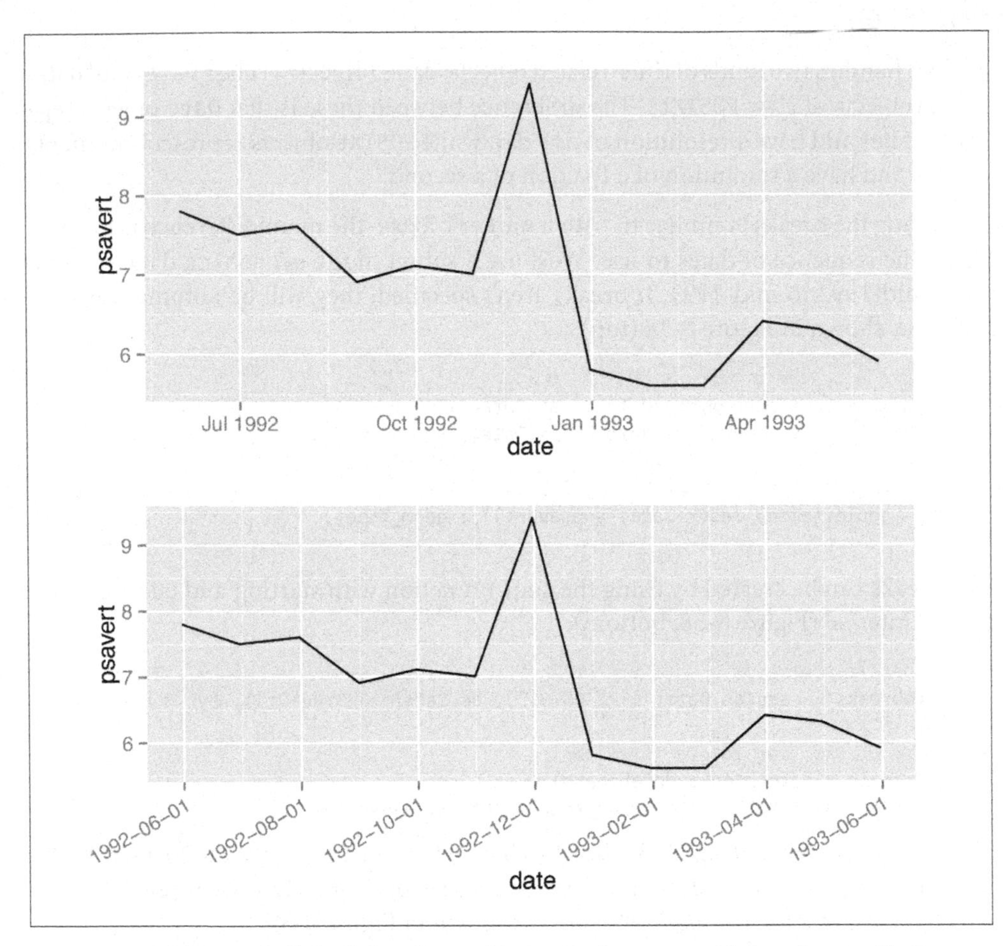

Figure 8-38. Top: with default breaks on the x-axis; bottom: with breaks specified

Option	Description
%b	Abbreviated month name in current locale (Aug)
%B	Full month name in current locale (August)
%d	Day of month as a decimal number (04)
%U	Week of the year as a decimal number, with Sunday as the first day of the week (00–53)
%W	Week of the year as a decimal number, with Monday as the first day of the week (00–53)
%w	Day of week (0–6, Sunday is 0)
%a	Abbreviated weekday name (Thu)
%A	Full weekday name (Thursday)

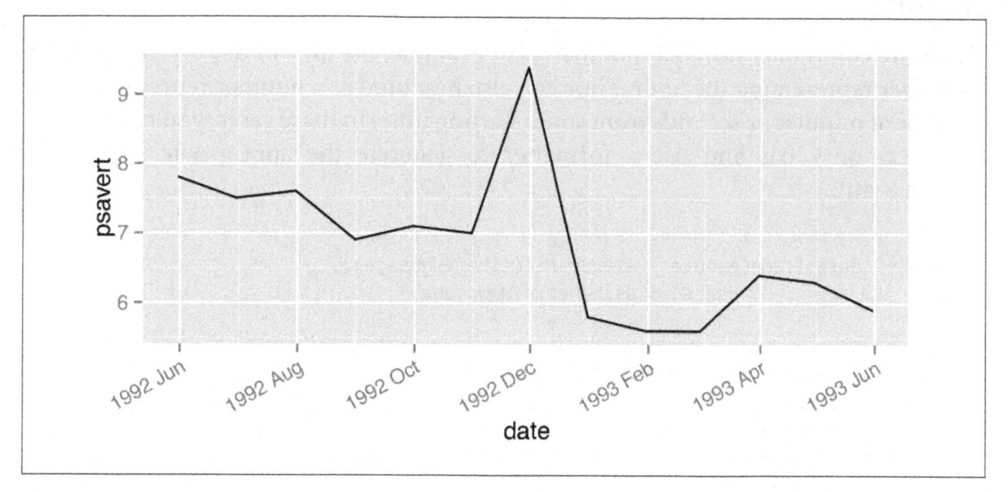

Figure 8-39. Line graph with date format specified

Some of these items are specific to the computer's locale. Months and days have different names in different languages (the examples here are generated with a US locale). You can change the locale with `Sys.setlocale()`. For example, this will change the date formatting to use an Italian locale:

```
# Mac and Linux
Sys.setlocale("LC_TIME", "it_IT.UTF-8")

# Windows
Sys.setlocale("LC_TIME", "italian")
```

Note that the locale names may differ between platforms, and your computer must have support for the locale installed at the operating system level.

See Also

See `?Sys.setlocale` for more about setting the locale.

See `?strptime` for information about converting strings to dates, and for information about formatting the date output.

8.18. Using Relative Times on an Axis

Problem

You want to use relative times on an axis.

Solution

Times are commonly stored as numbers. For example, the time of day can be stored as a number representing the hour. Time can also be stored as a number representing the number of minutes or seconds from some starting time. In these cases, you map a value to the x- or y-axis and use a formatter to generate the appropriate axis labels (Figure 8-40):

```
# Convert WWWusage time-series object to data frame
www <- data.frame(minute = as.numeric(time(WWWusage)),
                  users  = as.numeric(WWWusage))

# Define a formatter function - converts time in minutes to a string
timeHM_formatter <- function(x) {
    h <- floor(x/60)
    m <- floor(x %% 60)
    lab <- sprintf("%d:%02d", h, m) # Format the strings as HH:MM
    return(lab)
}

# Default x axis
ggplot(www, aes(x=minute, y=users)) + geom_line()

# With formatted times
ggplot(www, aes(x=minute, y=users)) + geom_line() +
    scale_x_continuous(name="time", breaks=seq(0, 100, by=10),
                       labels=timeHM_formatter)
```

Discussion

In some cases it might be simpler to specify the breaks and labels manually, with something like this:

```
scale_x_continuous(breaks=c(0, 20, 40, 60, 80, 100),
    labels=c("0:00", "0:20", "0:40", "1:00", "1:20", "1:40"))
```

In the preceding example, we used the `timeHM_formatter()` function to convert the numeric time (in minutes) to a string like `"1:10"`:

```
timeHM_formatter(c(0, 50, 51, 59, 60, 130, 604))

 "0:00" "0:50" "0:51" "0:59" "1:00" "2:10" "10:04"
```

To convert to HH:MM:SS format, you can use the following formatter function:

```
timeHMS_formatter <- function(x) {
    h <- floor(x/3600)
    m <- floor((x/60) %% 60)
    s <- round(x %% 60)                       # Round to nearest second
    lab <- sprintf("%02d:%02d:%02d", h, m, s) # Format the strings as HH:MM:SS
    lab <- sub("^00:", "", lab)               # Remove leading 00: if present
    lab <- sub("^0", "", lab)                 # Remove leading 0 if present
```

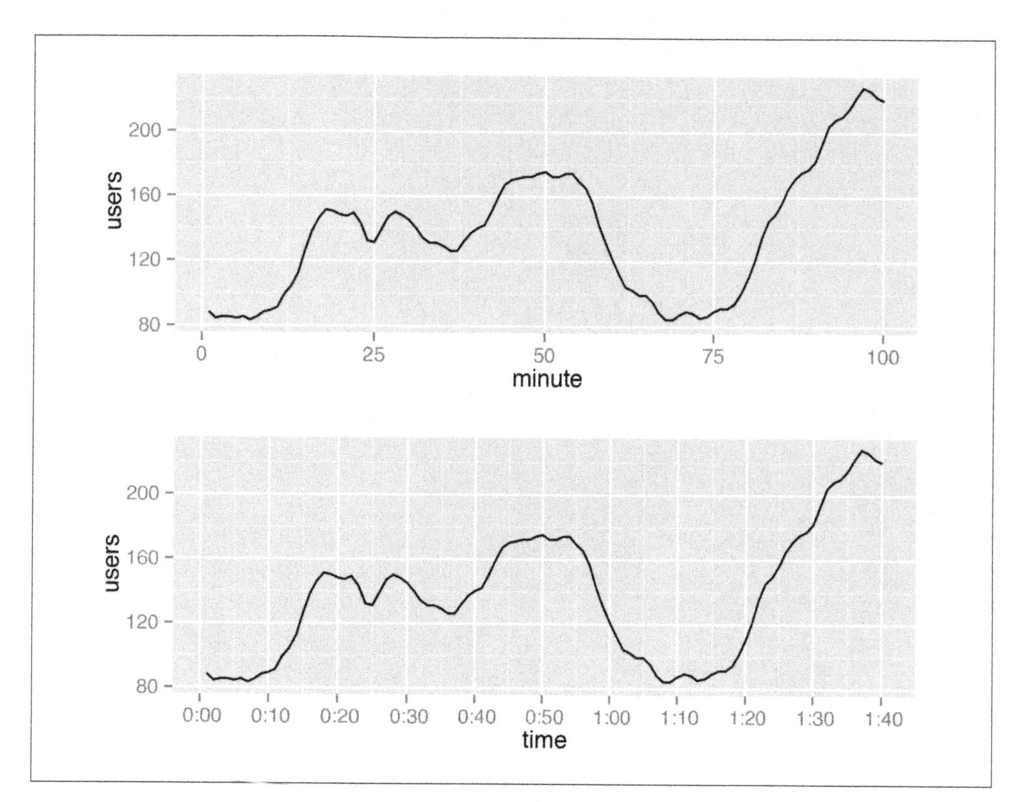

Figure 8-40. Top: relative times on x-axis; bottom: with formatted times

```
    return(lab)
}
```

Running it on some sample numbers yields:

```
timeHMS_formatter(c(20, 3000, 3075, 3559.2, 3600, 3606, 7813.8))

 "0:20"    "50:00"   "51:15"   "59:19"   "1:00:00" "1:00:06" "2:10:14"
```

See Also

See Recipe 15.21 for information about converting time series objects to data frames.

CHAPTER 9

Controlling the Overall Appearance of Graphs

In this chapter I'll discuss how to control the overall appearance of graphics made by ggplot2. The grammar of graphics that underlies ggplot2 is concerned with how data is processed and displayed—it's not concerned with things like fonts, background colors, and so on. When it comes to presenting your data, there's a good chance that you'll want to tune the appearance of these things. ggplot2's theming system provides control over the appearance of non-data elements. I touched on the theme system in the previous chapter, and here I'll explain a bit more about how it works.

9.1. Setting the Title of a Graph

Problem

You want to set the title of a graph.

Solution

Set `title` with `ggtitle()`, as shown in Figure 9-1:

```
library(gcookbook) # For the data set

p <- ggplot(heightweight, aes(x=ageYear, y=heightIn)) + geom_point()

p + ggtitle("Age and Height of Schoolchildren")

# Use \n for a newline
p + ggtitle("Age and Height\nof Schoolchildren")
```

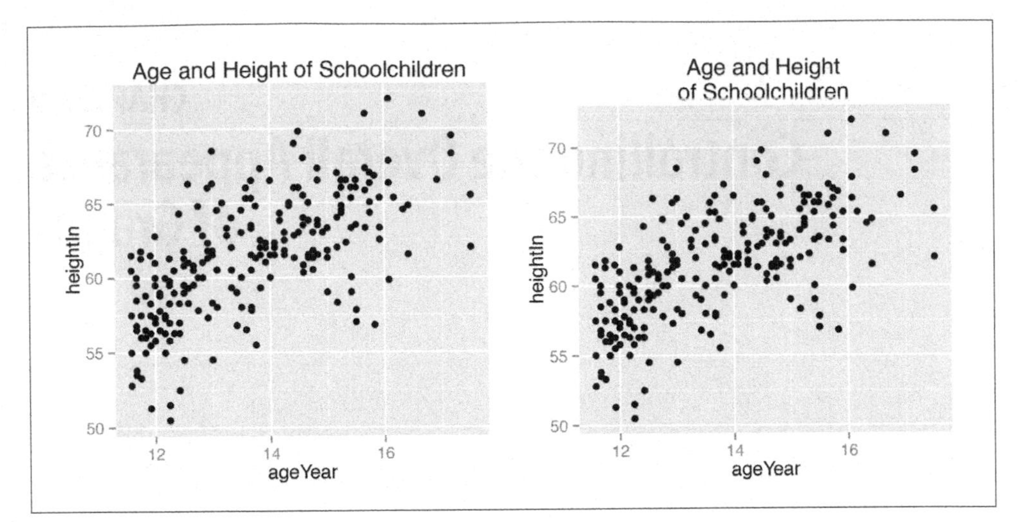

Figure 9-1. Left: scatter plot with a title added; right: with a /n for a newline

Discussion

`ggtitle()` is equivalent to using `labs(title = "Title text")`.

If you want to move the title inside the plotting area, you can use one of two methods, both of which are a little bit of a hack (Figure 9-2). The first method is to use `ggtitle()` with a negative `vjust` value. The drawback of this method is that it still reserves blank space above the plotting region for the title.

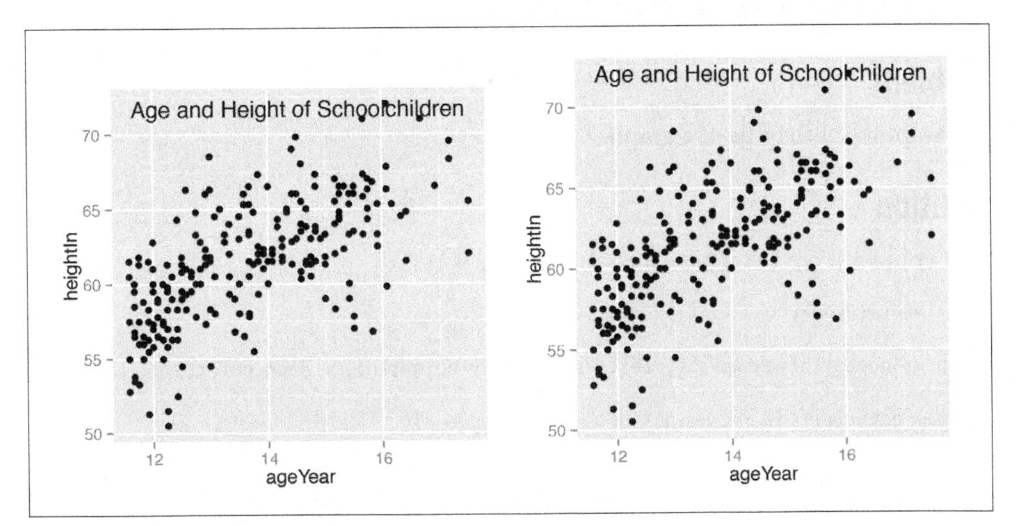

Figure 9-2. Left: title with ggtitle and a negative vjust() value (note the extra space above the plotting area); right: with a text annotation at the top of the figure

The second method is to instead use a text annotation, setting its *x* position to the middle of the *x* range and its *y* position to `Inf`, which places it at the top of the plotting region. This also requires a positive `vjust` value to bring the text fully inside the plotting region:

```
# Move the title inside
p + ggtitle("Age and Height of Schoolchildren") +
    theme(plot.title=element_text(vjust = -2.5))

# Use a text annotation instead
p + annotate("text", x=mean(range(heightweight$ageYear)), y=Inf,
             label="Age and Height of Schoolchildren", vjust=1.5, size=6)
```

9.2. Changing the Appearance of Text

Problem

You want to change the appearance of text in a plot.

Solution

To set the appearance of theme items such as the title, axis labels, and axis tick marks, use `theme()` and set the item with `element_text()`. For example, `axis.title.x` controls the appearance of the x-axis label and `plot.title` controls the appearance of the title text (Figure 9-3, left):

```
library(gcookbook) # For the data set

# Base plot
p <- ggplot(heightweight, aes(x=ageYear, y=heightIn)) + geom_point()

# Controlling appearance of theme items
p + theme(axis.title.x=element_text(size=16, lineheight=.9, family="Times",
                                    face="bold.italic", colour="red"))

p + ggtitle("Age and Height\nof Schoolchildren") +
    theme(plot.title=element_text(size=rel(1.5), lineheight=.9, family="Times",
                                  face="bold.italic", colour="red"))

# rel(1.5) means that the font will be 1.5 times the base font size of the theme.
# For theme elements, font size is in points.
```

To set the appearance of text geoms (text that's in the plot itself, with `geom_text()` or `annotate()`), set the text properties. For example (Figure 9-3, right):

```
p + annotate("text", x=15, y=53, label="Some text", size = 7, family="Times",
         fontface="bold.italic", colour="red")

p + geom_text(aes(label=weightLb), size=4, family="Times", colour="red")

# For text geoms, font size is in mm
```

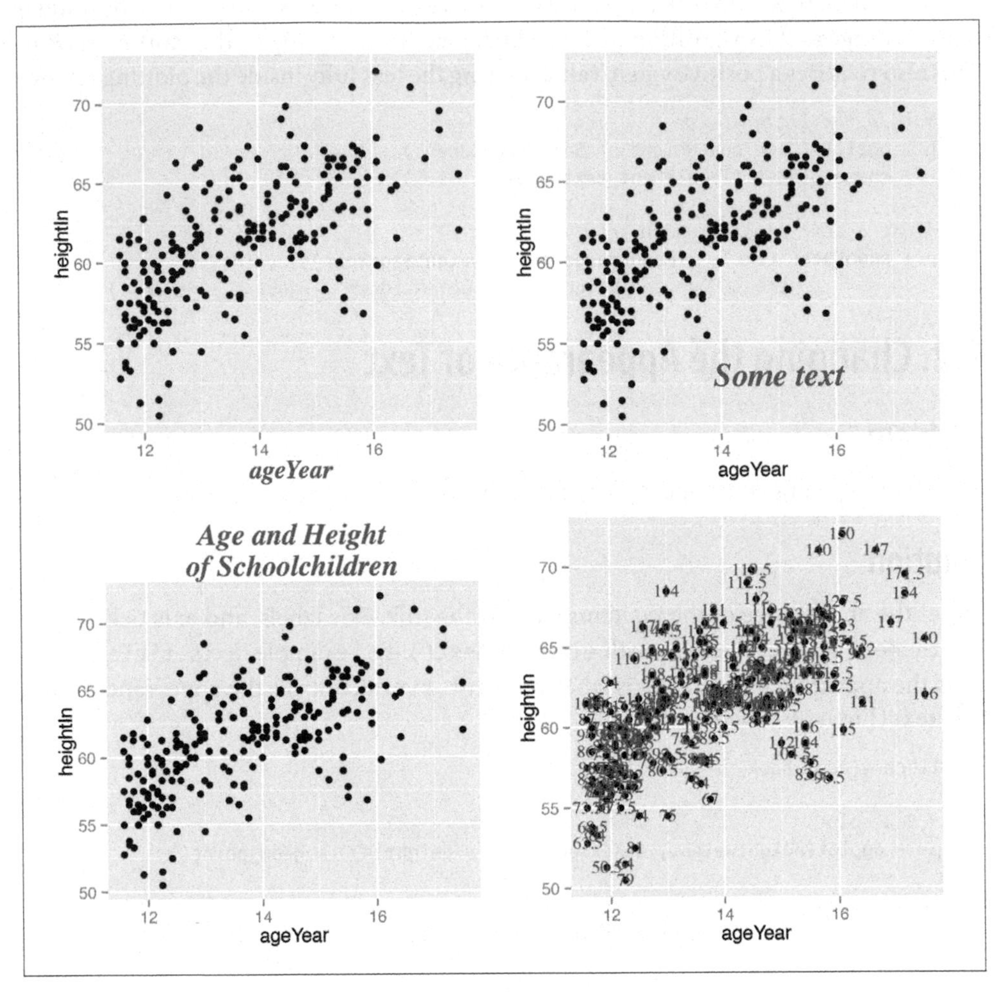

Figure 9-3. Counterclockwise from top left: axis.title.x, plot.title, geom_text(), and annotate("text")

Discussion

There are two kinds of text items in ggplot2: theme elements and text geoms. Theme elements are all the non-data elements in the plot: the title, legends, and axes. Text geoms are things that are part of the plot itself.

There are differences in the parameters, as shown in Table 9-1.

Table 9-1. Text properties of theme elements and text geoms

Theme elements	Text geoms	Description
`family`	`family`	`Helvetica`, `Times`, `Courier`
`face`	`fontface`	`plain`, `bold`, `italic`, `bold.italic`
`colour`	`colour`	Color (name or `"#RRGGBB"`)
`size`	`size`	Font size (in points for theme elements; in mm for geoms)
`hjust`	`hjust`	Horizontal alignment: 0=left, 0.5=center, 1=right
`vjust`	`vjust`	Vertical alignment: 0=bottom, 0.5=middle, 1=top
`angle`	`angle`	Angle in degrees
`lineheight`	`lineheight`	Line spacing multiplier

The theme elements are listed in Table 9-2. Most of them are straightforward. Some are shown in Figure 9-4.

Table 9-2. Theme items that control text appearance in theme()

Element name	Description
`axis.title`	Appearance of axis labels on both axes
`axis.title.x`	Appearance of x-axis label
`axis.title.y`	Appearance of y-axis label
`axis.ticks`	Appearance of tick labels on both axes
`axis.ticks.x`	Appearance of x tick labels
`axis.ticks.y`	Appearance of y tick labels
`legend.title`	Appearance of legend title
`legend.text`	Appearance of legend items
`plot.title`	Appearance of overall plot title
`strip.text`	Appearance of facet labels in both directions
`strip.text.x`	Appearance of horizontal facet labels
`strip.text.y`	Appearance of vertical facet labels

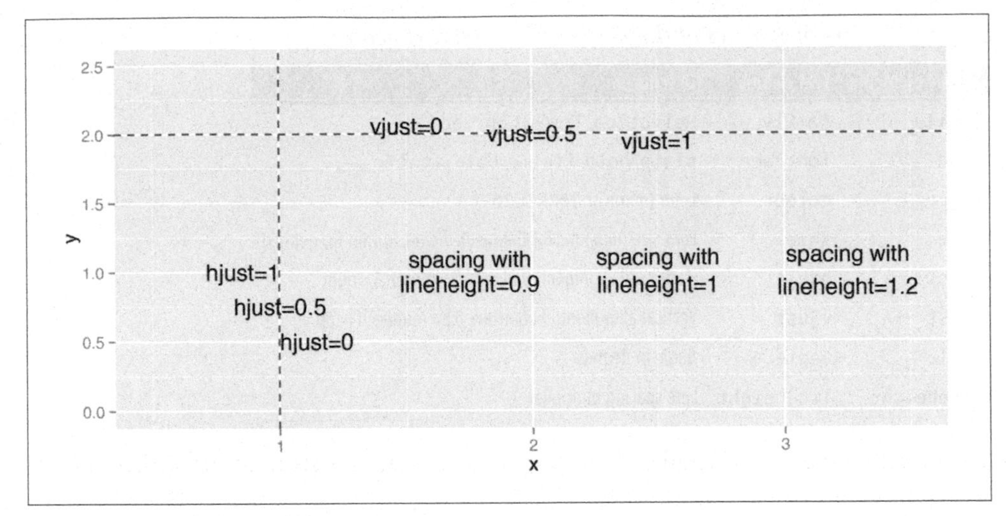

Figure 9-4. Aligning with hjust and vjust, and spacing with lineheight

9.3. Using Themes

Problem

You want to use premade themes to control the overall plot appearance.

Solution

To use a premade theme, add `theme_bw()` or `theme_grey()` (Figure 9-5):

```
library(gcookbook) # For the data set

# Base plot
p <- ggplot(heightweight, aes(x=ageYear, y=heightIn)) + geom_point()

# Grey theme (the default)
p + theme_grey()

# Black-and-white theme
p + theme_bw()
```

Discussion

Some commonly used properties of theme elements in ggplot2 are those things that are controlled by `theme()`. Most of these things, like the title, legend, and axes, are outside the plot area, but some of them are inside the plot area, such as grid lines and the background coloring.

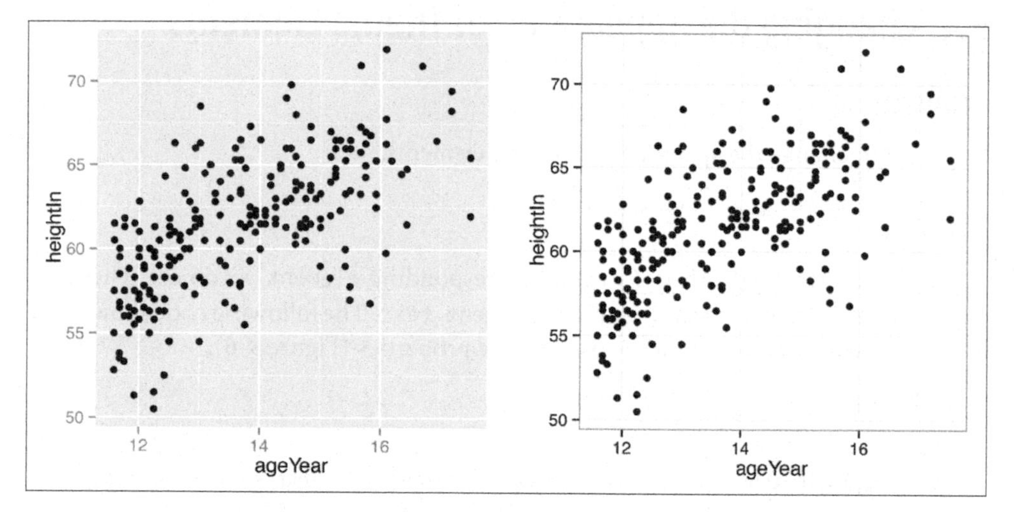

Figure 9-5. Left: scatter plot with theme_grey() (the default); right: with theme_bw()

The two included themes are `theme_grey()` and `theme_bw()`, but it is also possible to create your own.

You can set the base font family and size with either of the included themes (the default base font family is Helvetica, and the default size is 12):

```
p + theme_grey(base_size=16, base_family="Times")
```

You can set the default theme for the current R session with `theme_set()`:

```
# Set default theme for current session
theme_set(theme_bw())

# This will use theme_bw()
p

# Reset the default theme back to theme_grey()
theme_set(theme_grey())
```

See Also

To modify a theme, see Recipe 9.4.

To create your own themes, see Recipe 9.5.

See `?theme` to see all the available theme properties.

9.4. Changing the Appearance of Theme Elements

Problem

You want to change the appearance of theme elements.

Solution

To modify a theme, add `theme()` with a corresponding `element_xx` object. These include `element_line`, `element_rect`, and `element_text`. The following code shows how to modify many of the commonly used theme properties (Figure 9-6):

```
library(gcookbook) # For the data set

# Base plot
p <- ggplot(heightweight, aes(x=ageYear, y=heightIn, colour=sex)) + geom_point()

# Options for the plotting area
p + theme(
    panel.grid.major = element_line(colour="red"),
    panel.grid.minor = element_line(colour="red", linetype="dashed", size=0.2),
    panel.background = element_rect(fill="lightblue"),
    panel.border = element_rect(colour="blue", fill=NA, size=2))

# Options for text items
p + ggtitle("Plot title here") +
    theme(
    axis.title.x = element_text(colour="red", size=14),
    axis.text.x  = element_text(colour="blue"),
    axis.title.y = element_text(colour="red", size=14, angle = 90),
    axis.text.y  = element_text(colour="blue"),
    plot.title = element_text(colour="red", size=20, face="bold"))

# Options for the legend
p + theme(
    legend.background = element_rect(fill="grey85", colour="red", size=1),
    legend.title = element_text(colour="blue", face="bold", size=14),
    legend.text = element_text(colour="red"),
    legend.key = element_rect(colour="blue", size=0.25))

# Options for facets
p + facet_grid(sex ~ .) + theme(
    strip.background = element_rect(fill="pink"),
    strip.text.y = element_text(size=14, angle=-90, face="bold"))
    # strip.text.x is the same, but for horizontal facets
```

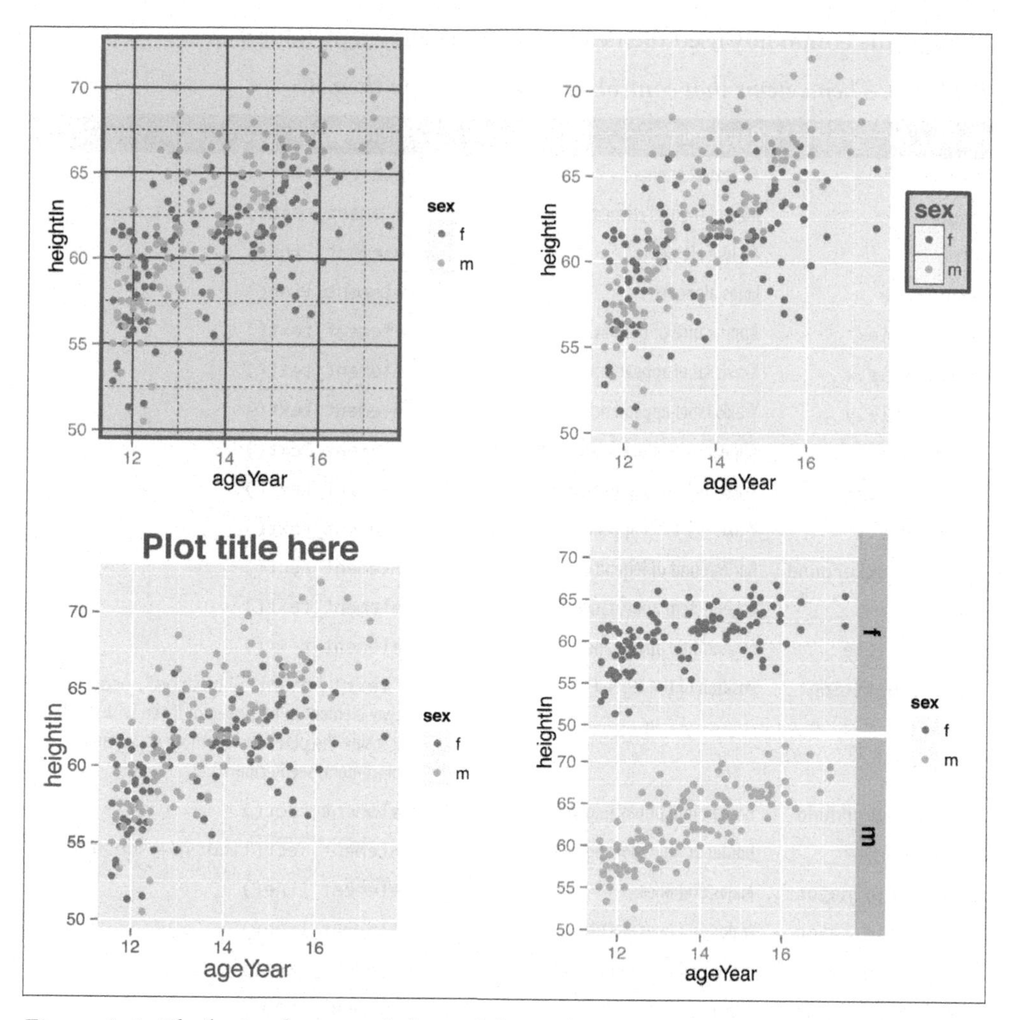

Figure 9-6. Clockwise from top left: modifying theme properties for the plotting area, the legend, the facets, and the text items

Discussion

If you want to use a saved theme and tweak a few parts of it with `theme()`, the `theme()` must come after the theme specification. Otherwise, anything set by `theme()` will be unset by the theme you add:

```
# theme() has no effect if before adding a complete theme
p + theme(axis.title.x = element_text(colour="red")) + theme_bw()

# theme() works if after a compete theme
p + theme_bw() + theme(axis.title.x = element_text(colour="red", size=12))
```

Many of the commonly used theme properties are shown in Table 9-3.

Table 9-3. Theme items that control text appearance in theme()

Name	Description	Element type
text	All text elements	element_text()
rect	All rectangular elements	element_rect()
line	All line elements	element_line()
axis.line	Lines along axes	element_line()
axis.title	Appearance of both axis labels	element_text()
axis.title.x	X-axis label appearance	element_text()
axis.title.y	Y-axis label appearance	element_text()
axis.text	Appearance of tick labels on both axes	element_text()
axis.text.x	X-axis tick label appearance	element_text()
axis.text.y	Y-axis tick label appearance	element_text()
legend.background	Background of legend	element_rect()
legend.text	Legend item appearance	element_text()
legend.title	Legend title appearance	element_text()
legend.position	Position of the legend	"left", "right", "bottom", "top", or two-element numeric vector if you wish to place it inside the plot area (for more on legend placement, see Recipe 10.2)
panel.background	Background of plotting area	element_rect()
panel.border	Border around plotting area	element_rect(linetype="dashed")
panel.grid.major	Major grid lines	element_line()
panel.grid.major.x	Major grid lines, vertical	element_line()
panel.grid.major.y	Major grid lines, horizontal	element_line()
panel.grid.minor	Minor grid lines	element_line()
panel.grid.minor.x	Minor grid lines, vertical	element_line()
panel.grid.minor.y	Minor grid lines, horizontal	element_line()
plot.background	Background of the entire plot	element_rect(fill = "white", colour = NA)
plot.title	Title text appearance	element_text()
strip.background	Background of facet labels	element_rect()
strip.text	Text appearance for vertical and horizontal facet labels	element_text()
strip.text.x	Text appearance for horizontal facet labels	element_text()
strip.text.y	Text appearance for vertical facet labels	element_text()

9.5. Creating Your Own Themes

Problem

You want to create your own theme.

Solution

You can create your own theme by adding elements to an existing theme (Figure 9-7):

```
library(gcookbook) # For the data set

# Start with theme_bw() and modify a few things
mytheme <- theme_bw() +
    theme(text       = element_text(colour="red"),
          axis.title = element_text(size = rel(1.25)))

# Base plot
p <- ggplot(heightweight, aes(x=ageYear, y=heightIn)) + geom_point()

# Plot with modified theme
p + mytheme
```

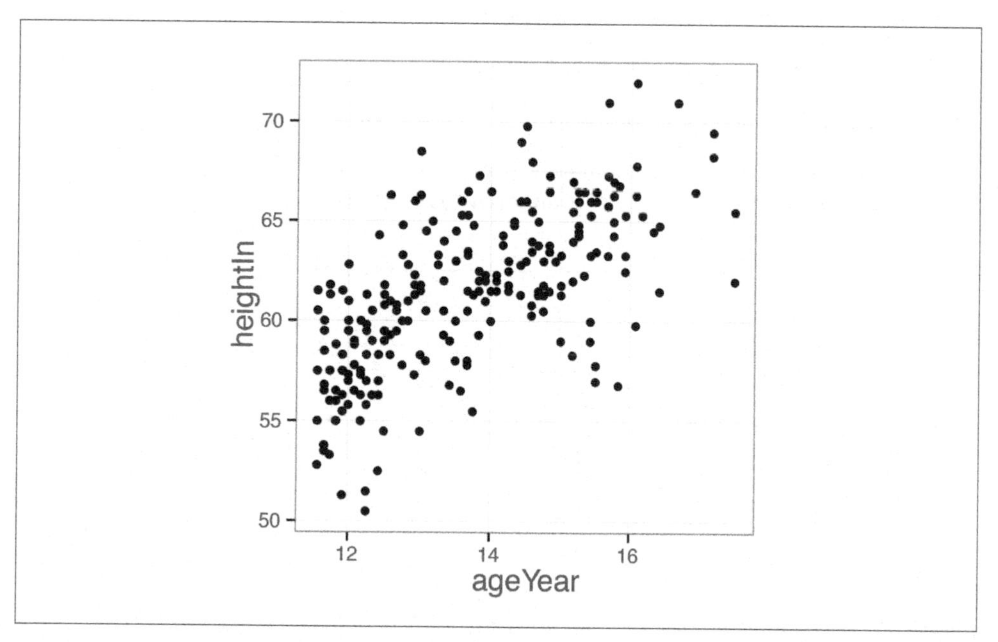

Figure 9-7. A modified default theme

Discussion

With ggplot2, you can not only make use of the default themes, but also modify these themes to suit your needs. You can add new theme elements or change the values of existing ones, and apply your changes globally or to a single plot.

See Also

The options for modifying themes are listed in Recipe 9.4.

9.6. Hiding Grid Lines

Problem

You want to hide the grid lines in a plot.

Solution

The major grid lines (those that align with the tick marks) are controlled with `panel.grid.major`. The minor grid lines (the ones between the major lines) are controlled with `panel.grid.minor`. This will hide them both, as shown in Figure 9-8 (left):

```
library(gcookbook) # For the data set

p <- ggplot(heightweight, aes(x=ageYear, y=heightIn)) + geom_point()

p + theme(panel.grid.major = element_blank(),
          panel.grid.minor = element_blank())
```

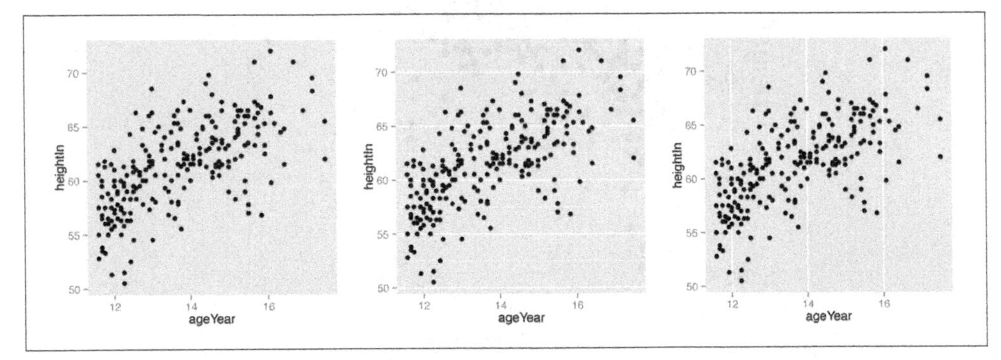

Figure 9-8. Left: no grid lines; middle: no vertical lines; right: no horizontal lines

Discussion

It's possible to hide just the vertical or horizontal grid lines, as shown in the middle and righthand graphs in Figure 9-8, with `panel.grid.major.x`, `panel.grid.major.y`, `panel.grid.minor.x`, and `panel.grid.minor.y`:

```
# Hide the vertical grid lines (which intersect with the x-axis)
p + theme(panel.grid.major.x = element_blank(),
          panel.grid.minor.x = element_blank())

# Hide the horizontal grid lines (which intersect with the y-axis)
p + theme(panel.grid.major.y = element_blank(),
          panel.grid.minor.y = element_blank())
```

CHAPTER 10
Legends

Like the x- or y-axis, a legend is a guide: it shows people how to map visual (aesthetic) properties back to data values.

10.1. Removing the Legend

Problem

You want to remove the legend from a graph.

Solution

Use `guides()`, and specify the scale that should have its legend removed (Figure 10-1):

```
# The base plot (with legend)
p <- ggplot(PlantGrowth, aes(x=group, y=weight, fill=group)) + geom_boxplot()
p

# Remove the legend for fill
p + guides(fill=FALSE)
```

Discussion

Another way to remove a legend is to set `guide=FALSE` in the scale. This will result in the exact same output as the preceding code:

```
# Remove the legend for fill
p + scale_fill_discrete(guide=FALSE)
```

Yet another way to remove the legend is to use the theming system. If you have more than one aesthetic mapping with a legend (`color` and `shape`, for example), this will remove legends for all of them:

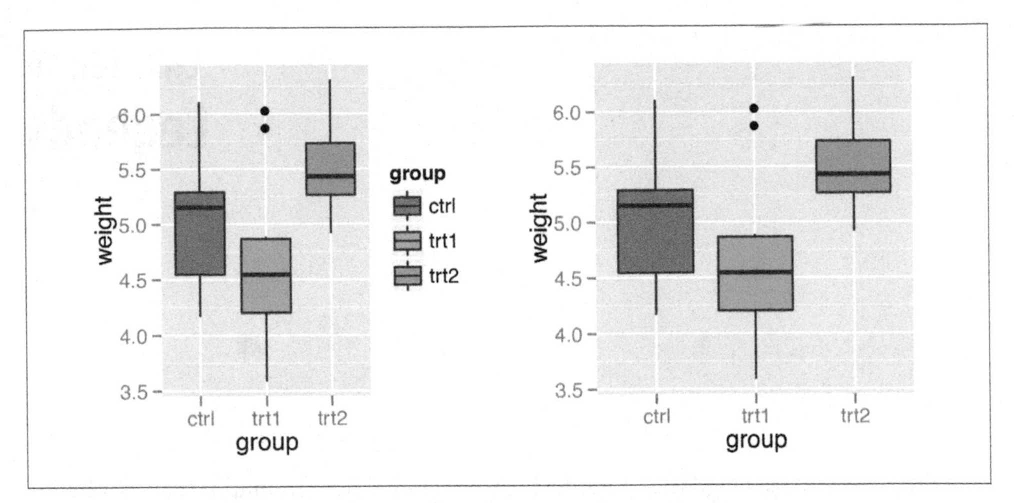

Figure 10-1. Left: default appearance; right: with legend removed

```
p + theme(legend.position="none")
```

Sometimes a legend is redundant, or it is supplied in another graph that will be displayed with the current one. In these cases, it can be useful to remove the legend from a graph.

In the example used here, the colors provide the same information that is on the x-axis, so the legend is unnecessary. Notice that with the legend removed, the area used for graphing the data is larger. If you want to achieve the same proportions in the graphing area, you will need to adjust the overall dimensions of the graph.

When a variable is mapped to `fill`, the default scale used is `scale_fill_discrete()` (equivalent to `scale_fill_hue()`), which maps the factor levels to colors that are equally spaced around the color wheel. There are other scales for `fill`, such as `scale_fill_manual()`. If you use scales for other aesthetics, such as `colour` (for lines and points) or `shape` (for points), you must use the appropriate scale. Commonly used scales include:

- `scale_fill_discrete()`
- `scale_fill_hue()`
- `scale_fill_manual()`
- `scale_fill_grey()`
- `scale_fill_brewer()`
- `scale_colour_discrete()`
- `scale_colour_hue()`

- `scale_colour_manual()`
- `scale_colour_grey()`
- `scale_colour_brewer()`
- `scale_shape_manual()`
- `scale_linetype()`

10.2. Changing the Position of a Legend

Problem

You want to move the legend from its default place on the right side.

Solution

Use `theme(legend.position=...)`. It can be put on the top, left, right, or bottom by using one of those strings as the position (Figure 10-2, left):

```
p <- ggplot(PlantGrowth, aes(x=group, y=weight, fill=group)) + geom_boxplot() +
    scale_fill_brewer(palette="Pastel2")

p + theme(legend.position="top")
```

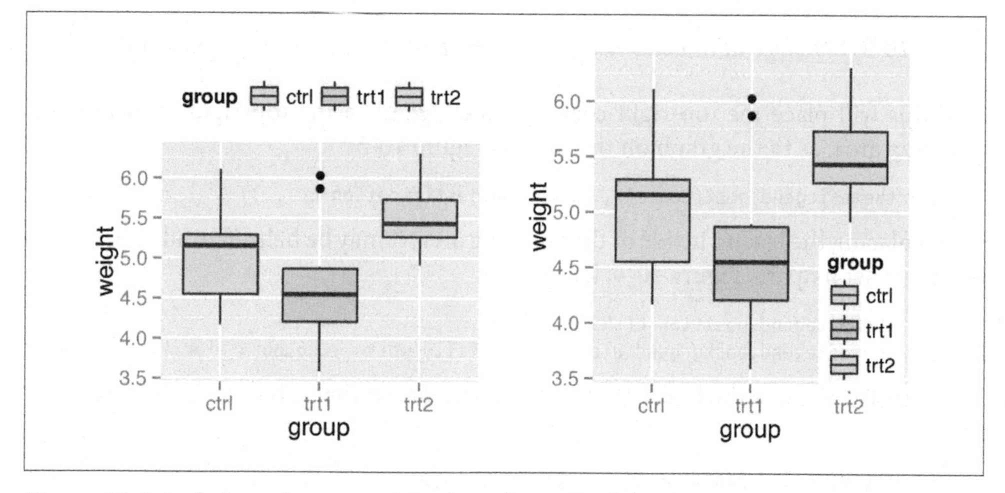

Figure 10-2. Left: legend on top; right: legend inside of graphing area

The legend can also be placed inside the graphing area by specifying a coordinate position, as in `legend.position=c(1,0)` (Figure 10-2, right). The coordinate space starts at (0, 0) in the bottom left and goes to (1, 1) in the top right.

Discussion

You can also use `legend.justification` to set which *part* of the legend box is set to the position at `legend.position`. By default, the center of the legend (.5, .5) is placed at the coordinate, but it is often useful to specify a different point.

For example, this will place the bottom-right corner of the legend (1,0) in the bottom-right corner of the graphing area (1,0):

```
p + theme(legend.position=c(1,0), legend.justification=c(1,0))
```

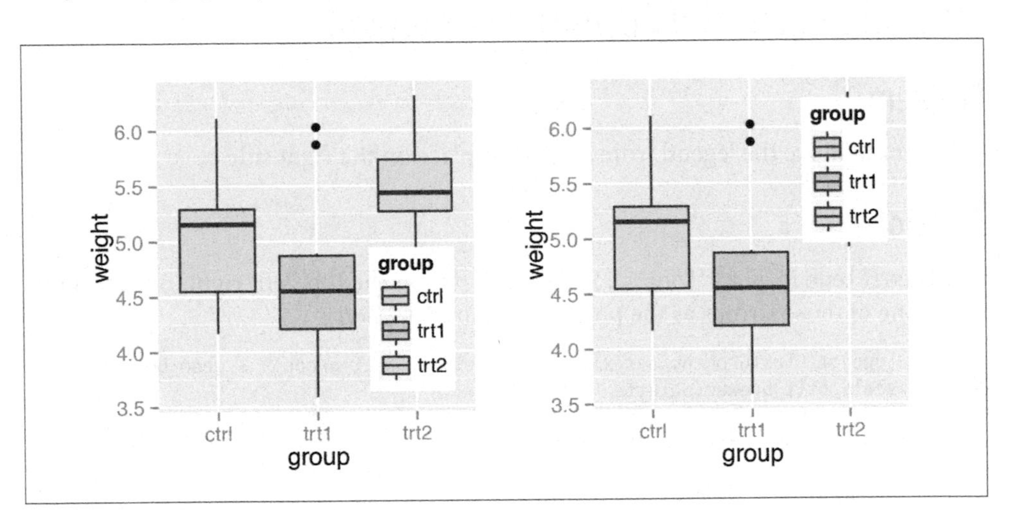

Figure 10-3. Left: legend in bottom-right corner; right: legend in top-right corner.

And this will place the top-right corner of the legend in the top-right corner of the graphing area, as in the graph on the right in Figure 10-3:

```
p + theme(legend.position=c(1,1), legend.justification=c(1,1))
```

When placing the legend inside of the graphing area, it may be helpful to add an opaque border to set it apart (Figure 10-4, left):

```
p + theme(legend.position=c(.85,.2)) +
    theme(legend.background=element_rect(fill="white", colour="black"))
```

You can also remove the border around its elements so that it blends in (Figure 10-4, right):

```
p + theme(legend.position=c(.85,.2)) +
    theme(legend.background=element_blank()) +  # Remove overall border
    theme(legend.key=element_blank())            # Remove border around each item
```

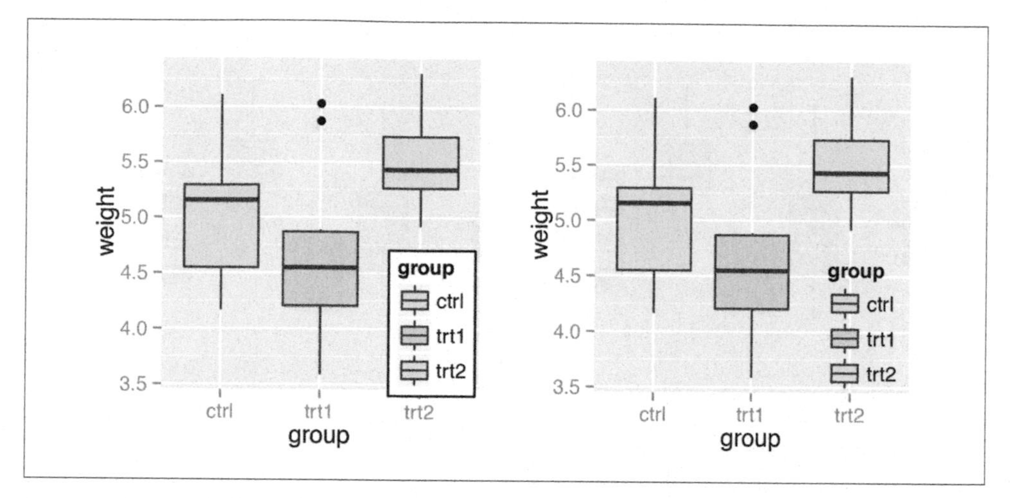

Figure 10-4. Left: legend with opaque background and outline; right: with no background or outlines

10.3. Changing the Order of Items in a Legend

Problem

You want to change the order of the items in a legend.

Solution

Set the `limits` in the scale to the desired order (Figure 10-5):

```
# The base plot
p <- ggplot(PlantGrowth, aes(x=group, y=weight, fill=group)) + geom_boxplot()
p

# Change the order of items
p + scale_fill_discrete(limits=c("trt1", "trt2", "ctrl"))
```

Discussion

Note that the order of the items on the x-axis did not change. To do that, you would have to set the `limits` of `scale_x_discrete()` (Recipe 8.4), or change the data to have a different factor level order (Recipe 15.8).

In the preceding example, `group` was mapped to the `fill` aesthetic. By default this uses `scale_fill_discrete()` (which is the same as `scale_fill_hue()`), which maps the factor levels to colors that are equally spaced around the color wheel. We could have

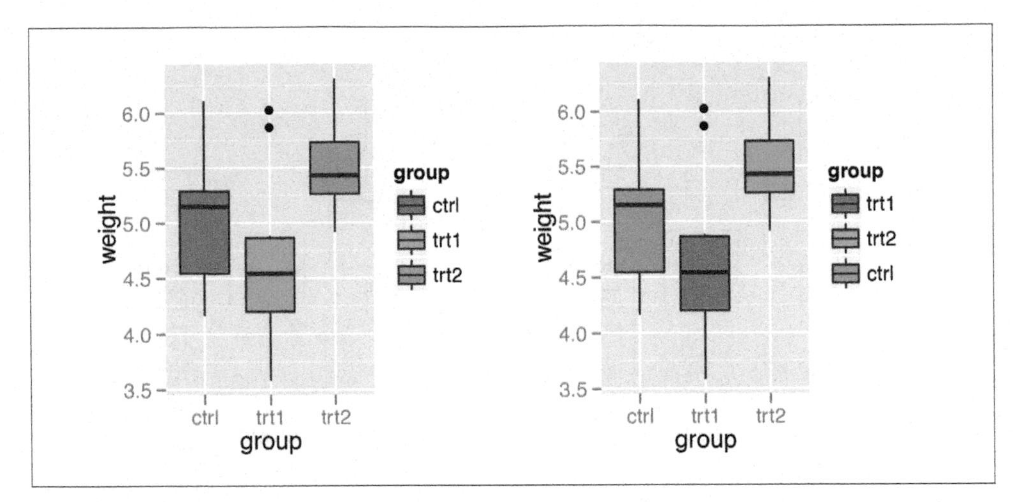

Figure 10-5. Left: default order for legend; right: modified order

used a different `scale_fill_xxx()`, though. For example, we could use a grey palette (Figure 10-6, left):

```
p + scale_fill_grey(start=.5, end=1, limits=c("trt1", "trt2", "ctrl"))
```

Or we could use a palette from RColorBrewer (Figure 10-6, right):

```
p + scale_fill_brewer(palette="Pastel2", limits=c("trt1", "trt2", "ctrl"))
```

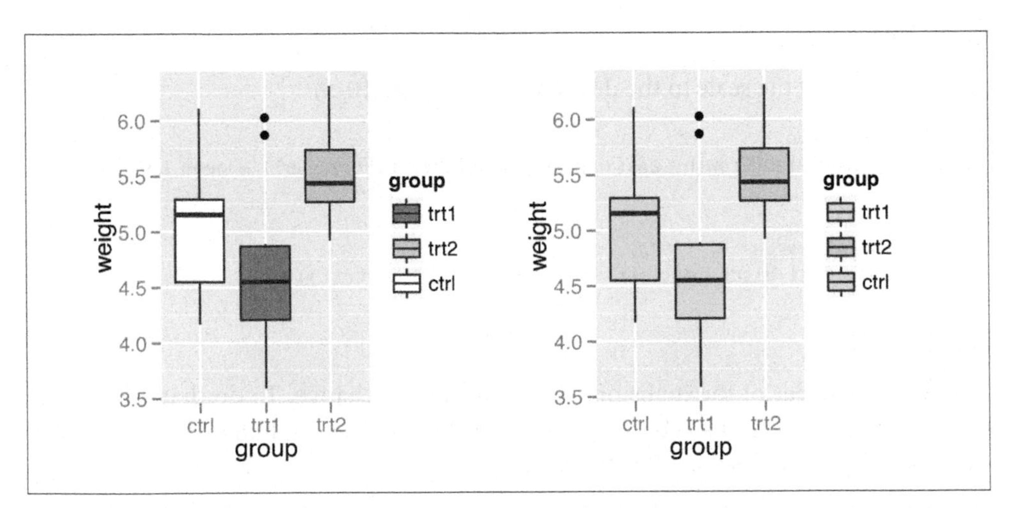

Figure 10-6. Left: modified order with a grey palette; right: with a palette from RColorBrewer

All the previous examples were for `fill`. If you use scales for other aesthetics, such as `colour` (for lines and points) or `shape` (for points), you must use the appropriate scale. Commonly used scales include:

- `scale_fill_discrete()`
- `scale_fill_hue()`
- `scale_fill_manual()`
- `scale_fill_grey()`
- `scale_fill_brewer()`
- `scale_colour_discrete()`
- `scale_colour_hue()`
- `scale_colour_manual()`
- `scale_colour_grey()`
- `scale_colour_brewer()`
- `scale_shape_manual()`
- `scale_linetype()`

By default, using `scale_fill_discrete()` is equivalent to using `scale_fill_hue()`; the same is true for color scales.

See Also

To reverse the order of the legend, see Recipe 10.4.

To change the order of factor levels, see Recipe 15.8. To order legend items based on values in another variable, see Recipe 15.9.

10.4. Reversing the Order of Items in a Legend

Problem

You want to reverse the order of items in a legend.

Solution

Add `guides(fill=guide_legend(reverse=TRUE))` to reverse the order of the legend, as in Figure 10-7 (for other aesthetics, replace `fill` with the name of the aesthetic, such as `colour` or `size`):

```
# The base plot
p <- ggplot(PlantGrowth, aes(x=group, y=weight, fill=group)) + geom_boxplot()
p

# Reverse the legend order
p + guides(fill=guide_legend(reverse=TRUE))
```

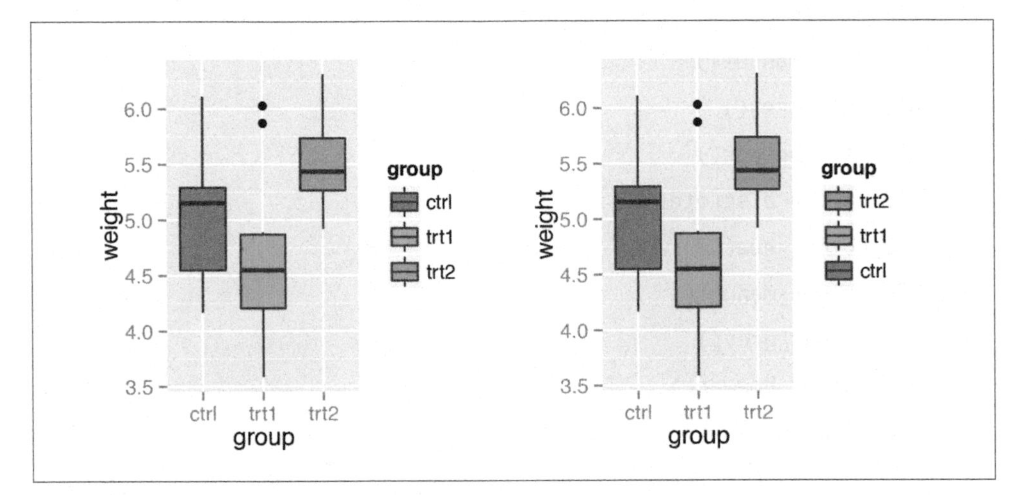

Figure 10-7. Left: default order for legend; right: reversed order

Discussion

It is also possible to control the legend when specifying the scale, as in the following:

```
scale_fill_hue(guide=guide_legend(reverse=TRUE))
```

10.5. Changing a Legend Title

Problem

You want to change the text of a legend title.

Solution

Use `labs()` and set the value of `fill`, `colour`, `shape`, or whatever aesthetic is appropriate for the legend (Figure 10-8):

```
# The base plot
p <- ggplot(PlantGrowth, aes(x=group, y=weight, fill=group)) + geom_boxplot()
p

# Set the legend title to "Condition"
p + labs(fill="Condition")
```

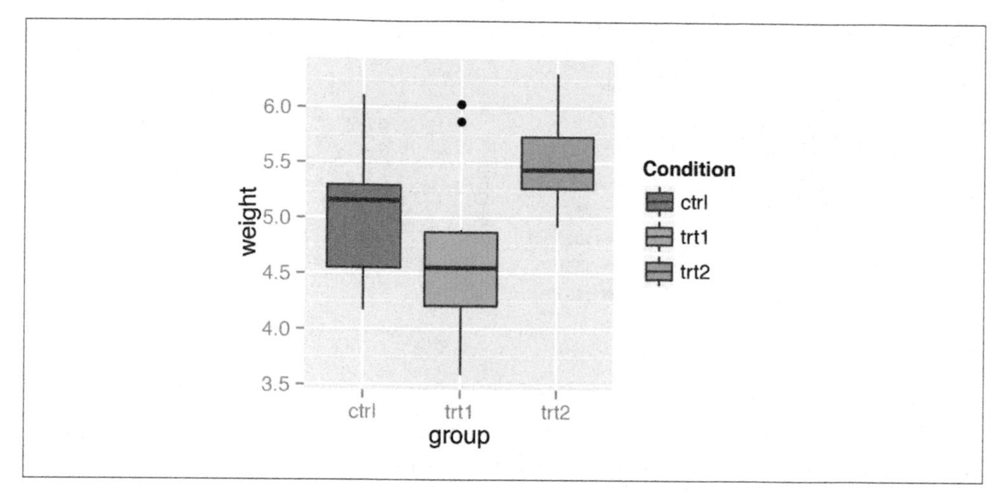

Figure 10-8. With the legend title set to "Condition"

Discussion

It's also possible to set the title of the legend in the scale specification. Since legends and axes are both guides, this works the same way as setting the title of the x- or y-axis.

This would have the same effect as the previous code:

```
p + scale_fill_discrete(name="Condition")
```

If there are multiple variables mapped to aesthetics with a legend (those other than x and y), you can set the title of each individually. In the example here we'll use \n to add a line break in one of the titles (Figure 10-9):

```
library(gcookbook) # For the data set

# Make the base plot
hw <- ggplot(heightweight, aes(x=ageYear, y=heightIn, colour=sex)) +
      geom_point(aes(size=weightLb)) + scale_size_continuous(range=c(1,4))

hw

# With new legend titles
hw + labs(colour="Male/Female", size="Weight\n(pounds)")
```

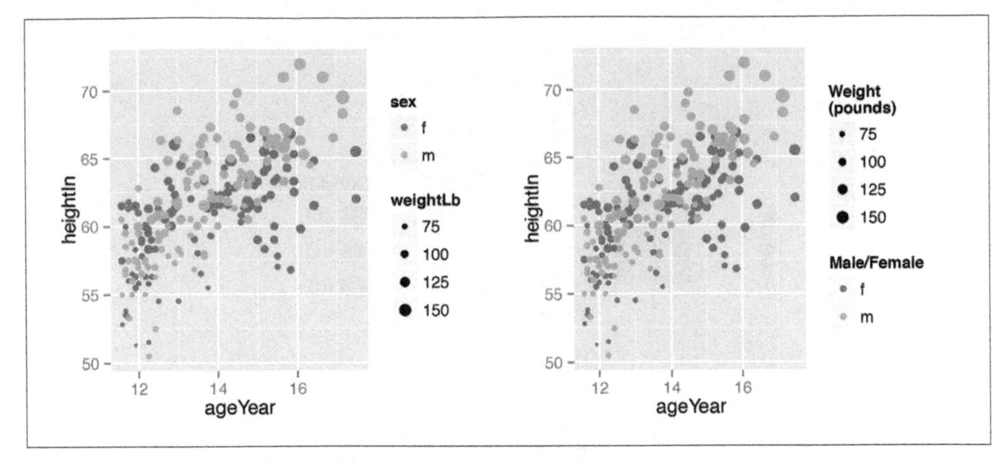

Figure 10-9. Left: two legends with original titles; right: with new titles

If you have one variable mapped to two separate aesthetics, the default is to have a single legend that combines both. For example, if we map sex to both shape and weight, there will be just one legend (Figure 10-10, left):

```
hw1 <- ggplot(heightweight, aes(x=ageYear, y=heightIn, shape=sex, colour=sex)) +
       geom_point()

hw1
```

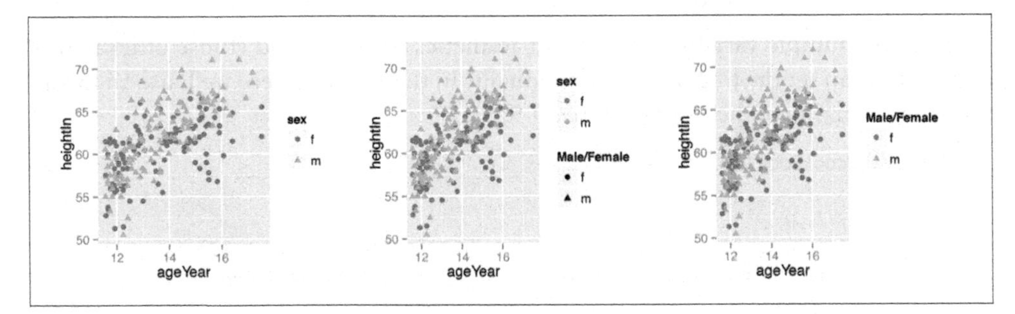

Figure 10-10. Left: default legend with a variable mapped to shape and colour; middle: with shape renamed; right: with both shape and colour renamed

To change the title (Figure 10-10, right), you need to set the name for both of them. If you change the name for just one, it will result in two separate legends (Figure 10-10, middle):

```
# Change just shape
hw1 + labs(shape="Male/Female")
```

```
# Change both shape and colour
hw1 + labs(shape="Male/Female", colour="Male/Female")
```

It is also possible to control the legend title with the `guides()` function. It's a little more verbose, but it can be useful when you're already using it to control other properties:

```
p + guides(fill=guide_legend(title="Condition"))
```

10.6. Changing the Appearance of a Legend Title

Problem

You want to change the appearance of a legend title's text.

Solution

Use `theme(legend.title=element_text())` (Figure 10-11):

```
p <- ggplot(PlantGrowth, aes(x=group, y=weight, fill=group)) + geom_boxplot()

p + theme(legend.title=element_text(face="italic", family="Times", colour="red",
                                   size=14))
```

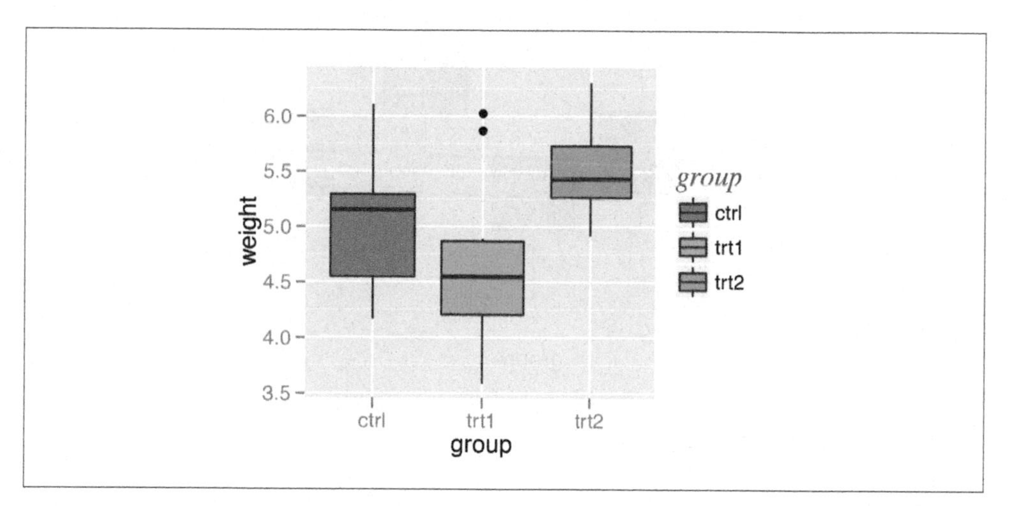

Figure 10-11. Customized legend title appearance

Discussion

It's also possible to specify the legend title's appearance via `guides()`, but this method can be a bit verbose. This has the same effect as the previous code:

```
p + guides(fill=guide_legend(title.theme=
           element_text(face="italic", family="times", colour="red", size=14)))
```

See Also

See Recipe 9.2 for more on controlling the appearance of text.

10.7. Removing a Legend Title

Problem

You want to remove a legend title.

Solution

Add `guides(fill=guide_legend(title=NULL))` to remove the title from a legend, as in Figure 10-12 (for other aesthetics, replace `fill` with the name of the aesthetic, such as `colour` or `size`):

```
ggplot(PlantGrowth, aes(x=group, y=weight, fill=group)) + geom_boxplot() +
    guides(fill=guide_legend(title=NULL))
```

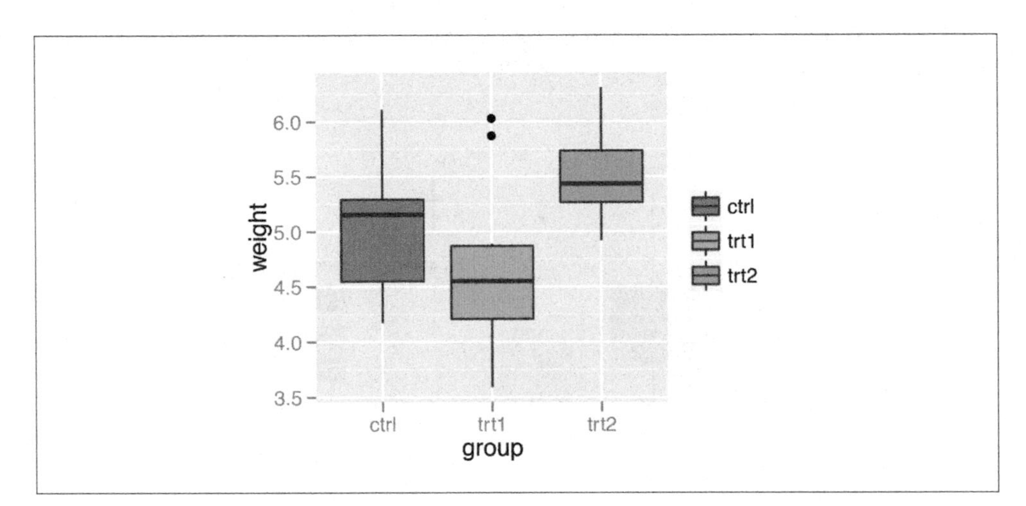

Figure 10-12. Box plot with no legend title

Discussion

It is also possible to control the legend title when specifying the scale. This has the same effect as the preceding code:

```
scale_fill_hue(guide = guide_legend(title=NULL))
```

10.8. Changing the Labels in a Legend

Problem

You want to change the text of labels in a legend.

Solution

Set the `labels` in the scale (Figure 10-13, left):

```
library(gcookbook) # For the data set

# The base plot
p <- ggplot(PlantGrowth, aes(x=group, y=weight, fill=group)) + geom_boxplot()

# Change the legend labels
p + scale_fill_discrete(labels=c("Control", "Treatment 1", "Treatment 2"))
```

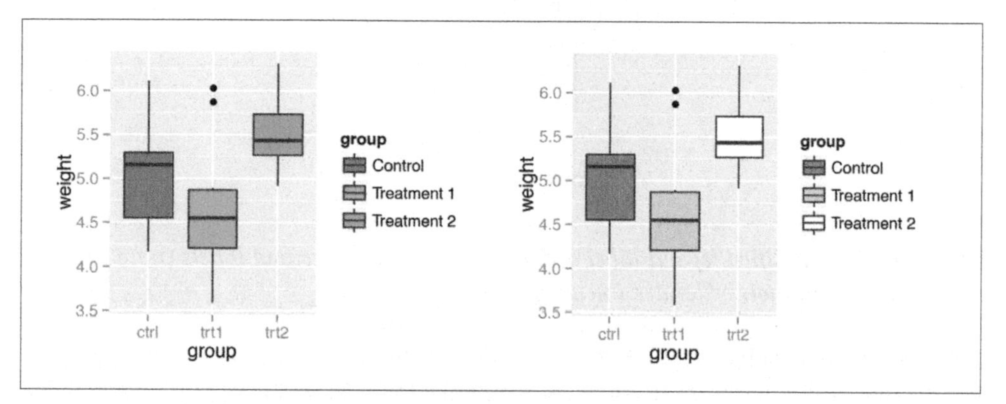

Figure 10-13. Left: manually specified legend labels with the default discrete scale; right: manually specified labels with a different scale

Discussion

Note that the labels on the x-axis did not change. To do that, you would have to set the `labels` of `scale_x_discrete()` (Recipe 8.10), or change the data to have different factor level names (Recipe 15.10).

In the preceding example, `group` was mapped to the `fill` aesthetic. By default this uses `scale_fill_discrete()`, which maps the factor levels to colors that are equally spaced around the color wheel (the same as `scale_fill_hue()`). There are other `fill` scales we could use, and setting the labels works the same way. For example, to produce the graph on the right in Figure 10-13:

```
p + scale_fill_grey(start=.5, end=1,
                    labels=c("Control", "Treatment 1", "Treatment 2"))
```

If you are also changing the order of items in the legend, the labels are matched to the items by position. In this example we'll change the item order, and make sure to set the labels in the same order (Figure 10-14):

```
p + scale_fill_discrete(limits=c("trt1", "trt2", "ctrl"),
                        labels=c("Treatment 1", "Treatment 2", "Control"))
```

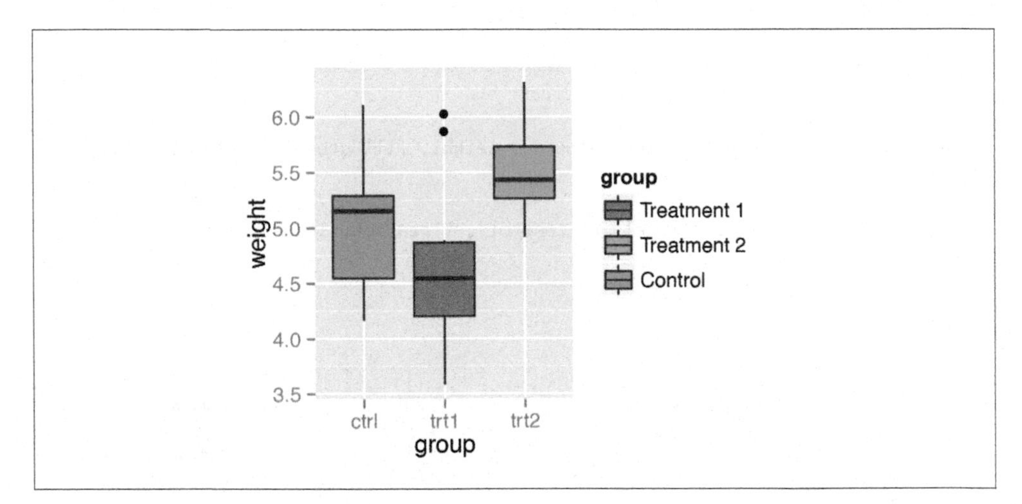

Figure 10-14. Modified legend label order and manually specified labels (note that the x-axis labels and their order are unchanged)

If you have one variable mapped to two separate aesthetics, the default is to have a single legend that combines both. If you want to change the legend labels, you must change them for both scales; otherwise you will end up with two separate legends, as shown in Figure 10-15:

```
# The base plot
p <- ggplot(heightweight, aes(x=ageYear, y=heightIn, shape=sex, colour=sex)) +
     geom_point()
p

# Change the labels for one scale
p + scale_shape_discrete(labels=c("Female", "Male"))

# Change the labels for both scales
p + scale_shape_discrete(labels=c("Female", "Male")) +
    scale_colour_discrete(labels=c("Female", "Male"))
```

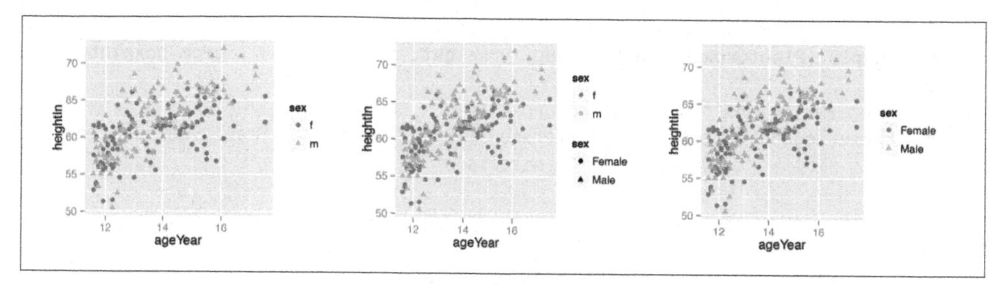

Figure 10-15. Left: a variable mapped to shape and colour; middle: with new labels for shape; right: with new labels for both shape and colour

Other commonly used scales with legends include:

- `scale_fill_discrete()`
- `scale_fill_hue()`
- `scale_fill_manual()`
- `scale_fill_grey()`
- `scale_fill_brewer()`
- `scale_colour_discrete()`
- `scale_colour_hue()`
- `scale_colour_manual()`
- `scale_colour_grey()`
- `scale_colour_brewer()`
- `scale_shape_manual()`
- `scale_linetype()`

By default, using `scale_fill_discrete()` is equivalent to using `scale_fill_hue()`; the same is true for color scales.

10.9. Changing the Appearance of Legend Labels

Problem

You want to change the appearance of labels in a legend.

Solution

Use `theme(legend.text=element_text())` (Figure 10-16):

```
# The base plot
p <- ggplot(PlantGrowth, aes(x=group, y=weight, fill=group)) + geom_boxplot()

# Change the legend label appearance
p + theme(legend.text=element_text(face="italic", family="Times", colour="red",
                                   size=14))
```

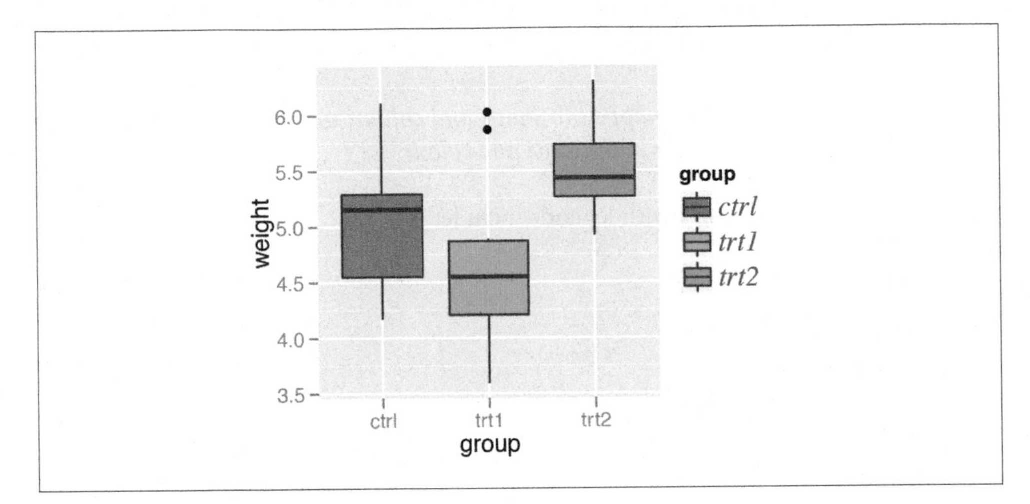

Figure 10-16. Customized legend label appearance

Discussion

It's also possible to specify the legend label appearance via `guides()`, although this method is a bit unwieldy. This has the same effect as the previous code:

```
# Changes the legend title text for the fill legend
p + guides(fill=guide_legend(label.theme=
           element_text(face="italic", family="Times", colour="red", size=14)))
```

See Also

See Recipe 9.2 for more on controlling the appearance of text.

10.10. Using Labels with Multiple Lines of Text

Problem

You want to use legend labels that have more than one line of text.

Solution

Set the labels in the scale, using \n to represent a newline. In this example, we'll use scale_fill_discrete() to control the legend for the fill scale (Figure 10-17, left):

```
p <- ggplot(PlantGrowth, aes(x=group, y=weight, fill=group)) + geom_boxplot()

# Labels that have more than one line
p + scale_fill_discrete(labels=c("Control", "Type 1\ntreatment",
                                 "Type 2\ntreatment"))
```

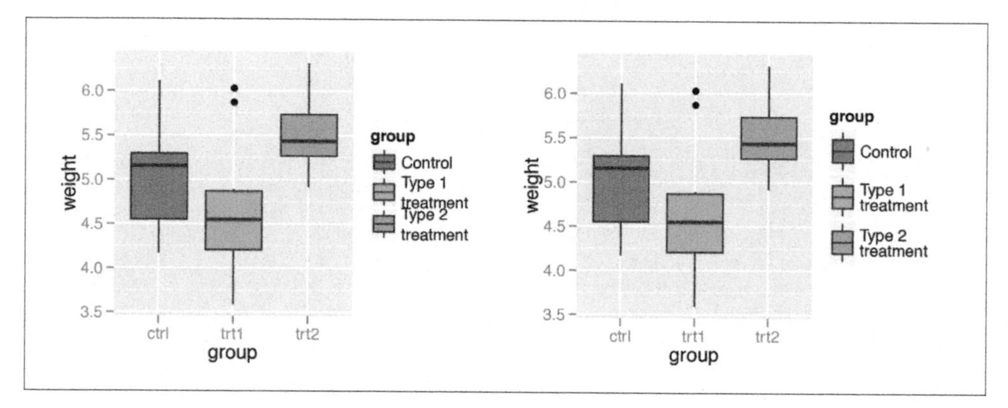

Figure 10-17. Left: multiline legend labels; right: with increased key height and reduced line spacing

Discussion

As you can see in the version on the left in Figure 10-17, with the default settings the lines of text will run into each other when you use labels that have more than one line. To deal with this problem, you can increase the height of the legend keys and decrease the spacing between lines, using theme() (Figure 10-17, right). To do this, you will need to specify the height using the unit() function from the grid package:

```
library(grid)
p + scale_fill_discrete(labels=c("Control", "Type 1\ntreatment",
                                 "Type 2\ntreatment")) +
    theme(legend.text=element_text(lineheight=.8),
          legend.key.height=unit(1, "cm"))
```

CHAPTER 11

Facets

One of the most useful techniques in data visualization is rendering groups of data alongside each other, making it easy to compare the groups. With ggplot2, one way to do this is by mapping a discrete variable to an aesthetic, like *x* position, color, or shape. Another way of doing this is to create a subplot for each group and draw the subplots side by side.

These kinds of plots are known as *Trellis* displays. They're implemented in the lattice package as well as in the ggplot2 package. In ggplot2, they're called *facets*. In this chapter I'll explain how to use them.

11.1. Splitting Data into Subplots with Facets

Problem

You want to plot subsets of your data in separate panels.

Solution

Use `facet_grid()` or `facet_wrap()`, and specify the variables on which to split.

With `facet_grid()`, you can specify a variable to split the data into vertical subpanels, and another variable to split it into horizontal subpanels (Figure 11-1):

```
# The base plot
p <- ggplot(mpg, aes(x=displ, y=hwy)) + geom_point()

# Faceted by drv, in vertically arranged subpanels
p + facet_grid(drv ~ .)

# Faceted by cyl, in horizontally arranged subpanels
p + facet_grid(. ~ cyl)
```

```
# Split by drv (vertical) and cyl (horizontal)
p + facet_grid(drv ~ cyl)
```

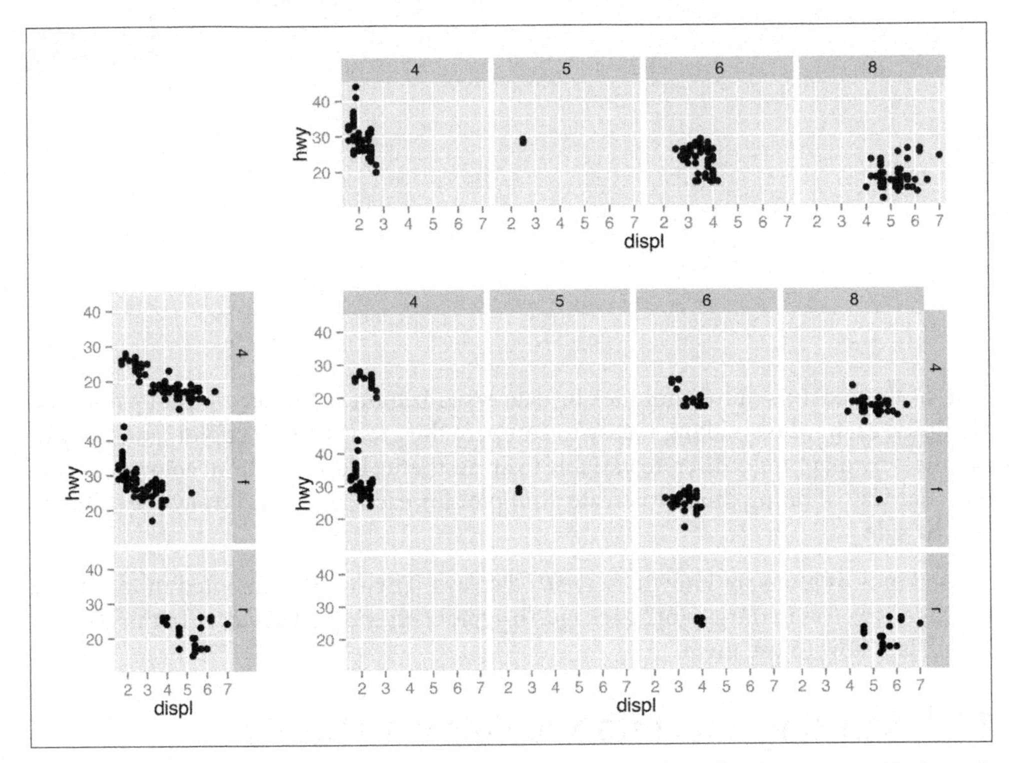

Figure 11-1. Top: faceting horizontally by drv; bottom left: faceting vertically by cyl; bottom right: faceting in both directions, with both variables

With `facet_wrap()`, the subplots are laid out horizontally and wrap around, like words on a page, as in Figure 11-2:

```
# Facet on class
# Note there is nothing before the tilde
p + facet_wrap( ~ class)
```

Discussion

With `facet_wrap()`, the default is to use the same number of rows and columns. In Figure 11-2, there were seven facets, and they fit into a 3×3 "square." To change this, you can pass a value for `nrow` or `ncol`:

```
# These will have the same result: 2 rows and 4 cols
p + facet_wrap( ~ class, nrow=2)
p + facet_wrap( ~ class, ncol=4)
```

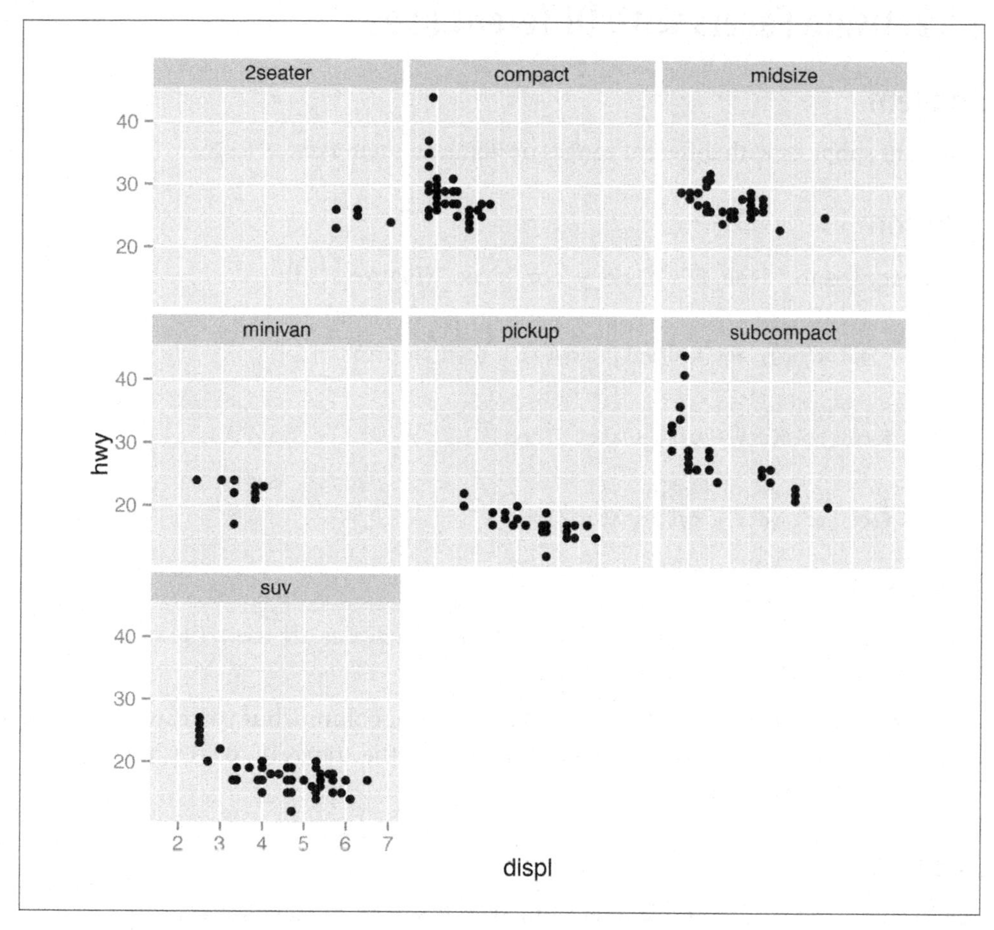

Figure 11-2. A scatter plot with facet_wrap() on class

The choice of faceting direction depends on the kind of comparison you would like to encourage. For example, if you want to compare heights of bars, it's useful to make the facets go horizontally. If, on the other hand, you want to compare the horizontal distribution of histograms, it makes sense to make the facets go vertically.

Sometimes both kinds of comparisons are important—there may not be a clear answer as to which faceting direction is best. It may turn out that displaying the groups in a single plot by mapping the grouping variable to an aesthetic like color works better than using facets. In these situations, you'll have to rely on your judgment.

11.2. Using Facets with Different Axes

Problem

You want subplots with different ranges or items on their axes.

Solution

Set the `scales` to `"free_x"`, `"free_y"`, or `"free"` (Figure 11-3):

```
# The base plot
p <- ggplot(mpg, aes(x=displ, y=hwy)) + geom_point()

# With free y scales
p + facet_grid(drv ~ cyl, scales="free_y")

# With free x and y scales
p + facet_grid(drv ~ cyl, scales="free")
```

Discussion

Each row of subplots has its own *y* range when free *y* scales are used; the same applies to columns when free *x* scales are used.

It's not possible to directly set the range of each row or column, but you can control the ranges by dropping unwanted data (to reduce the ranges), or by adding `geom_blank()` (to expand the ranges).

See Also

See Recipe 3.10 for an example of faceting with free scales and a discrete axis.

11.3. Changing the Text of Facet Labels

Problem

You want to change the text of facet labels.

Solution

Change the names of the factor levels (Figure 11-4):

```
mpg2 <- mpg  # Make a copy of the original data

# Rename 4 to 4wd, f to Front, r to Rear
levels(mpg2$drv)[levels(mpg2$drv)=="4"]  <- "4wd"
levels(mpg2$drv)[levels(mpg2$drv)=="f"]  <- "Front"
```

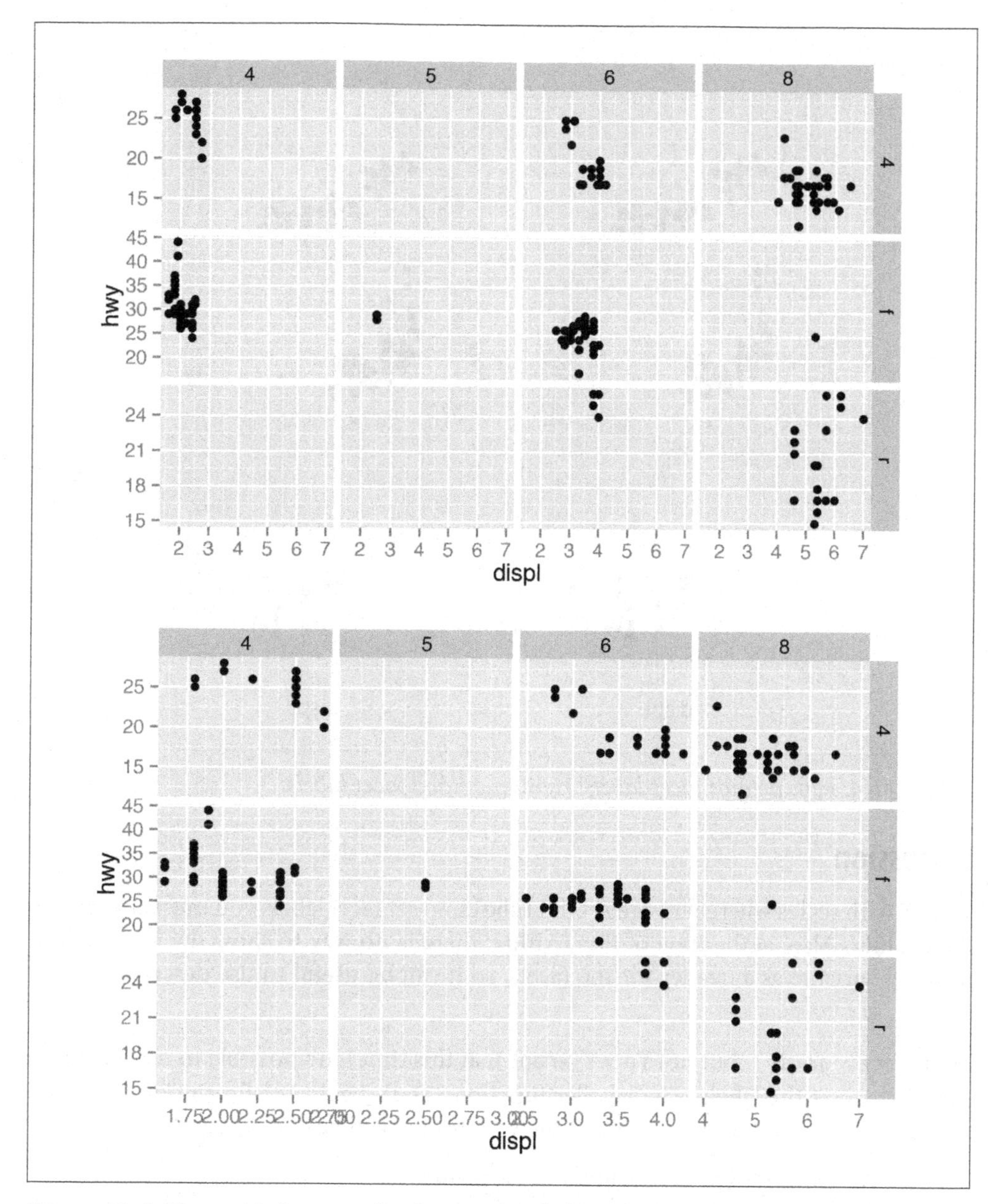

Figure 11-3. Top: with free y scales; bottom: with free x and y scales

```
levels(mpg2$drv)[levels(mpg2$drv)=="r"]  <- "Rear"

# Plot the new data
ggplot(mpg2, aes(x=displ, y=hwy)) + geom_point() + facet_grid(drv ~ .)
```

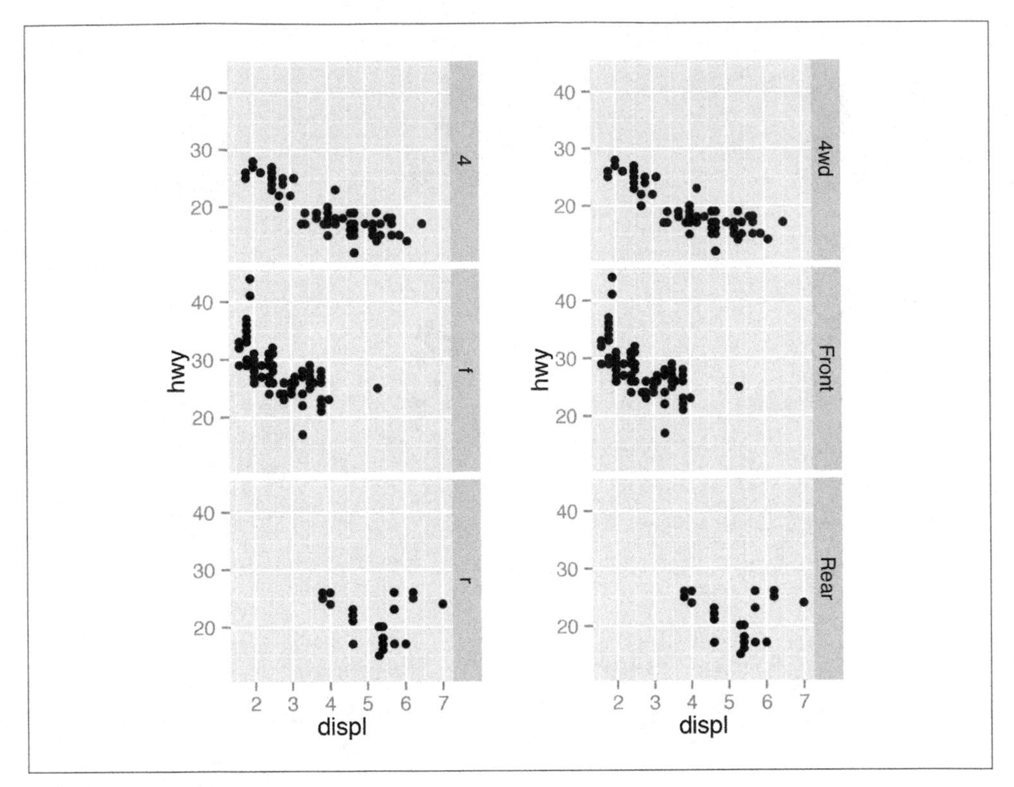

Figure 11-4. Left: default facet labels; right: modified facet labels

Discussion

Unlike with scales where you can set the `labels`, to set facet labels you must change the data values. Also, at the time of this writing, there is no way to show the name of the faceting variable as a header for the facets, so it can be useful to use descriptive facet labels.

With `facet_grid()` (but not `facet_wrap()`, at this time), it's possible to use a labeller function to set the labels. The labeller function `label_both()` will print out both the name of the variable and the value of the variable in each facet (Figure 11-5, left):

```
ggplot(mpg2, aes(x=displ, y=hwy)) + geom_point() +
    facet_grid(drv ~ ., labeller = label_both)
```

Another useful labeller is `label_parsed()`, which takes strings and treats them as R math expressions (Figure 11-5, right):

```
mpg3 <- mpg

levels(mpg3$drv)[levels(mpg3$drv)=="4"]  <- "4^{wd}"
```

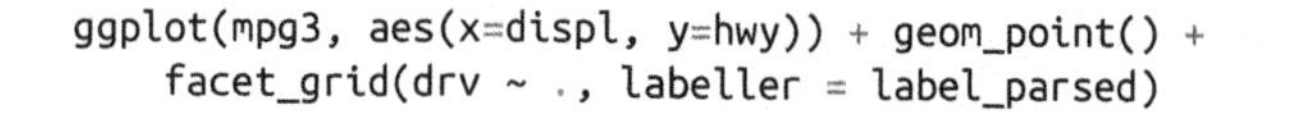

```
levels(mpg3$drv)[levels(mpg3$drv)=="f"]  <- "- Front %.% e^{pi * i}"
levels(mpg3$drv)[levels(mpg3$drv)=="r"]  <- "4^{wd} - Front"
```

```
ggplot(mpg3, aes(x=displ, y=hwy)) + geom_point() +
    facet_grid(drv ~ ., labeller = label_parsed)
```

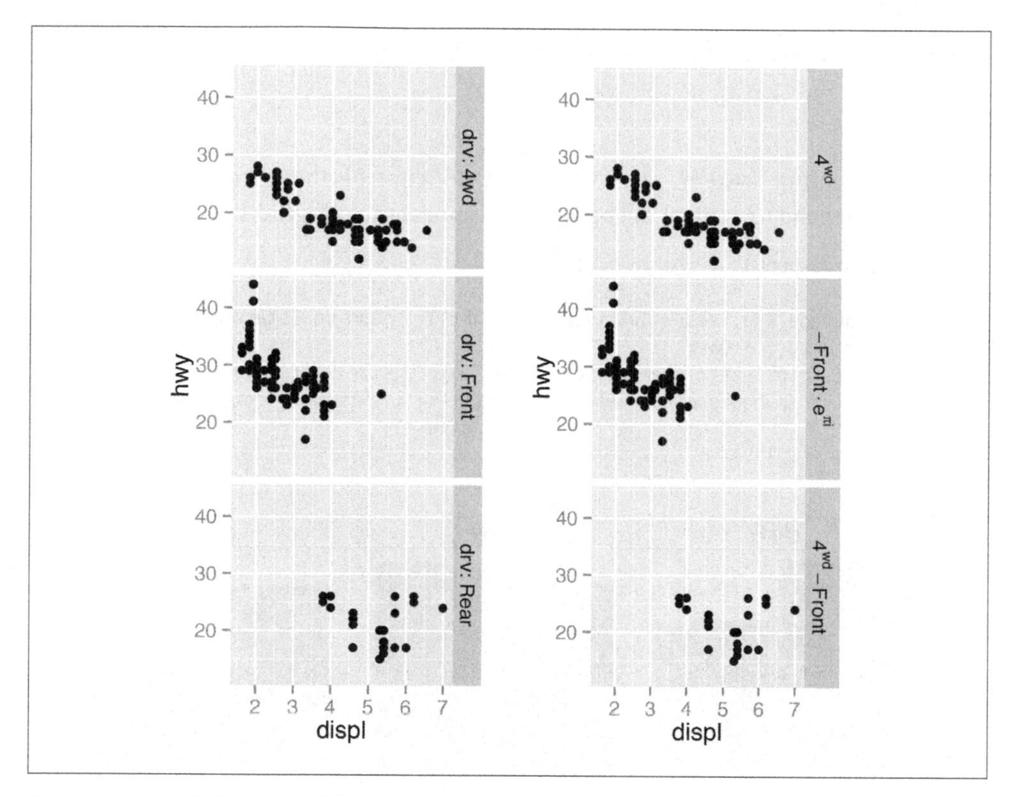

Figure 11-5. Left: with label_both(); right: with label_parsed() for mathematical expressions

See Also

See Recipe 15.10 for more on renaming factor levels. If the faceting variable is not a factor but a character vector, changing the values is somewhat different. See Recipe 15.12 for information on renaming items in character vectors.

11.4. Changing the Appearance of Facet Labels and Headers

Problem

You want to change the appearance of facet labels and headers.

Solution

With the theming system, set `strip.text` to control the text appearance and `strip.background` to control the background appearance (Figure 11-6):

```
library(gcookbook) # For the data set

ggplot(cabbage_exp, aes(x=Cultivar, y=Weight)) + geom_bar(stat="identity") +
    facet_grid(. ~ Date) +
    theme(strip.text = element_text(face="bold", size=rel(1.5)),
          strip.background = element_rect(fill="lightblue", colour="black",
                                          size=1))
```

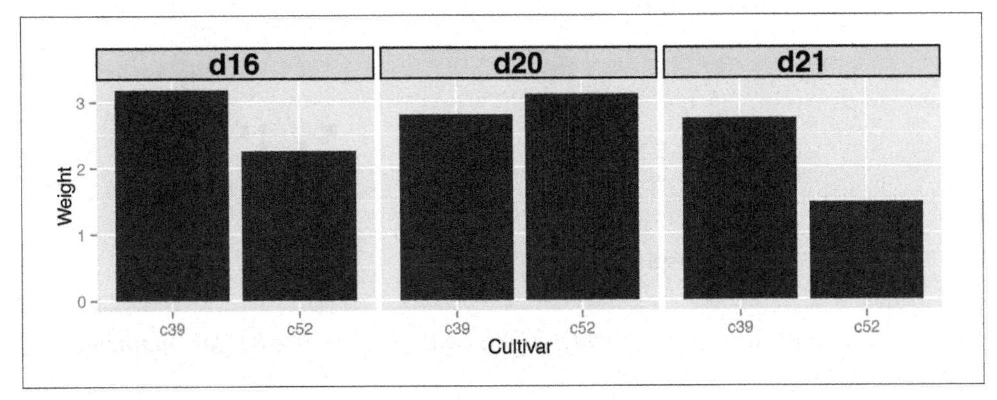

Figure 11-6. Customized appearance for facet labels

Discussion

Using `rel(1.5)` makes the label text 1.5 times the size of the base text size for the theme. Using `size=1` for the background makes the outline of the facets 1 mm thick.

See Also

For more on how the theme system works, see Recipes 9.3 and 9.4.

CHAPTER 12

Using Colors in Plots

In ggplot2's implementation of the grammar of graphics, color is an aesthetic, just like *x* position, *y* position, and size. If color is just another aesthetic, why does it deserve its own chapter? The reason is that color is a more complicated aesthetic than the others. Instead of simply moving geoms left and right or making them larger and smaller, when you use color, there are many degrees of freedom and many more choices to make. What palette should you use for discrete values? Should you use a gradient with several different hues? How do you choose colors that can be interpreted accurately by those with color-vision deficiencies? In this chapter, I'll address these issues.

12.1. Setting the Colors of Objects

Problem

You want to set the color of some geoms in your graph.

Solution

In the call to the geom, set the values of `colour` or `fill` (Figure 12-1):

```
ggplot(mtcars, aes(x=wt, y=mpg)) + geom_point(colour="red")

library(MASS) # For the data set
ggplot(birthwt, aes(x=bwt)) + geom_histogram(fill="red", colour="black")
```

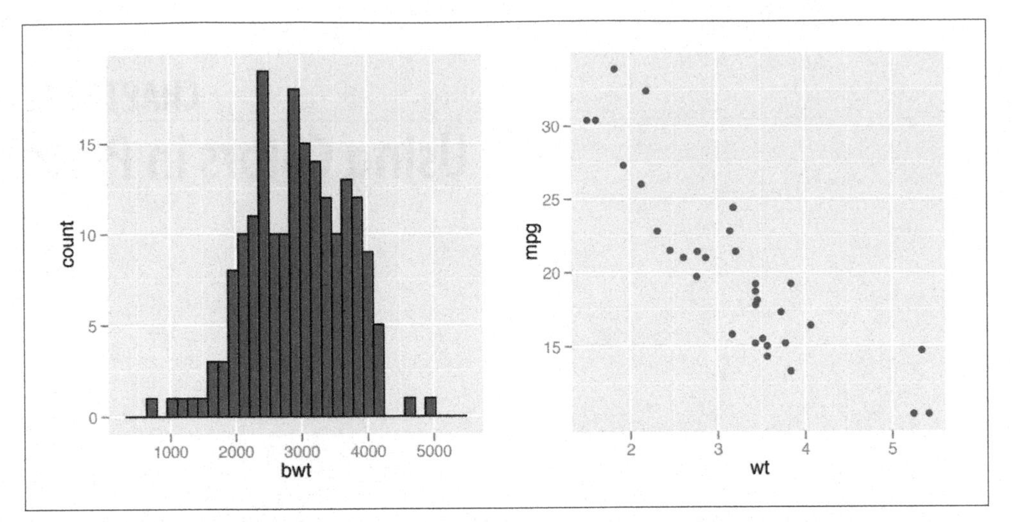

Figure 12-1. Left: setting fill and colour; right: setting colour for points

Discussion

In ggplot2, there's an important difference between *setting* and *mapping* aesthetic properties. In the preceding example, we set the color of the objects to `"red"`.

Generally speaking, `colour` controls the color of lines and of the outlines of polygons, while `fill` controls the color of the fill area of polygons. However, point shapes are sometimes a little different. For most point shapes, the color of the entire point is controlled by `colour`, not `fill`. The exception is the point shapes (21–25) that have both a fill and an outline.

See Also

For more information about point shapes, see Recipe 4.5.

See Recipe 12.4 for more on specifying colors.

12.2. Mapping Variables to Colors

Problem

You want to use a variable (column from a data frame) to control the color of geoms.

Solution

In the call to the geom, set the value of `colour` or `fill` to the name of one of the columns in the data (Figure 12-2):

```
library(gcookbook) # For the data set

# These both have the same effect
ggplot(cabbage_exp, aes(x=Date, y=Weight, fill=Cultivar)) +
    geom_bar(colour="black", position="dodge")

ggplot(cabbage_exp, aes(x=Date, y=Weight)) +
    geom_bar(aes(fill=Cultivar), colour="black", position="dodge")

# These both have the same effect
ggplot(mtcars, aes(x=wt, y=mpg, colour=cyl)) + geom_point()

ggplot(mtcars, aes(x=wt, y=mpg)) + geom_point(aes(colour=cyl))
```

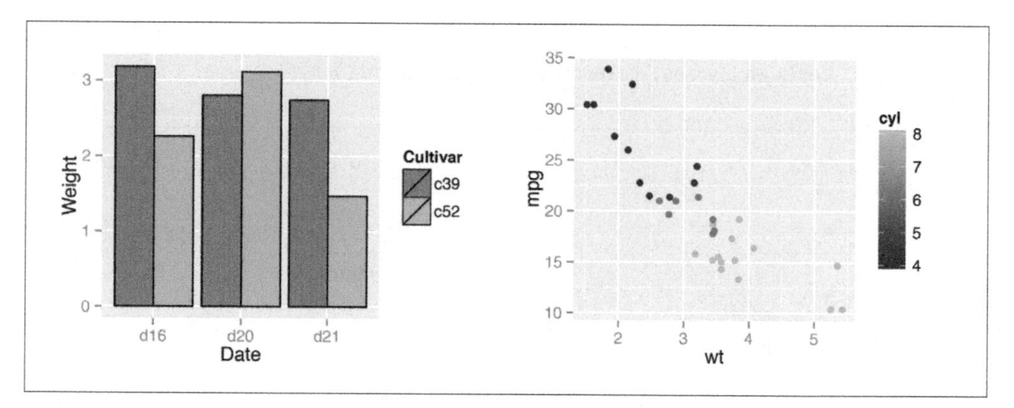

Figure 12-2. Left: mapping a variable to fill; right: mapping a variable to colour for points

When the mapping is specified in `ggplot()` it sets the default mapping, which is inherited by all the geoms. The default mappings can be overridden by specifying mappings within a geom.

Discussion

In the `cabbage_exp` example, the variable `Cultivar` is mapped to `fill`. The `Cultivar` column in `cabbage_exp` is a factor, so ggplot2 treats it as a discrete variable. You can check the type using `str()`:

```
str(cabbage_exp)

'data.frame':    6 obs. of  6 variables:
 $ Cultivar: Factor w/ 2 levels "c39","c52": 1 1 1 2 2 2
 $ Date    : Factor w/ 3 levels "d16","d20","d21": 1 2 3 1 2 3
 $ Weight  : num  3.18 2.8 2.74 2.26 3.11 1.47
 $ sd      : num  0.957 0.279 0.983 0.445 0.791 ...
```

```
 $ n      : int  10 10 10 10 10 10
 $ se     : num  0.3025 0.0882 0.311 0.1408 0.2501 ...
```

In the mtcars example, cyl is numeric, so it is treated as a continuous variable. Because of this, even though the actual values of cyl include only 4, 6, and 8, the legend has entries for the intermediate values 5 and 7. To make ggplot() treat cyl as a categorical variable, you can convert it to a factor in the call to ggplot(), or you can modify the data so that the column is a character vector or factor (Figure 12-3):

```
# Convert to factor in call to ggplot()
ggplot(mtcars, aes(x=wt, y=mpg, colour=factor(cyl))) + geom_point()

# Another method: Convert to factor in the data
m <- mtcars                   # Make a copy of mtcars
m$cyl <- factor(m$cyl)        # Convert cyl to a factor
ggplot(m, aes(x=wt, y=mpg, colour=cyl)) + geom_point()
```

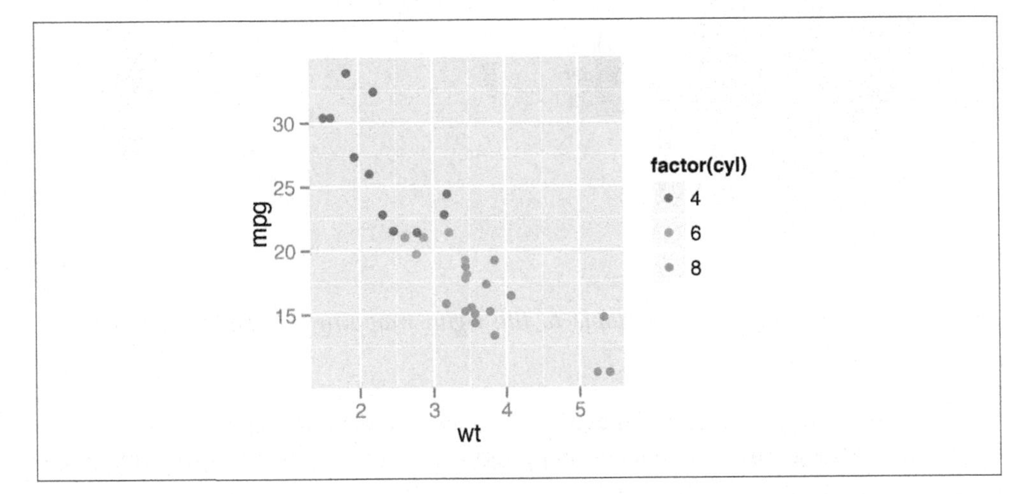

Figure 12-3. Mapping to colour with a continuous variable converted to a factor

See Also

You may also want to change the colors that are used in the scale. For continuous data, see Recipe 12.6. For discrete data, see Recipes 12.3 and 12.4.

12.3. Using a Different Palette for a Discrete Variable

Problem

You want to use different colors for a discrete mapped variable.

Solution

Use one of the scales listed in Table 12-1.

Table 12-1. Discrete fill and color scales

Fill scale	Color scale	Description
`scale_fill_discrete()`	`scale_colour_discrete()`	Colors evenly spaced around the color wheel (same as hue)
`scale_fill_hue()`	`scale_colour_hue()`	Colors evenly spaced around the color wheel (same as discrete)
`scale_fill_grey()`	`scale_colour_grey()`	Greyscale palette
`scale_fill_brewer()`	`scale_colour_brewer()`	ColorBrewer palettes
`scale_fill_manual()`	`scale_colour_manual()`	Manually specified colors

In the example here we'll use the default palette (hue), and a ColorBrewer palette (Figure 12-4):

```
library(gcookbook) # For the data set

# Base plot
p <- ggplot(uspopage, aes(x=Year, y=Thousands, fill=AgeGroup)) + geom_area()

# These three have the same effect
p
p + scale_fill_discrete()
p + scale_fill_hue()

# ColorBrewer palette
p + scale_fill_brewer()
```

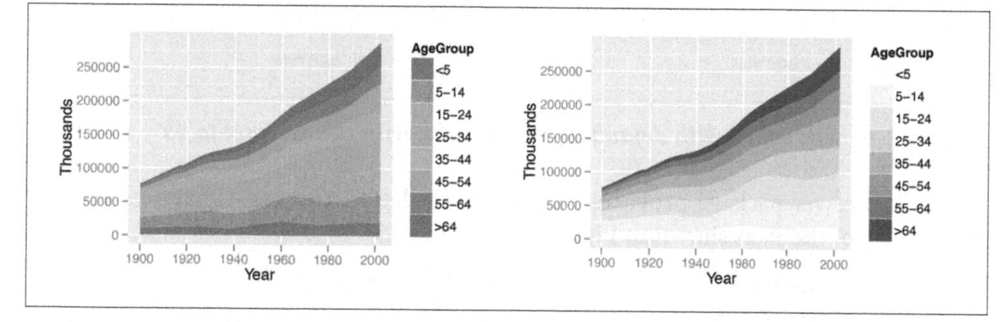

Figure 12-4. Left: default palette (using hue); right: a ColorBrewer palette

Discussion

Changing a palette is a modification of the color (or fill) scale: it involves a change in the mapping from numeric or categorical values to aesthetic attributes. There are two types of scales that use colors: *fill* scales and *color* scales.

With `scale_fill_hue()`, the colors are taken from around the color wheel in the HCL (hue-chroma-lightness) color space. The default lightness value is 65 on a scale from 0–100. This is good for filled areas, but it's a bit light for points and lines. To make the colors darker for points and lines, as in Figure 12-5 (right), set the value of `l` (luminance/lightness):

```
# Basic scatter plot
h <- ggplot(heightweight, aes(x=ageYear, y=heightIn, colour=sex)) +
    geom_point()

# Default lightness = 65
h

# Slightly darker
h + scale_colour_hue(l=45)
```

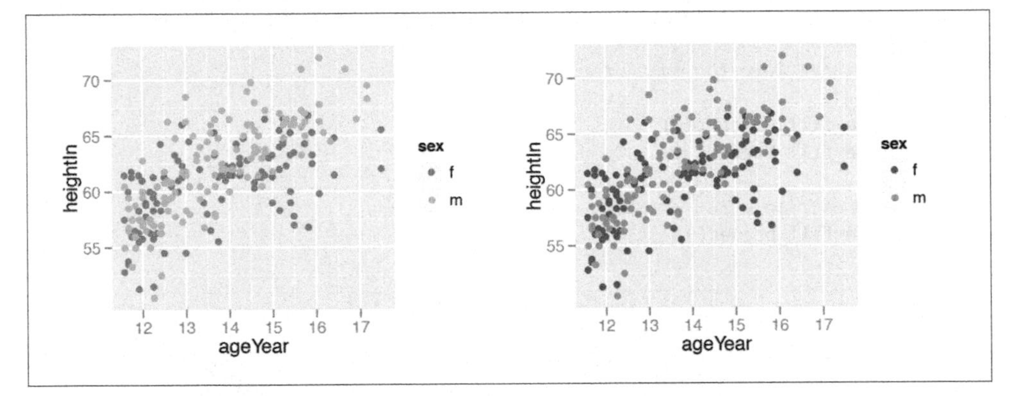

Figure 12-5. Left: points with default lightness; right: with lightness set to 45

The ColorBrewer package provides a number of palettes. You can generate a graphic showing all of them, as shown in Figure 12-6:

```
library(RColorBrewer)
display.brewer.all()
```

The ColorBrewer palettes can be selected by name. For example, this will use the `Oranges` palette (Figure 12-7):

```
p + scale_fill_brewer(palette="Oranges")
```

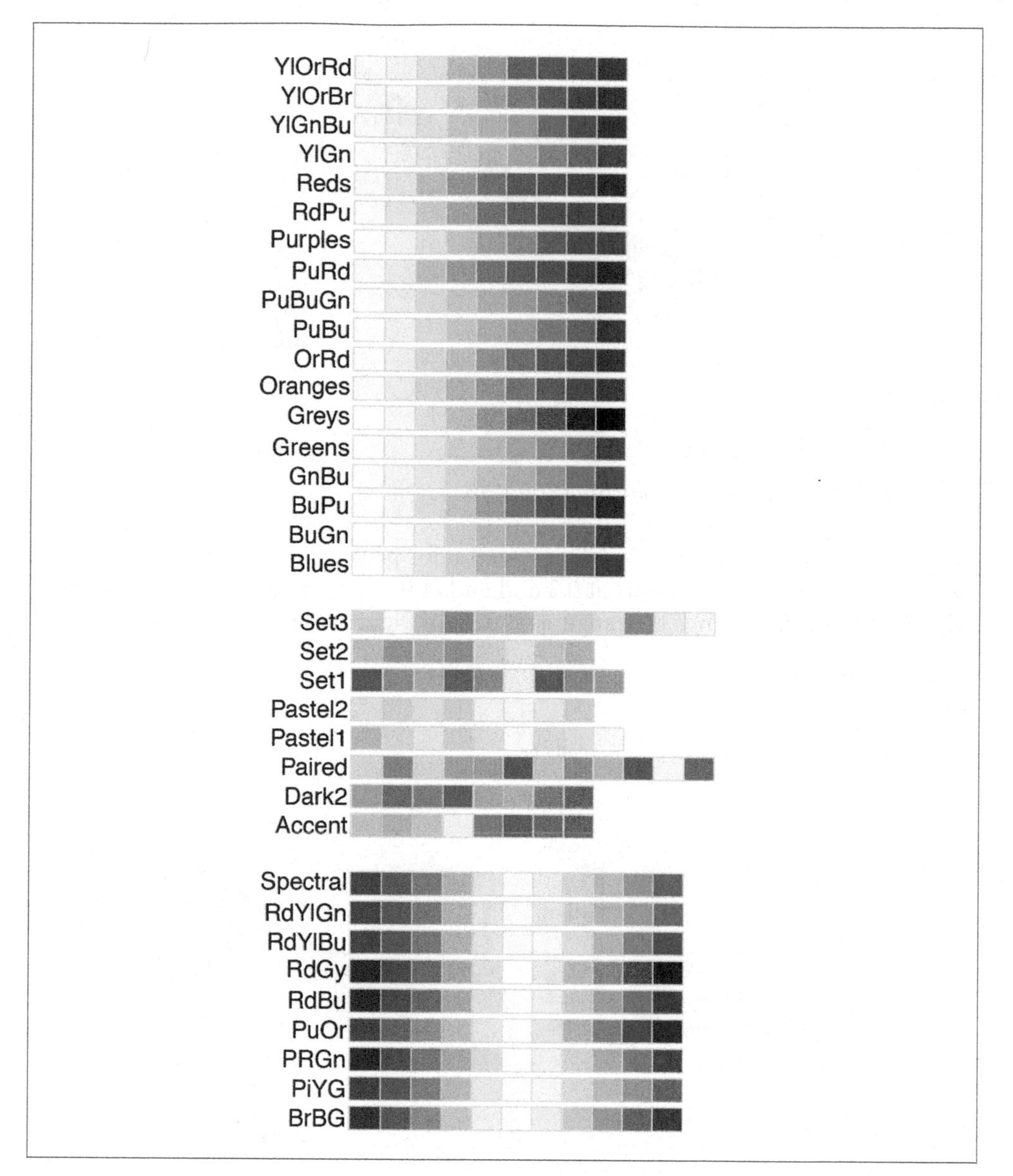

Figure 12-6. All the ColorBrewer palettes

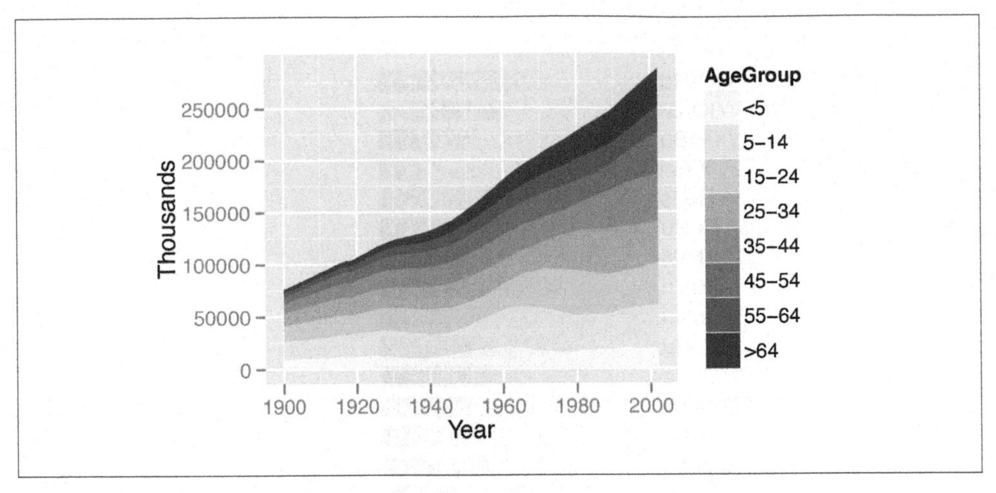

Figure 12-7. Using a named ColorBrewer palette

You can also use a palette of greys. This is useful for print when the output is in black and white. The default is to start at 0.2 and end at 0.8, on a scale from 0 (black) to 1 (white), but you can change the range, as shown in Figure 12-8.

```
p + scale_fill_grey()

# Reverse the direction and use a different range of greys
p + scale_fill_grey(start=0.7, end=0)
```

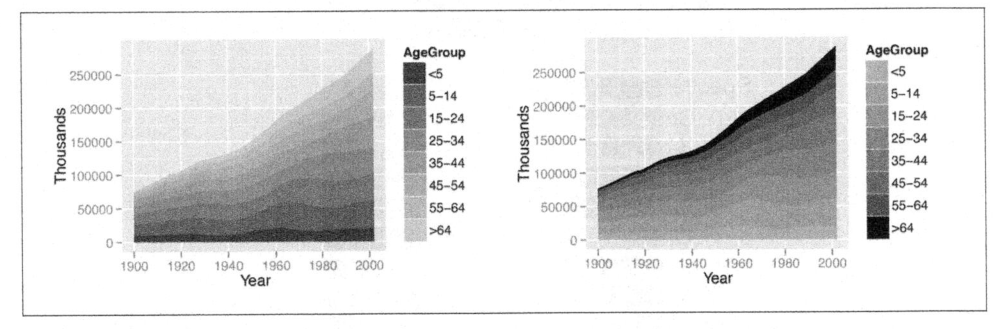

Figure 12-8. Left: using the default grey palette; right: a different grey palette

See Also

See Recipe 10.4 for more information about reversing the legend.

To select colors manually, see Recipe 12.4.

For more about ColorBrewer, see *http://colorbrewer.org*.

12.4. Using a Manually Defined Palette for a Discrete Variable

Problem

You want to use different colors for a discrete mapped variable.

Solution

In the example here, we'll manually define colors by specifying `values` with `scale_colour_manual()` (Figure 12-9). The colors can be named, or they can be specified with RGB values:

```
library(gcookbook) # For the data set

# Base plot
h <- ggplot(heightweight, aes(x=ageYear, y=heightIn, colour=sex)) + geom_point()

# Using color names
h + scale_colour_manual(values=c("red", "blue"))

# Using RGB values
h + scale_colour_manual(values=c("#CC6666", "#7777DD"))
```

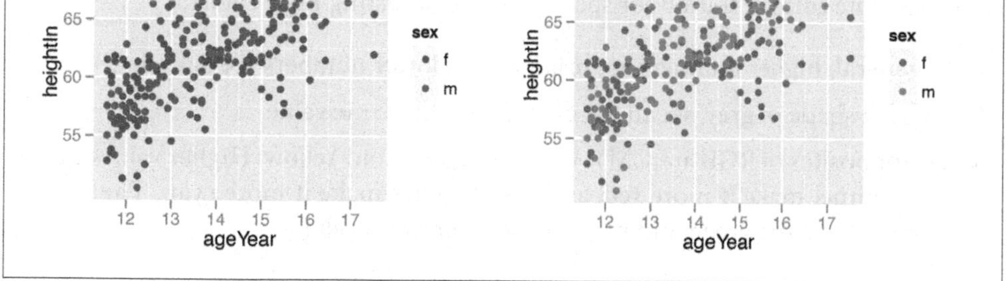

Figure 12-9. Left: scatter plot with named colors; right: with slightly different RGB colors

For fill scales, use `scale_fill_manual()` instead.

Discussion

The order of the items in the `values` vector matches the order of the factor levels for the discrete scale. In the preceding example, the order of `sex` is `f`, then `m`, so the first item in `values` goes with `f` and the second goes with `m`. Here's how to see the order of factor levels:

```
levels(heightweight$sex)

"f" "m"
```

If the variable is a character vector, not a factor, it will automatically be converted to a factor, and by default the levels will appear in alphabetical order.

It's possible to specify the colors in a different order by using a named vector:

```
h + scale_colour_manual(values=c(m="blue", f="red"))
```

There is a large set of named colors in R, which you can see by running `color()`. Some basic color names are useful: `"white"`, `"black"`, `"grey80"`, `"red"`, `"blue"`, `"darkred"`, and so on. There are many other named colors, but their names are generally not very informative (I certainly have no idea what `"thistle3"` and `"seashell"` look like), so it's often easier to use numeric RGB values for specifying colors.

RGB colors are specified as six-digit hexadecimal (base-16) numbers of the form `"#RRGGBB"`. In hexadecimal, the digits go from 0 to 9, and then continue with A (10 in base 10) to F (15 in base 10). Each color is represented by two digits and can range from `00` to `FF` (255 in base 10). So, for example, the color `"#FF0099"` has a value of 255 for red, 0 for green, and 153 for blue, resulting in a shade of magenta. The hexadecimal numbers for each color channel often repeat the same digit because it makes them a little easier to read, and because the precise value of the second digit has a relatively insignificant effect on appearance.

Here are some rules of thumb for specifying and adjusting RGB colors:

- In general, higher numbers are brighter and lower numbers are darker.
- To get a shade of grey, set all the channels to the same value.
- The opposites of RGB are CMY: Cyan, Magenta, and Yellow. Higher values for the red channel make it more red, and lower values make it more cyan. The same is true for the pairs green and magenta, and blue and yellow.

See Also

A chart of RGB color codes (*http://html-color-codes.com*).

12.5. Using a Colorblind-Friendly Palette

Problem

You want to use colors that can be distinguished by colorblind viewers.

Solution

Use the palette defined here (cb_palette) with scale_fill_manual() (Figure 12-10):

```
library(gcookbook) # For the data set

# Base plot
p <- ggplot(uspopage, aes(x=Year, y=Thousands, fill=AgeGroup)) + geom_area()

# The palette with grey:
cb_palette <- c("#999999", "#E69F00", "#56B4E9", "#009E73", "#F0E442",
                "#0072B2", "#D55E00", "#CC79A7")

# Add it to the plot
p + scale_fill_manual(values=cb_palette)
```

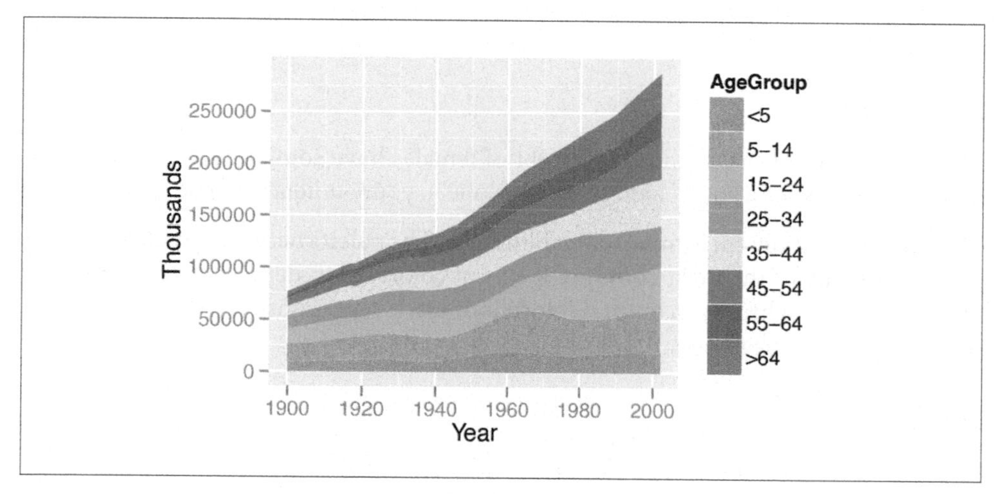

Figure 12-10. A graph with the colorblind-friendly palette

A chart of the colors is shown in Figure 12-11.

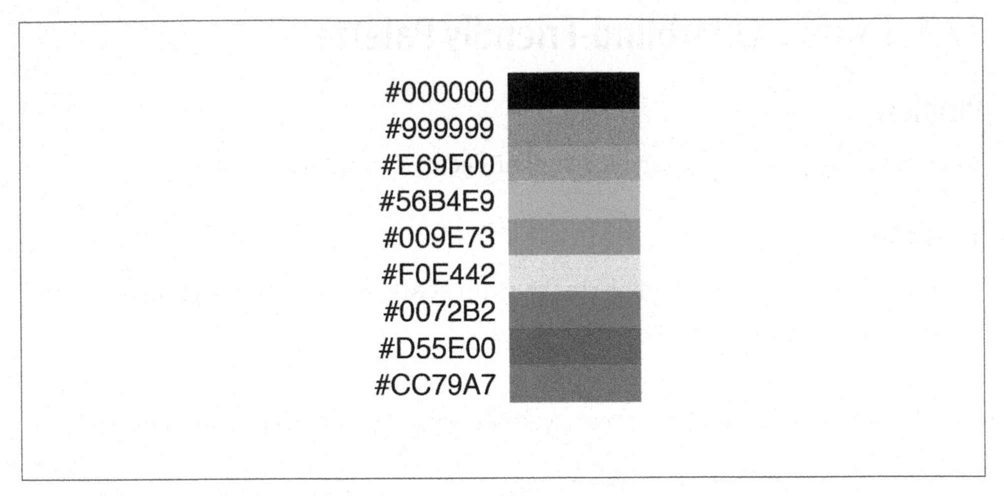

Figure 12-11. Colorblind palette with RGB values

In some cases it may be better to use black instead of grey. To do this, replace the `"#999999"` with `"#000000"` or `"black"`:

```
c("#000000", "#E69F00", "#56B4E9", "#009E73", "#F0E442", "#0072B2", "#D55E00",
  "#CC79A7")
```

Discussion

About 8 percent of males and 0.5 percent of females have some form of color-vision deficiency, so there's a good chance that someone in your audience will be among them.

There are many different forms of color blindness. The palette here is designed to enable people with any of the most common forms of color-vision deficiency to distinguish the colors. (Monochromacy, or total colorblindness, is rare. Those who have it can only see differences in brightness.)

See Also

The source of this palette (*http://jfly.iam.u-tokyo.ac.jp/color/*).

The Color Oracle program (*http://colororacle.org*) can simulate how things on your screen appear to someone with color vision deficiency, but keep in mind that the simulation isn't perfect. In my informal testing, I viewed an image with simulated red-green deficiency, and I could distinguish the colors just fine—but others with actual red-green deficiency viewed the same image and couldn't tell the colors apart!

12.6. Using a Manually Defined Palette for a Continuous Variable

Problem

You want to use different colors for a continuous variable.

Solution

In the example here, we'll specify the colors for a continuous variable using various gradient scales (Figure 12-12). The colors can be named, or they can be specified with RGB values:

```
library(gcookbook) # For the data set

# Base plot
p <- ggplot(heightweight, aes(x=ageYear, y=heightIn, colour=weightLb)) +
       geom_point(size=3)

p

# With a gradient between two colors
p + scale_colour_gradient(low="black", high="white")

# A gradient with a white midpoint
library(scales)
p + scale_colour_gradient2(low=muted("red"), mid="white", high=muted("blue"),
     midpoint=110)

# A gradient of n colors
p + scale_colour_gradientn(colours = c("darkred", "orange", "yellow", "white"))
```

For fill scales, use `scale_fill_xxx()` versions instead, where `xxx` is one of `gradient`, `gradient2`, or `gradientn`.

Discussion

Mapping continuous values to a color scale requires a continuously changing palette of colors. Table 12-2 lists the continuous color and fill scales.

Table 12-2. Continuous fill and color scales

Fill scale	Color scale	Description
`scale_fill_gradient()`	`scale_colour_gradient()`	Two-color gradient
`scale_fill_gradient2()`	`scale_colour_gradient2()`	Gradient with a middle color and two colors that diverge from it
`scale_fill_gradientn()`	`scale_colour_gradientn()`	Gradient with *n* colors, equally spaced

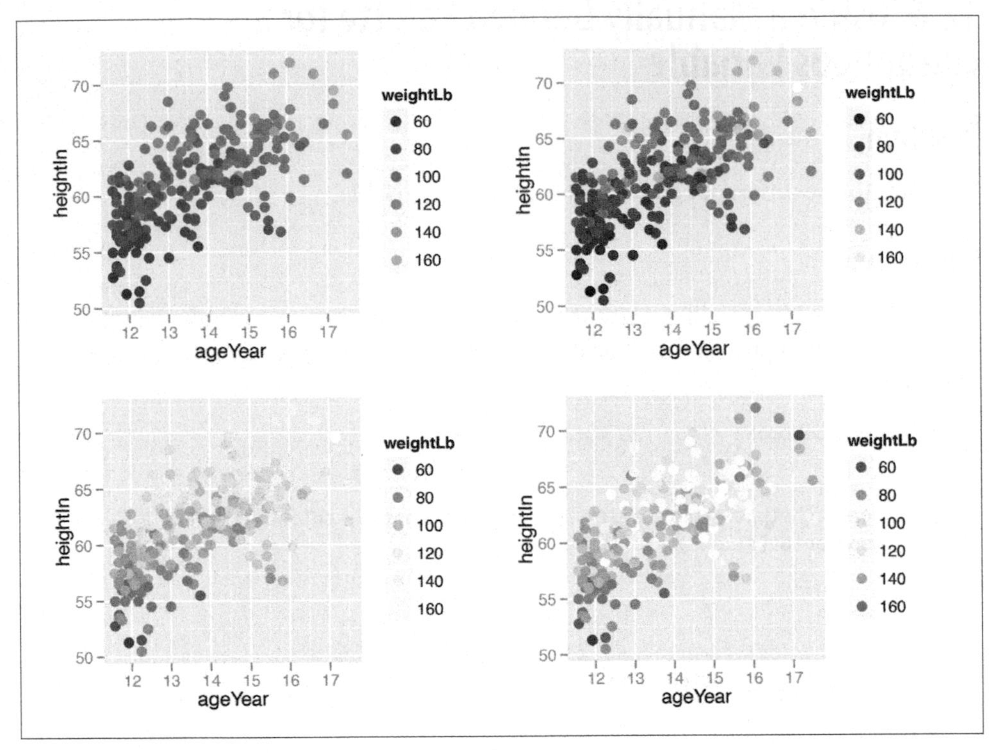

Figure 12-12. Clockwise from top left: default colors, two-color gradient with scale_colour_gradient(), three-color gradient with midpoint with scale_colour_gradient2(), four-color gradient with scale_colour_gradientn()

Notice that we used the `muted()` function in the examples. This is a function from the scales package that returns an RGB value that is a less-saturated version of the color chosen.

See Also

If you want use a discrete (categorical) scale instead of a continuous one, you can recode your data into categorical values. See Recipe 15.14.

12.7. Coloring a Shaded Region Based on Value

Problem

You want to set the color of a shaded region based on the *y* value.

Solution

Add a column that categorizes the *y* values, then map that column to `fill`. In this example, we'll first categorize the values as positive or negative:

```
library(gcookbook) # For the data set

cb <- subset(climate, Source=="Berkeley")

cb$valence[cb$Anomaly10y >= 0] <- "pos"
cb$valence[cb$Anomaly10y < 0]  <- "neg"

cb
```

```
   Source Year Anomaly1y Anomaly5y Anomaly10y Unc10y valence
 Berkeley 1800        NA        NA     -0.435  0.505     neg
 Berkeley 1801        NA        NA     -0.453  0.493     neg
 Berkeley 1802        NA        NA     -0.460  0.486     neg
 ...
 Berkeley 2002        NA        NA      0.856  0.028     pos
 Berkeley 2003        NA        NA      0.869  0.028     pos
 Berkeley 2004        NA        NA      0.884  0.029     pos
```

Once we've categorized the values as positive or negative, we can make the plot, mapping `valence` to the fill color, as shown in Figure 12-13:

```
ggplot(cb, aes(x=Year, y=Anomaly10y)) +
    geom_area(aes(fill=valence)) +
    geom_line() +
    geom_hline(yintercept=0)
```

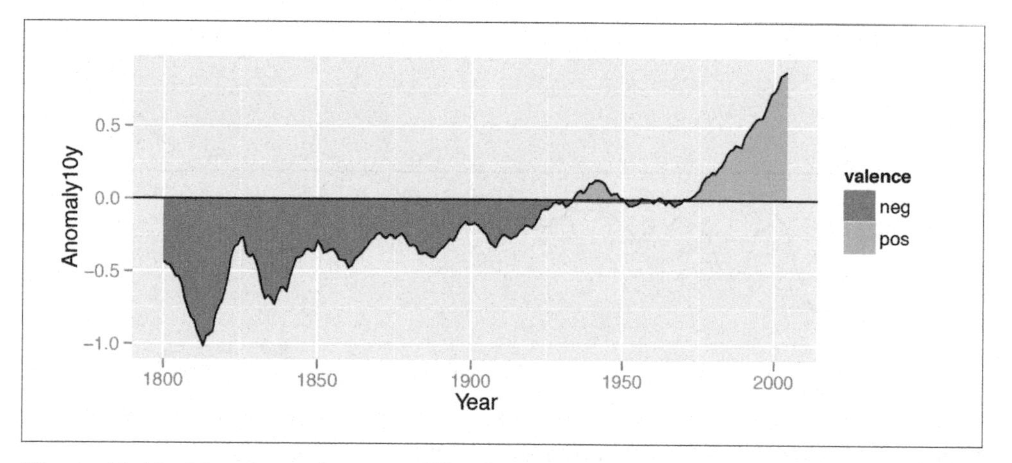

Figure 12-13. Mapping valence to fill color—notice the red area under the zero line around 1950

Discussion

If you look closely at the figure, you'll notice that there are some stray shaded areas near the zero line. This is because each of the two colored areas is a single polygon bounded by the data points, and the data points are not actually at zero. To solve this problem, we can interpolate the data to 1,000 points by using `approx()`:

```
# approx() returns a list with x and y vectors
interp <- approx(cb$Year, cb$Anomaly10y, n=1000)

# Put in a data frame and recalculate valence
cbi <- data.frame(Year=interp$x, Anomaly10y=interp$y)
cbi$valence[cbi$Anomaly10y >= 0] <- "pos"
cbi$valence[cbi$Anomaly10y < 0]  <- "neg"
```

It would be more precise (and more complicated) to interpolate exactly where the line crosses zero, but `approx()` works fine for the purposes here.

Now we can plot the interpolated data (Figure 12-14). This time we'll make a few adjustments—we'll make the shaded regions partially transparent, change the colors, remove the legend, and remove the padding on the left and right sides:

```
ggplot(cbi, aes(x=Year, y=Anomaly10y)) +
    geom_area(aes(fill=valence), alpha = .4) +
    geom_line() +
    geom_hline(yintercept=0) +
    scale_fill_manual(values=c("#CCEEFF", "#FFDDDD"), guide=FALSE) +
    scale_x_continuous(expand=c(0, 0))
```

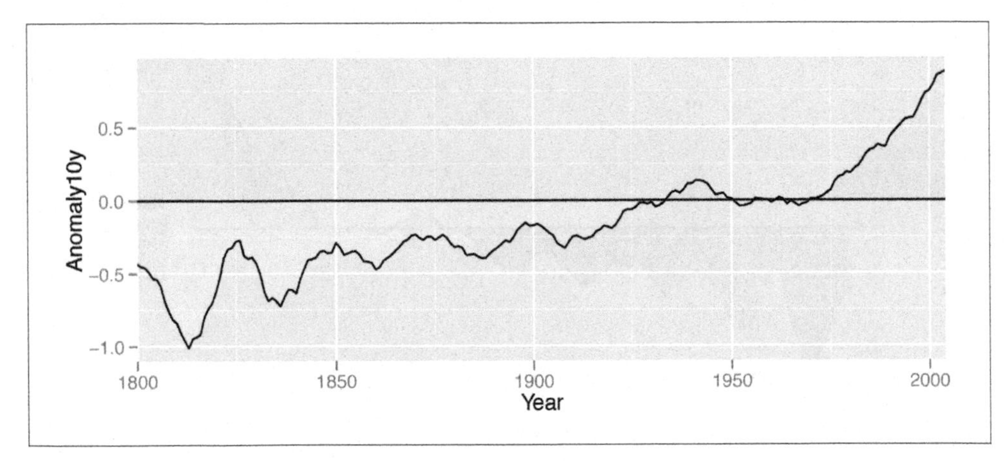

Figure 12-14. Shaded regions with interpolated data

CHAPTER 13
Miscellaneous Graphs

There are many, many ways of visualizing data, and sometimes things don't fit into nice, tidy categories. This chapter shows how to make some of these other visualizations.

13.1. Making a Correlation Matrix

Problem

You want to make a graphical correlation matrix.

Solution

We'll look at the `mtcars` data set:

```
mtcars
```

```
                   mpg cyl  disp  hp drat    wt  qsec vs am gear carb
Mazda RX4         21.0   6 160.0 110 3.90 2.620 16.46  0  1    4    4
Mazda RX4 Wag     21.0   6 160.0 110 3.90 2.875 17.02  0  1    4    4
Datsun 710        22.8   4 108.0  93 3.85 2.320 18.61  1  1    4    1
...
Ferrari Dino      19.7   6 145.0 175 3.62 2.770 15.50  0  1    5    6
Maserati Bora     15.0   8 301.0 335 3.54 3.570 14.60  0  1    5    8
Volvo 142E        21.4   4 121.0 109 4.11 2.780 18.60  1  1    4    2
```

First, generate the numerical correlation matrix using `cor`. This will generate correlation coefficients for each pair of columns:

```
mcor <- cor(mtcars)

# Print mcor and round to 2 digits
round(mcor, digits=2)
```

```
       mpg   cyl  disp    hp  drat    wt  qsec    vs    am  gear  carb
```

```
mpg   1.00 -0.85 -0.85 -0.78  0.68 -0.87  0.42  0.66  0.60  0.48 -0.55
cyl  -0.85  1.00  0.90  0.83 -0.70  0.78 -0.59 -0.81 -0.52 -0.49  0.53
disp -0.85  0.90  1.00  0.79 -0.71  0.89 -0.43 -0.71 -0.59 -0.56  0.39
hp   -0.78  0.83  0.79  1.00 -0.45  0.66 -0.71 -0.72 -0.24 -0.13  0.75
drat  0.68 -0.70 -0.71 -0.45  1.00 -0.71  0.09  0.44  0.71  0.70 -0.09
wt   -0.87  0.78  0.89  0.66 -0.71  1.00 -0.17 -0.55 -0.69 -0.58  0.43
qsec  0.42 -0.59 -0.43 -0.71  0.09 -0.17  1.00  0.74 -0.23 -0.21 -0.66
vs    0.66 -0.81 -0.71 -0.72  0.44 -0.55  0.74  1.00  0.17  0.21 -0.57
am    0.60 -0.52 -0.59 -0.24  0.71 -0.69 -0.23  0.17  1.00  0.79  0.06
gear  0.48 -0.49 -0.56 -0.13  0.70 -0.58 -0.21  0.21  0.79  1.00  0.27
carb -0.55  0.53  0.39  0.75 -0.09  0.43 -0.66 -0.57  0.06  0.27  1.00
```

If there are any columns that you don't want used for correlations (for example, a column of names), you should exclude them. If there are any `NA` cells in the original data, the resulting correlation matrix will have `NA` values. To deal with this, you will probably want to use the option `use="complete.obs"` or `use="pairwise.complete.obs"`.

To graph the correlation matrix (Figure 13-1), we'll use the `corrplot` package, which first must be installed with `install.packages("corrplot")`:

```
library(corrplot)

corrplot(mcor)
```

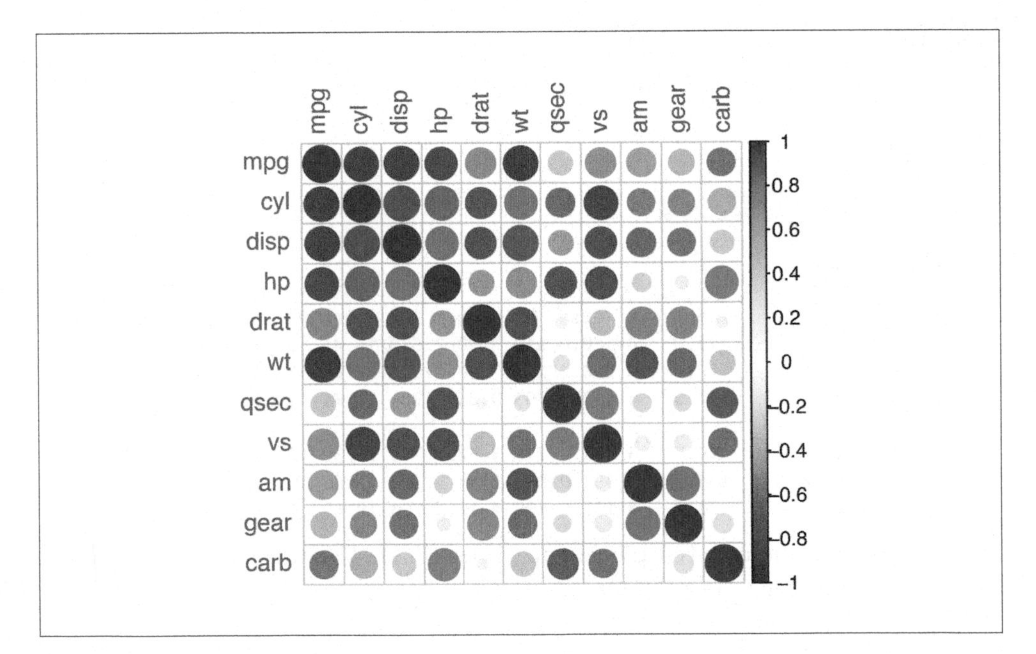

Figure 13-1. A correlation matrix

Discussion

The `corrplot()` function has many, many options. Here is an example of how to make a correlation matrix with colored squares and black labels, rotated 45 degrees along the top (Figure 13-2):

```
corrplot(mcor, method="shade", shade.col=NA, tl.col="black", tl.srt=45)
```

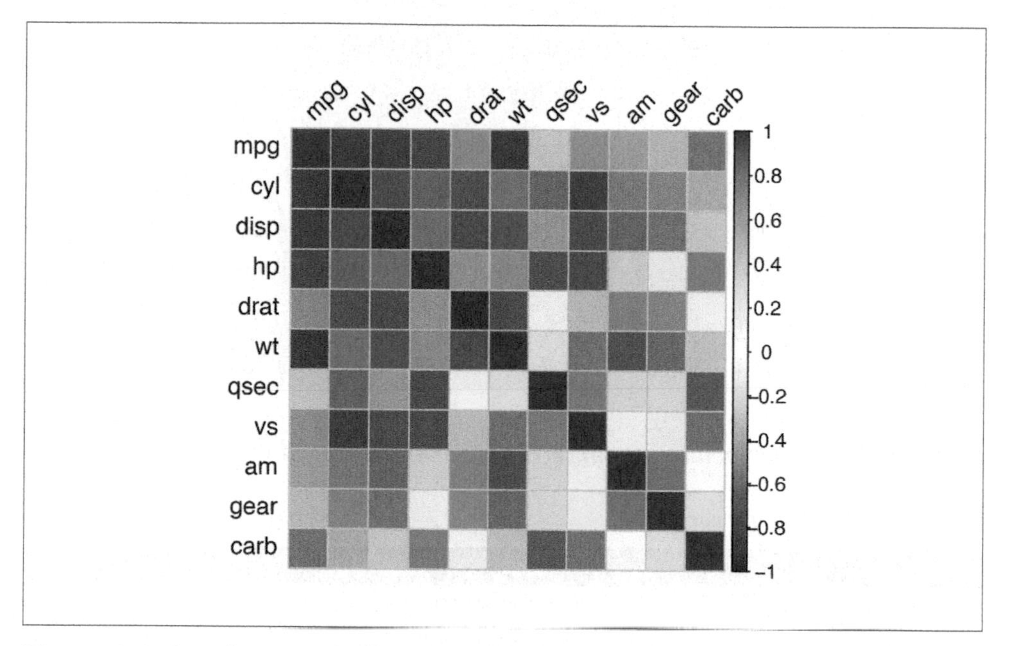

Figure 13-2. Correlation matrix with colored squares and black, rotated labels

It may also be helpful to display labels representing the correlation coefficient on each square in the matrix. In this example we'll make a lighter palette so that the text is readable, and we'll remove the color legend, since it's redundant. We'll also order the items so that correlated items are closer together, using the `order="AOE"` (angular order of eigenvectors) option. The result is shown in Figure 13-3:

```
# Generate a lighter palette
col <- colorRampPalette(c("#BB4444", "#EE9988", "#FFFFFF", "#77AADD", "#4477AA"))

corrplot(mcor, method="shade", shade.col=NA, tl.col="black", tl.srt=45,
         col=col(200), addCoef.col="black", addcolorlabel="no", order="AOE")
```

Like many other standalone graphing functions, `corrplot()` has its own menagerie of options, which can't all be illustrated here. Table 13-1 lists some useful options.

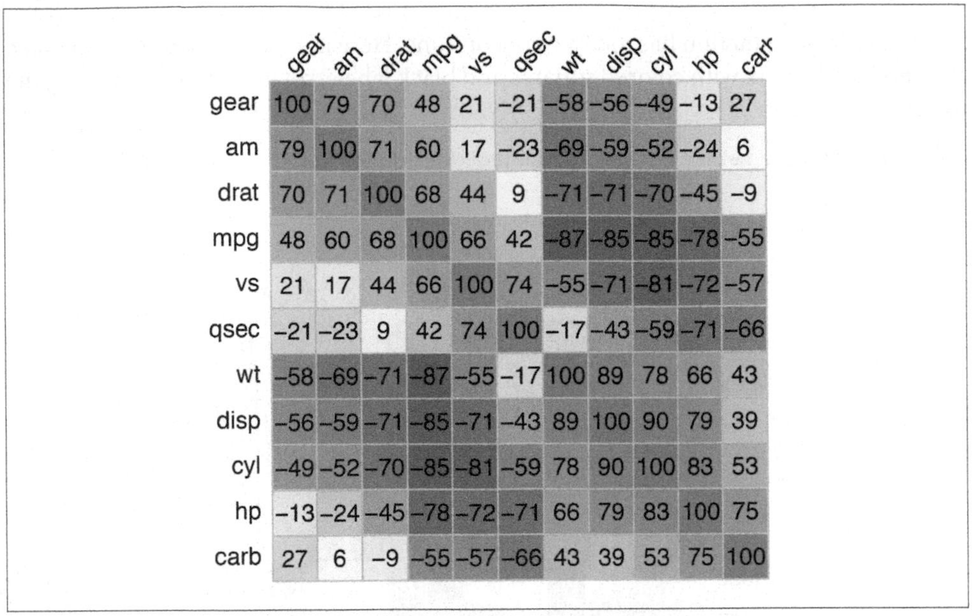

	gear	am	drat	mpg	vs	qsec	wt	disp	cyl	hp	carb
gear	100	79	70	48	21	−21	−58	−56	−49	−13	27
am	79	100	71	60	17	−23	−69	−59	−52	−24	6
drat	70	71	100	68	44	9	−71	−71	−70	−45	−9
mpg	48	60	68	100	66	42	−87	−85	−85	−78	−55
vs	21	17	44	66	100	74	−55	−71	−81	−72	−57
qsec	−21	−23	9	42	74	100	−17	−43	−59	−71	−66
wt	−58	−69	−71	−87	−55	−17	100	89	78	66	43
disp	−56	−59	−71	−85	−71	−43	89	100	90	79	39
cyl	−49	−52	−70	−85	−81	−59	78	90	100	83	53
hp	−13	−24	−45	−78	−72	−71	66	79	83	100	75
carb	27	6	−9	−55	−57	−66	43	39	53	75	100

Figure 13-3. Correlation matrix with correlation coefficients and no legend

Table 13-1. Options for corrplot()

Option	Description
`type={"lower" | "upper"}`	Only use the lower or upper triangle
`diag=FALSE`	Don't show values on the diagonal
`addshade="all"`	Add lines indicating the direction of the correlation
`shade.col=NA`	Hide correlation direction lines
`method="shade"`	Use colored squares
`method="ellipse"`	Use ellipses
`addCoef.col="color"`	Add correlation coefficients, in *color*
`tl.srt="number"`	Specify the rotation angle for top labels
`tl.col="color"`	Specify the label color
`order={"AOE" | "FPC" | "hclust"}`	Sort labels using angular order of eigenvectors, first principal component, or hierarchical clustering

See Also

To create a scatter plot matrix, see Recipe 5.13.

For more on subsetting data, see Recipe 15.7.

13.2. Plotting a Function

Problem

You want to plot a function.

Solution

Use `stat_function()`. It's also necessary to give `ggplot()` a dummy data frame so that it will get the proper *x* range. In this example we'll use `dnorm()`, which gives the density of the normal distribution (Figure 13-4, left):

```
# The data frame is only used for setting the range
p <- ggplot(data.frame(x=c(-3,3)), aes(x=x))

p + stat_function(fun = dnorm)
```

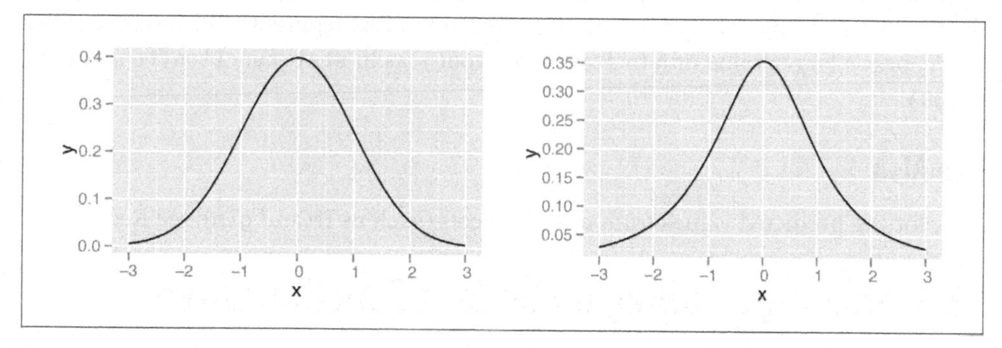

Figure 13-4. Left: the normal distribution; right: the t-distribution with df=2

Discussion

Some functions take additional arguments. For example, `dt()`, the function for the density of the *t*-distribution, takes a parameter for degrees of freedom (Figure 13-4, right). These additional arguments can be passed to the function by putting them in a list and giving the list to `args`:

```
p + stat_function(fun=dt, args=list(df=2))
```

It's also possible to define your own functions. It should take an *x* value for its first argument, and it should return a *y* value. In this example, we'll define a sigmoid function (Figure 13-5):

```
myfun <- function(xvar) {
    1/(1 + exp(-xvar + 10))
}

ggplot(data.frame(x=c(0, 20)), aes(x=x)) + stat_function(fun=myfun)
```

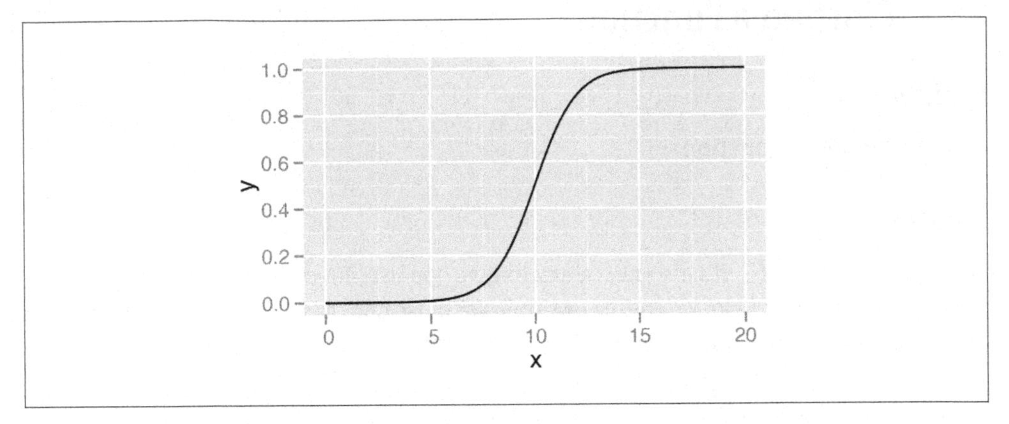

Figure 13-5. A user-defined function

By default, the function is calculated at 101 points along the *x* range. If you have a rapidly fluctuating function, you may be able to see the individual segments. To smooth out the curve, pass a larger value of `n` to `stat_function()`, as in `stat_function(fun=myfun, n=200)`.

See Also

For plotting predicted values from model objects (such as `lm` and `glm`), see Recipe 5.7.

13.3. Shading a Subregion Under a Function Curve

Problem

You want to shade part of the area under a function curve.

Solution

Define a new wrapper function around your curve function, and replace out-of-range values with `NA` (), as shown in Figure 13-6:

```
# Return dnorm(x) for 0 < x < 2, and NA for all other x
dnorm_limit <- function(x) {
    y <- dnorm(x)
    y[x < 0  |  x > 2] <- NA
    return(y)
}

# ggplot() with dummy data
p <- ggplot(data.frame(x=c(-3, 3)), aes(x=x))
```

```
p + stat_function(fun=dnorm_limit, geom="area", fill="blue", alpha=0.2) +
    stat_function(fun=dnorm)
```

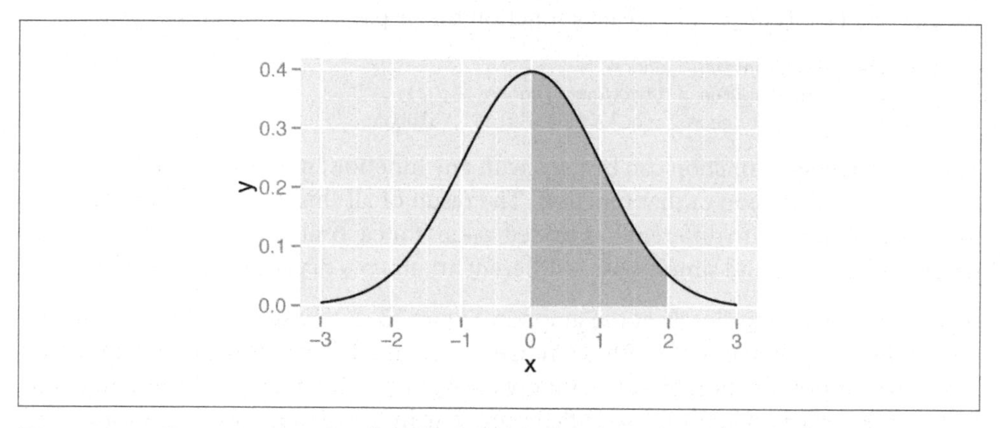

Figure 13-6. Function curve with a shaded region

Remember that what gets passed to this function is a vector, not individual values. If this function operated on single elements at a time, it might make sense to use an `if/else` statement to decide what to return, conditional on the value of `x`. But that won't work here, since `x` is a vector with many values.

Discussion

R has first-class functions, and we can write a function that returns a *closure*—that is, we can program a function to program another function.

This function will allow you to pass in a function, a minimum value, and a maximum value. Values outside the range will again be returned with `NA`:

```
limitRange <- function(fun, min, max) {
    function(x) {
        y <- fun(x)
        y[x < min  |  x > max] <- NA
        return(y)
    }
}
```

Now we can call this function to create another function—one that is effectively the same as the `dnorm_limit()` function used earlier:

```
# This returns a function
dlimit <- limitRange(dnorm, 0, 2)

# Now we'll try out the new function -- it only returns values for inputs
# between 0 and 2
dlimit(-2:4)
```

```
[1]         NA         NA 0.39894228 0.24197072 0.05399097         NA         NA
```

We can use `limitRange()` to create a function that is passed to `stat_function()`:

```
p + stat_function(fun = dnorm) +
    stat_function(fun = limitRange(dnorm, 0, 2),
                  geom="area", fill="blue", alpha=0.2)
```

The `limitRange()` function can be used with any function, not just `dnorm()`, to create a range-limited version of that function. The result of all this is that instead of having to write functions with different hardcoded values for each situation that arises, we can write one function and simply pass it different arguments depending on the situation.

If you look very, very closely at the graph in Figure 13-6, you may see that the shaded region does not align exactly with the range we specified. This is because ggplot2 does a numeric approximation by calculating values at fixed intervals, and these intervals may not fall exactly within the specified range. As in Recipe 13.2, we can improve the approximation by increasing the number of interpolated values with `stat_function(n=200)`.

13.4. Creating a Network Graph

Problem

You want to create a network graph.

Solution

Use the igraph package. To create a graph, pass a vector containing pairs of items to `graph()`, then plot the resulting object (Figure 13-7):

```
# May need to install first, with install.packages("igraph")
library(igraph)

# Specify edges for a directed graph
gd <- graph(c(1,2, 2,3, 2,4, 1,4, 5,5, 3,6))
plot(gd)

# For an undirected graph
gu <- graph(c(1,2, 2,3, 2,4, 1,4, 5,5, 3,6), directed=FALSE)
# No labels
plot(gu, vertex.label=NA)
```

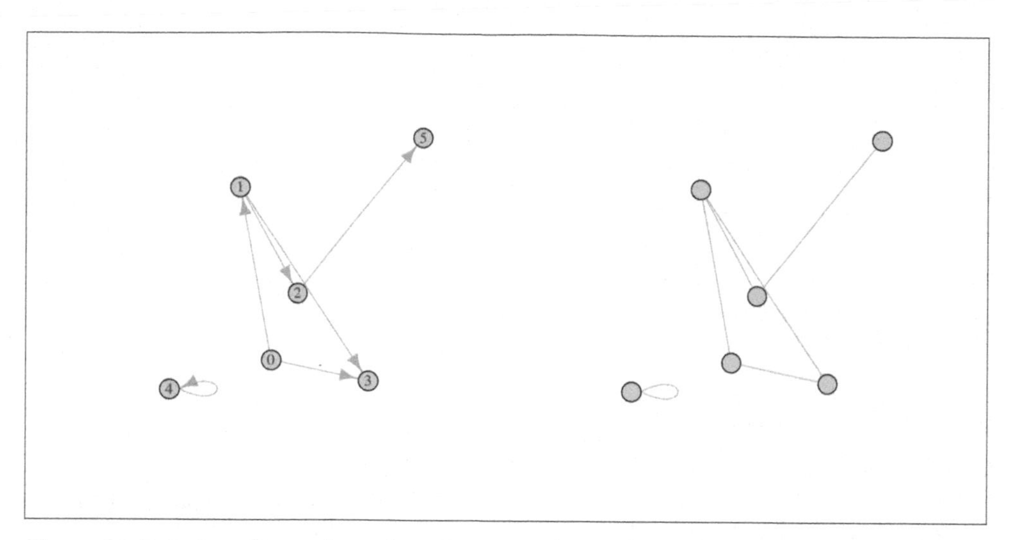

Figure 13-7. Left: a directed graph; right: an undirected graph, with no vertex labels

This is the structure of each of the graph objects:

```
str(gd)

IGRAPH D--- 6 6 --
+ edges:
[1] 1->2 2->3 2->4 1->4 5->5 3->6
```

```
str(gu)

IGRAPH U--- 6 6 --
+ edges:
[1] 1--2 2--3 2--4 1--4 5--5 3--6
```

Discussion

In a network graph, the position of the nodes is unspecified by the data, and they're placed randomly. To make the output repeatable, you can set the random seed before making the plot. You can try different random numbers until you get a result that you like:

```
set.seed(229)
plot(gu)
```

It's also possible to create a graph from a data frame. The first two columns of the data frame are used, and each row specifies a connection between two nodes. In the next example (Figure 13-8), we'll use the `madmen2` data set, which has this structure. We'll also use the Fruchterman-Reingold layout algorithm. The idea is that all the nodes have a magnetic repulsion from one another, but the edges between nodes act as springs, pulling the nodes together:

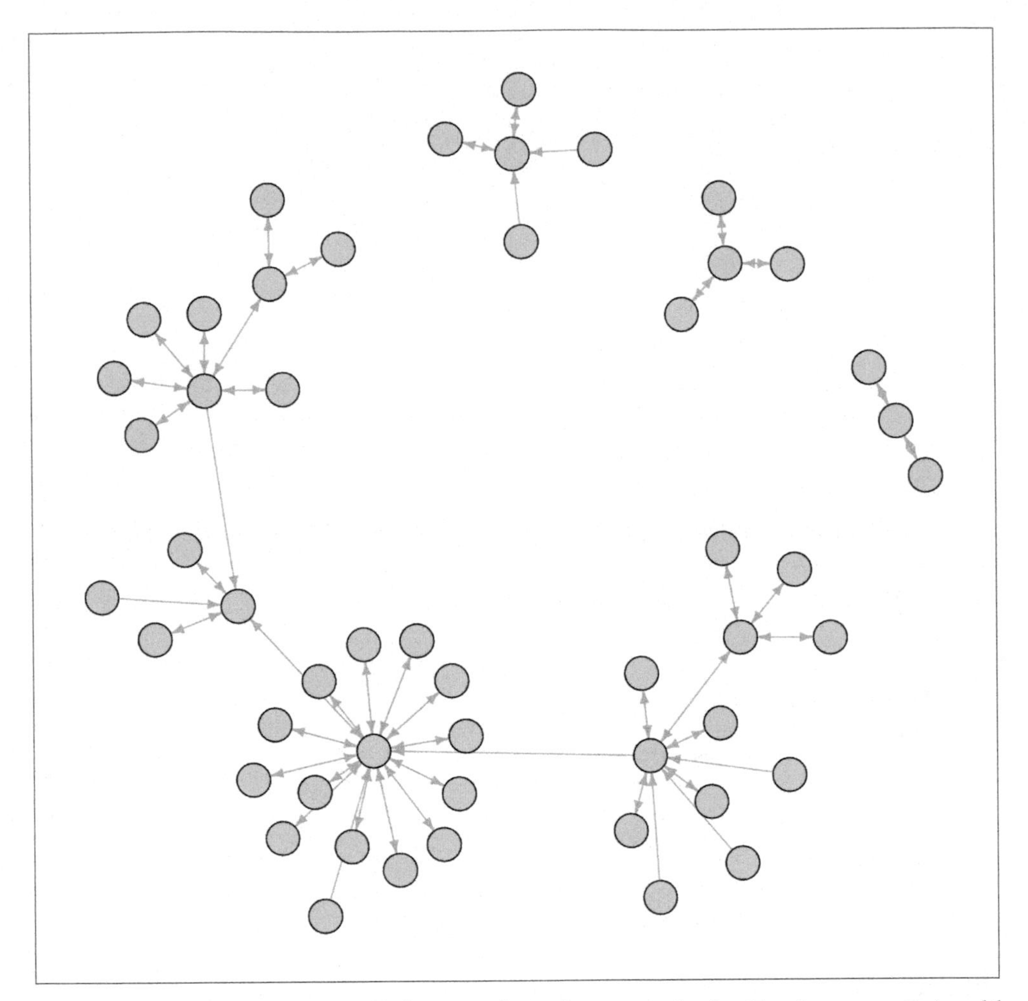

Figure 13-8. A directed graph from a data frame, with the Fruchterman-Reingold algorithm

```
library(gcookbook) # For the data set
madmen2

       Name1                 Name2
 Abe Drexler           Peggy Olson
     Allison            Don Draper
 Arthur Case          Betty Draper
 ...

# Create a graph object from the data set
g <- graph.data.frame(madmen2, directed=TRUE)
```

```
# Remove unnecessary margins
par(mar=c(0,0,0,0))

plot(g, layout=layout.fruchterman.reingold, vertex.size=8, edge.arrow.size=0.5,
     vertex.label=NA)
```

It's also possible to make a directed graph from a data frame. The madmen data set has only one row for each pairing, since direction doesn't matter for an undirected graph. This time we'll use a circle layout (Figure 13-9):

```
g <- graph.data.frame(madmen, directed=FALSE)
par(mar=c(0,0,0,0))  # Remove unnecessary margins
plot(g, layout=layout.circle, vertex.size=8, vertex.label=NA)
```

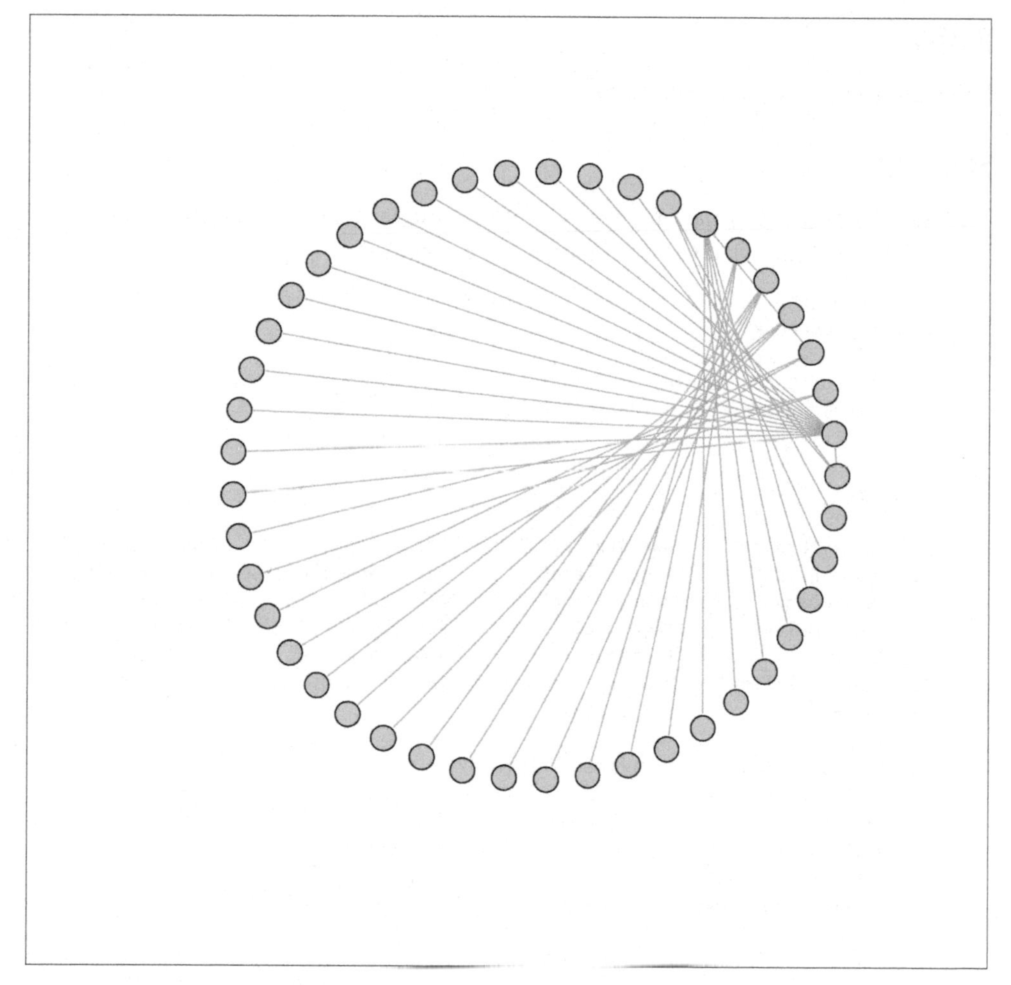

Figure 13-9. A circular undirected graph from a data frame

See Also

For more information about the available output options, see `?plot.igraph`. Also see `?igraph::layout` for layout options.

An alternative to igraph is Rgraphviz, which a frontend for Graphviz, an open-source library for visualizing graphs. It works better with labels and makes it easier to create graphs with a controlled layout, but it can be a bit challenging to install. Rgraphviz is available through the Bioconductor repository system.

13.5. Using Text Labels in a Network Graph

Problem

You want to use text labels in a network graph.

Solution

The vertices/nodes may have names, but these names are not used as labels by default. To set the labels, pass in a vector of names to `vertex.label` (Figure 13-10):

```
library(igraph)
library(gcookbook) # For the data set
# Copy madmen and drop every other row
m <- madmen[1:nrow(madmen) %% 2 == 1, ]
g <- graph.data.frame(m, directed=FALSE)

# Print out the names of each vertex
V(g)$name

 [1] "Betty Draper"    "Don Draper"          "Harry Crane"        "Joan Holloway"
 [5] "Lane Pryce"      "Peggy Olson"         "Pete Campbell"      "Roger Sterling"
 [9] "Sal Romano"      "Henry Francis"       "Allison"            "Candace"
[13] "Faye Miller"     "Megan Calvet"        "Rachel Menken"      "Suzanne Farrell"
[17] "Hildy"           "Franklin"            "Rebecca Pryce"      "Abe Drexler"
[21] "Duck Phillips"   "Playtex bra model" "Ida Blankenship"    "Mirabelle Ames"
[25] "Vicky"           "Kitty Romano"

plot(g, layout=layout.fruchterman.reingold,
     vertex.size        = 4,          # Smaller nodes
     vertex.label       = V(g)$name,  # Set the labels
     vertex.label.cex   = 0.8,        # Slightly smaller font
     vertex.label.dist  = 0.4,        # Offset the labels
     vertex.label.color = "black")
```

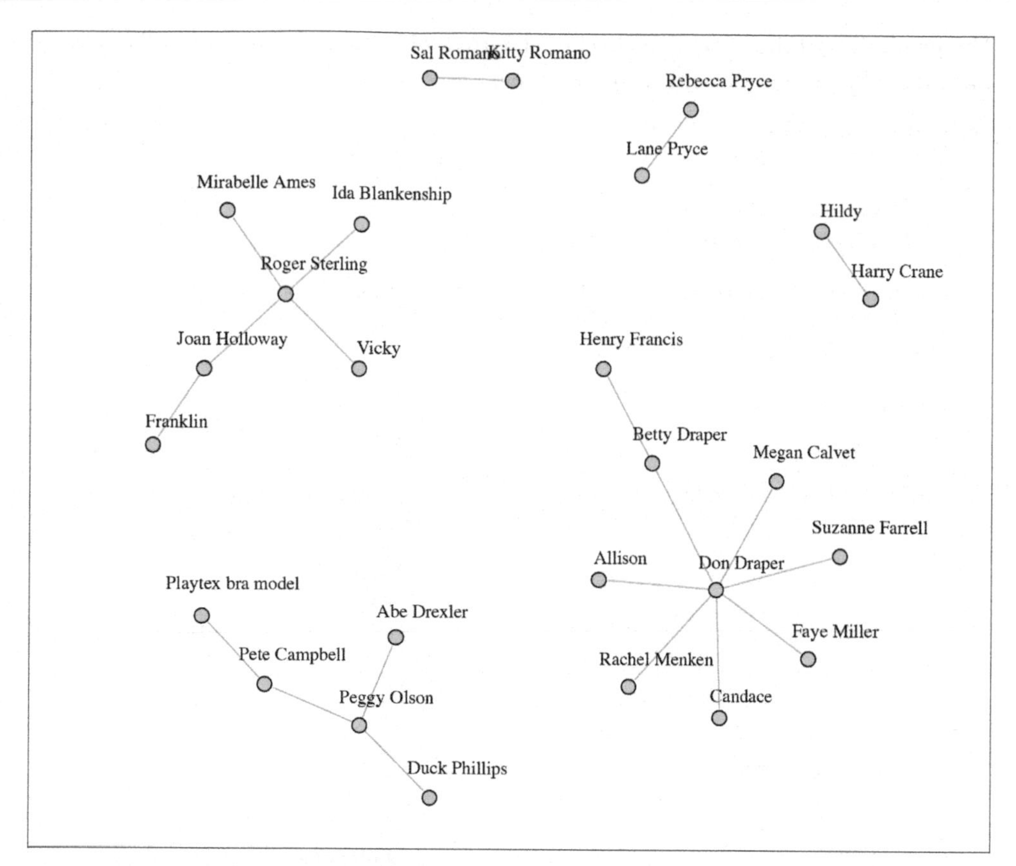

Figure 13-10. A network graph with labels

Discussion

Another way to achieve the same effect is to modify the `plot` object, instead of passing in the values as arguments to `plot()`. To do this, use `V()$xxx <-` instead of passing a value to a `vertex.xxx` argument. For example, this will result in the same output as the previous code:

```
# This is equivalent to the preceding code
V(g)$size        <- 4
V(g)$label       <- V(g)$name
V(g)$label.cex   <- 0.8
V(g)$label.dist  <- 0.4
V(g)$label.color <- "black"

# Set a property of the entire graph
g$layout <- layout.fruchterman.reingold

plot(g)
```

The properties of the edges can also be set, either with the `E()` function or by passing values to `edge.xxx` arguments (Figure 13-11):

```
# View the edges
E(g)

# Set some of the labels to "M"
E(g)[c(2,11,19)]$label <- "M"

# Set color of all to grey, and then color a few red
E(g)$color              <- "grey70"
E(g)[c(2,11,19)]$color <- "red"

plot(g)
```

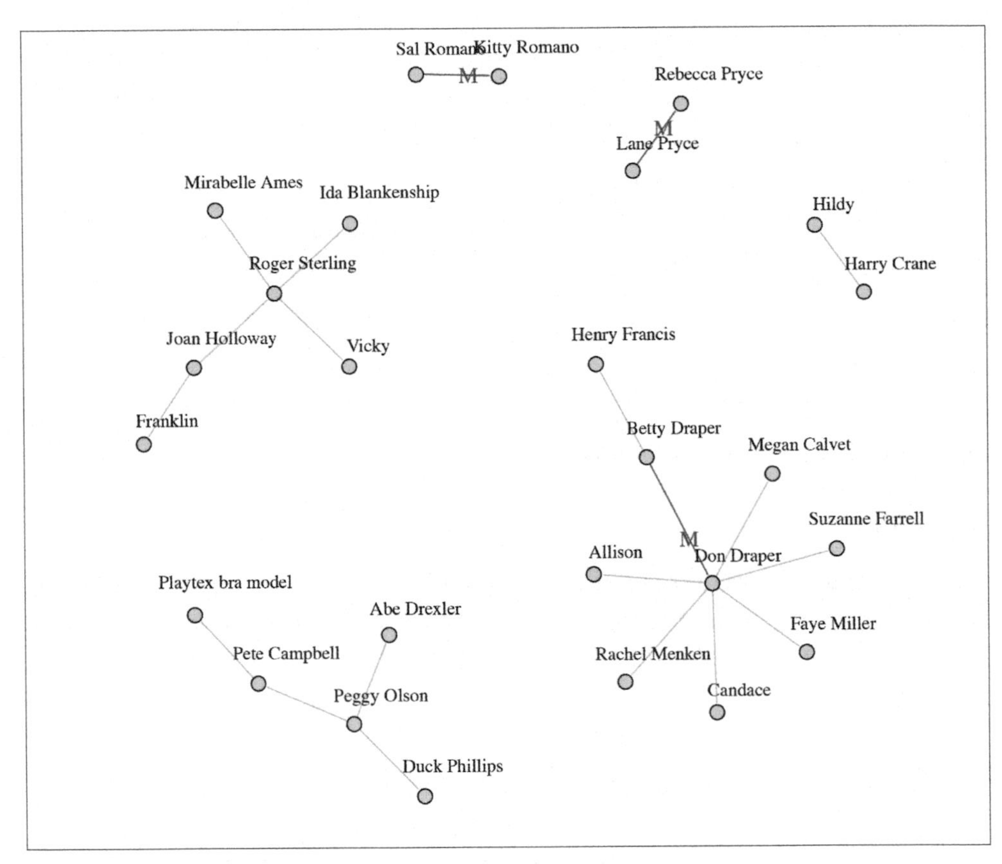

Figure 13-11. A network graph with labeled and colored edges

See Also

See `?igraph.plotting` for more information about graphical parameters in igraph.

13.6. Creating a Heat Map

Problem

You want to make a heat map.

Solution

Use `geom_tile()` or `geom_raster()` and map a continuous variable to `fill`. We'll use the `presidents` data set, which is a time series object rather than a data frame:

```
presidents

     Qtr1 Qtr2 Qtr3 Qtr4
1945   NA   87   82   75
1946   63   50   43   32
 ...
1973   68   44   40   27
1974   28   25   24   24

str(presidents)

Time-Series [1:120] from 1945 to 1975: NA 87 82 75 63 50 43 32 35 60 ...
```

We'll first convert it to a format that is usable by `ggplot()`—a data frame with columns that are numeric:

```
pres_rating <- data.frame(
    rating  = as.numeric(presidents),
    year    = as.numeric(floor(time(presidents))),
    quarter = as.numeric(cycle(presidents))
)

pres_rating

 rating year quarter
     NA 1945       1
     87 1945       2
     82 1945       3
 ...
     25 1974       2
     24 1974       3
     24 1974       4
```

Now we can make the plot using `geom_tile()` or `geom_raster()` (Figure 13-12). Simply map one variable to `x`, one to `y`, and one to `fill`:

```
# Base plot
p <- ggplot(pres_rating, aes(x=year, y=quarter, fill=rating))

# Using geom_tile()
```

```
p + geom_tile()

# Using geom_raster() - looks the same, but a little more efficient
p + geom_raster()
```

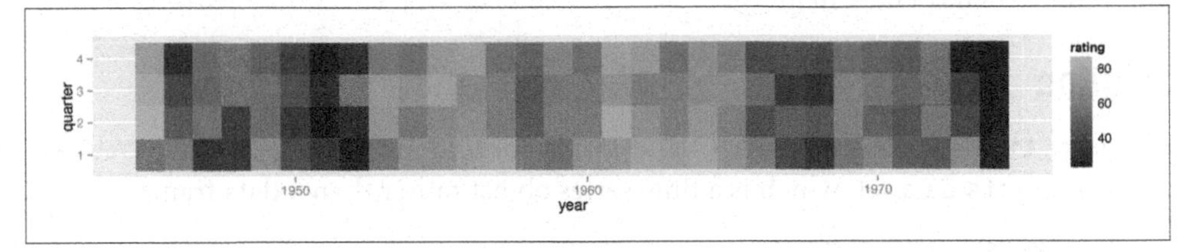

Figure 13-12. A heat map—the grey squares represent NAs in the data

The results with `geom_tile()` and `geom_raster()` *should* look the same, but in practice they might appear different. See Recipe 6.12 for more information about this issue.

Discussion

To better convey useful information, you may want to customize the appearance of the heat map. With this example, we'll reverse the y-axis so that it progresses from top to bottom, and we'll add tick marks every four years along the x-axis, to correspond with each presidential term. We'll also change the color scale using `scale_fill_gradient2()`, which lets you specify a midpoint color and the two colors at the low and high ends (Figure 13-13):

```
p + geom_tile() +
    scale_x_continuous(breaks = seq(1940, 1976, by = 4)) +
    scale_y_reverse() +
    scale_fill_gradient2(midpoint=50, mid="grey70", limits=c(0,100))
```

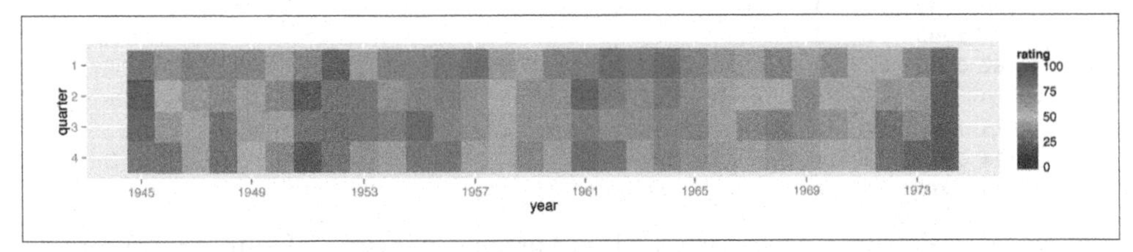

Figure 13-13. A heat map with customized appearance

See Also

If you want to use a different color palette, see Recipe 12.6.

13.7. Creating a Three-Dimensional Scatter Plot

Problem

You want to create a three-dimensional (3D) scatter plot.

Solution

We'll use the `rgl` package, which provides an interface to the OpenGL graphics library for 3D graphics. To create a 3D scatter plot, as in Figure 13-14, use `plot3d()` and pass in a data frame where the first three columns represent *x*, *y*, and *z* coordinates, or pass in three vectors representing the *x*, *y*, and *z* coordinates.

```
# You may need to install first, with install.packages("rgl")
library(rgl)
plot3d(mtcars$wt, mtcars$disp, mtcars$mpg, type="s", size=0.75, lit=FALSE)
```

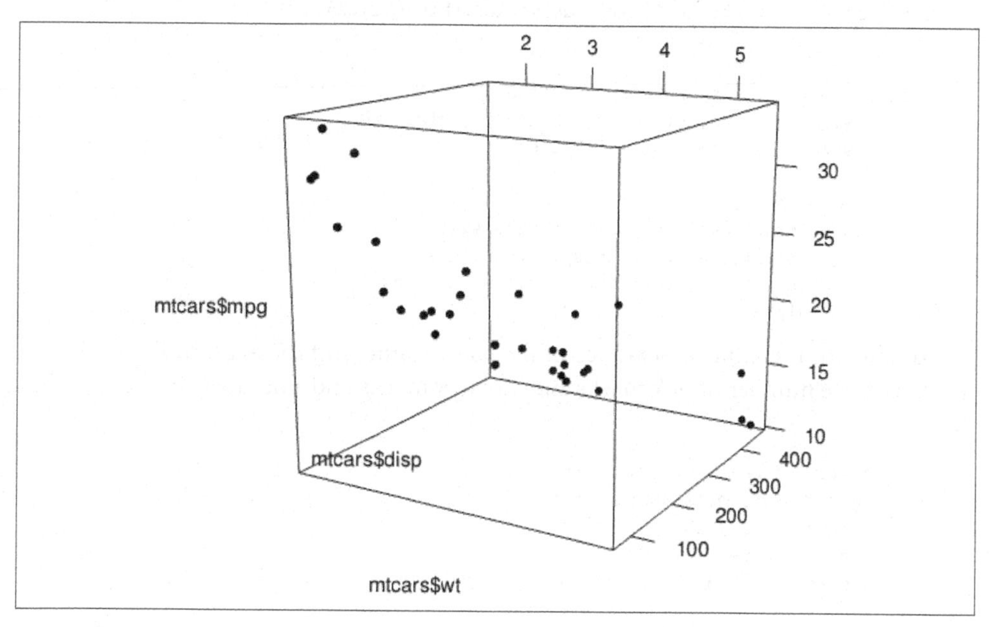

Figure 13-14. A 3D scatter plot

Viewers can rotate the image by clicking and dragging with the mouse, and zoom in and out with the scroll wheel.

By default, `plot3d()` uses square points, which do not appear properly when saving to a PDF. For improved appearance, we used `type="s"` for spherical points, made them smaller with `size=0.75`, and turned off the 3D lighting with `lit=FALSE` (otherwise they look like shiny spheres).

Discussion

Three-dimensional scatter plots can be difficult to interpret, so it's often better to use a two-dimensional representation of the data. That said, there are things that can help make a 3D scatter plot easier to understand.

In Figure 13-15, we'll add vertical segments to help give a sense of the spatial positions of the points:

```
# Function to interleave the elements of two vectors
interleave <- function(v1, v2)  as.vector(rbind(v1,v2))

# Plot the points
plot3d(mtcars$wt, mtcars$disp, mtcars$mpg,
       xlab="Weight", ylab="Displacement", zlab="MPG",
       size=.75, type="s", lit=FALSE)

# Add the segments
segments3d(interleave(mtcars$wt,   mtcars$wt),
           interleave(mtcars$disp, mtcars$disp),
           interleave(mtcars$mpg,  min(mtcars$mpg)),
           alpha=0.4, col="blue")
```

It's possible to tweak the appearance of the background and the axes. In Figure 13-16, we change the number of tick marks and add tick marks and axis labels to the specified sides:

```
# Make plot without axis ticks or labels
plot3d(mtcars$wt, mtcars$disp, mtcars$mpg,
       xlab = "", ylab = "", zlab = "",
       axes = FALSE,
       size=.75, type="s", lit=FALSE)

segments3d(interleave(mtcars$wt,   mtcars$wt),
           interleave(mtcars$disp, mtcars$disp),
           interleave(mtcars$mpg,  min(mtcars$mpg)),
           alpha = 0.4, col = "blue")

# Draw the box.
rgl.bbox(color="grey50",          # grey60 surface and black text
         emission="grey50",       # emission color is grey50
         xlen=0, ylen=0, zlen=0)  # Don't add tick marks

# Set default color of future objects to black
```

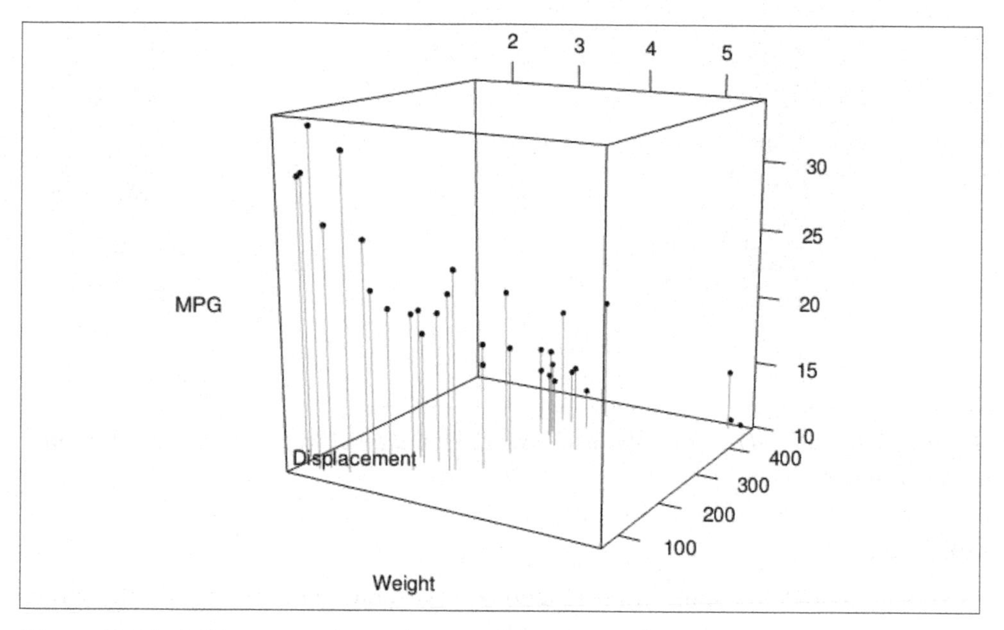

Figure 13-15. A 3D scatter plot with vertical lines for each point

```
rgl.material(color="black")

# Add axes to specific sides. Possible values are "x--", "x-+", "x+-", and "x++".
axes3d(edges=c("x--", "y+-", "z--"),
       ntick=6,                       # Attempt 6 tick marks on each side
       cex=.75)                       # Smaller font

# Add axis labels. 'line' specifies how far to set the label from the axis.
mtext3d("Weight",       edge="x--", line=2)
mtext3d("Displacement", edge="y+-", line=3)
mtext3d("MPG",          edge="z--", line=3)
```

See Also

See `?plot3d` for more options for controlling the output.

13.8. Adding a Prediction Surface to a Three-Dimensional Plot

Problem

You want to add a surface of predicted value to a three-dimensional scatter plot.

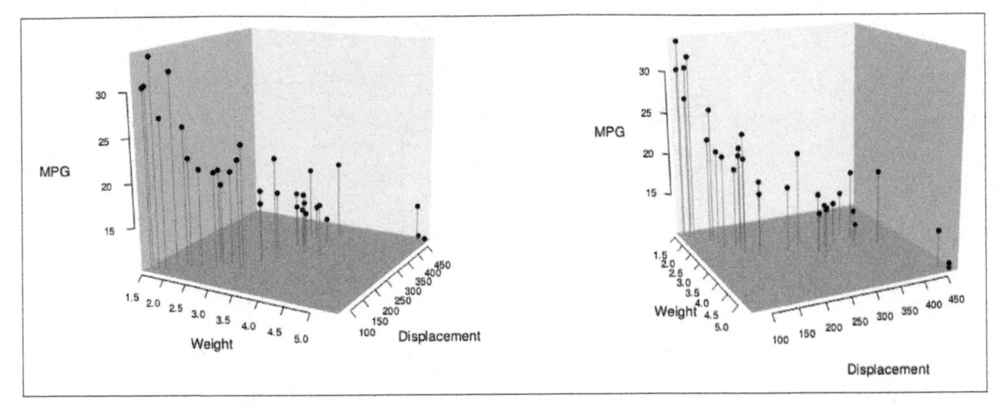

Figure 13-16. Left: 3D scatter plot with axis ticks and labels repositioned; right: from a different point of view

Solution

First we need to define some utility functions for generating the predicted values from a model object:

```
# Given a model, predict zvar from xvar and yvar
# Defaults to range of x and y variables, and a 16x16 grid
predictgrid <- function(model, xvar, yvar, zvar, res = 16, type = NULL) {
  # Find the range of the predictor variable. This works for lm and glm
  # and some others, but may require customization for others.
  xrange <- range(model$model[[xvar]])
  yrange <- range(model$model[[yvar]])

  newdata <- expand.grid(x = seq(xrange[1], xrange[2], length.out = res),
                         y = seq(yrange[1], yrange[2], length.out = res))
  names(newdata) <- c(xvar, yvar)
  newdata[[zvar]] <- predict(model, newdata = newdata, type = type)
  newdata
}


# Convert long-style data frame with x, y, and z vars into a list
# with x and y as row/column values, and z as a matrix.
df2mat <- function(p, xvar = NULL, yvar = NULL, zvar = NULL) {
  if (is.null(xvar)) xvar <- names(p)[1]
  if (is.null(yvar)) yvar <- names(p)[2]
  if (is.null(zvar)) zvar <- names(p)[3]

  x <- unique(p[[xvar]])
  y <- unique(p[[yvar]])
  z <- matrix(p[[zvar]], nrow = length(y), ncol = length(x))

  m <- list(x, y, z)
```

```
  names(m) <- c(xvar, yvar, zvar)
  m
}

# Function to interleave the elements of two vectors
interleave <- function(v1, v2)  as.vector(rbind(v1,v2))
```

With these utility functions defined, we can make a linear model from the data and plot it as a mesh along with the data, using the `surface3d()` function, as shown in Figure 13-17:

```
library(rgl)

# Make a copy of the data set
m <- mtcars

# Generate a linear model
mod <- lm(mpg ~ wt + disp + wt:disp, data = m)

# Get predicted values of mpg from wt and disp
m$pred_mpg <- predict(mod)

# Get predicted mpg from a grid of wt and disp
mpgrid_df <- predictgrid(mod, "wt", "disp", "mpg")
mpgrid_list <- df2mat(mpgrid_df)

# Make the plot with the data points
plot3d(m$wt, m$disp, m$mpg, type="s", size=0.5, lit=FALSE)

# Add the corresponding predicted points (smaller)
spheres3d(m$wt, m$disp, m$pred_mpg, alpha=0.4, type="s", size=0.5, lit=FALSE)

# Add line segments showing the error
segments3d(interleave(m$wt,   m$wt),
           interleave(m$disp, m$disp),
           interleave(m$mpg,  m$pred_mpg),
           alpha=0.4, col="red")

# Add the mesh of predicted values
surface3d(mpgrid_list$wt, mpgrid_list$disp, mpgrid_list$mpg,
          alpha=0.4, front="lines", back="lines")
```

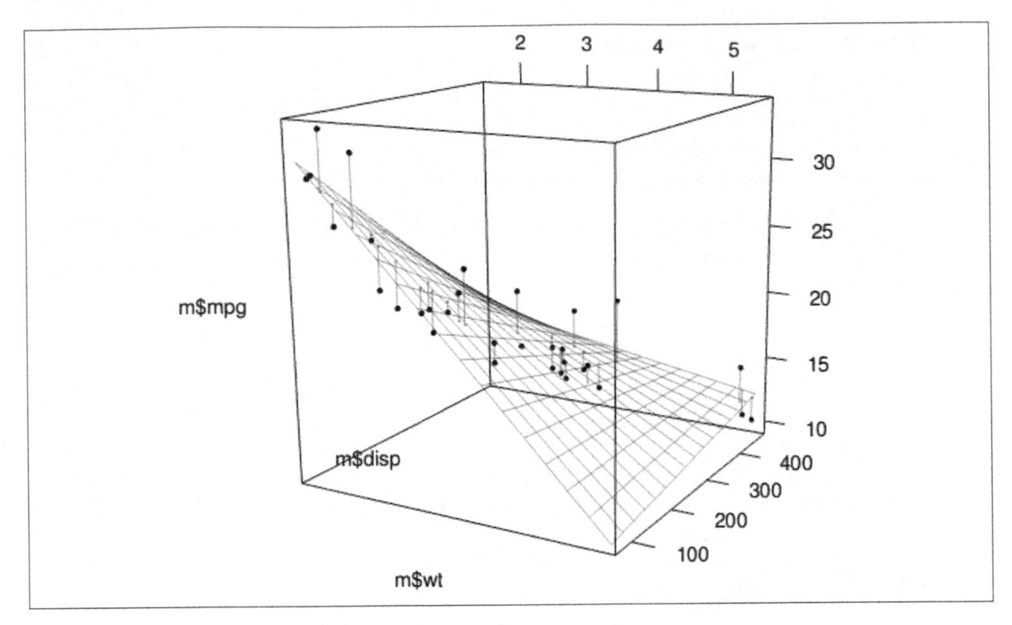

Figure 13-17. A 3D scatter plot with a prediction surface

Discussion

We can tweak the appearance of the graph, as shown in Figure 13-18. We'll add each of the components of the graph separately:

```
plot3d(mtcars$wt, mtcars$disp, mtcars$mpg,
       xlab = "", ylab = "", zlab = "",
       axes = FALSE,
       size=.5, type="s", lit=FALSE)

# Add the corresponding predicted points (smaller)
spheres3d(m$wt, m$disp, m$pred_mpg, alpha=0.4, type="s", size=0.5, lit=FALSE)

# Add line segments showing the error
segments3d(interleave(m$wt,   m$wt),
           interleave(m$disp, m$disp),
           interleave(m$mpg,  m$pred_mpg),
           alpha=0.4, col="red")

# Add the mesh of predicted values
surface3d(mpgrid_list$wt, mpgrid_list$disp, mpgrid_list$mpg,
          alpha=0.4, front="lines", back="lines")

# Draw the box
rgl.bbox(color="grey50",          # grey60 surface and black text
         emission="grey50",       # emission color is grey50
         xlen=0, ylen=0, zlen=0)  # Don't add tick marks
```

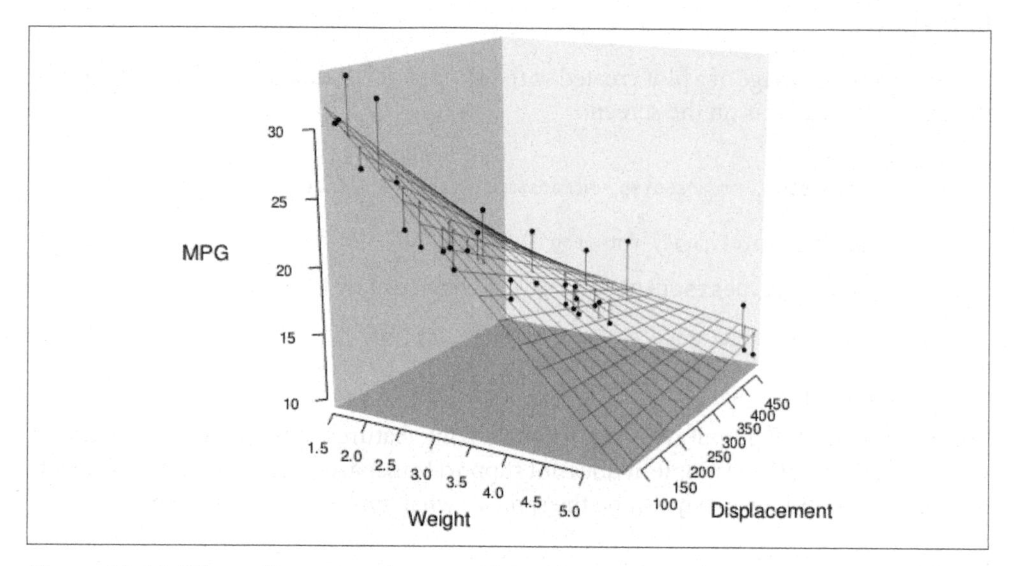

Figure 13-18. Three-dimensional scatter plot with customized appearance

```
# Set default color of future objects to black
rgl.material(color="black")

# Add axes to specific sides. Possible values are "x--", "x-+", "x+-", and "x++".
axes3d(edges=c("x--", "y+-", "z--"),
       ntick=6,                       # Attempt 6 tick marks on each side
       cex=.75)                       # Smaller font

# Add axis labels. 'line' specifies how far to set the label from the axis.
mtext3d("Weight",       edge="x--", line=2)
mtext3d("Displacement", edge="y+-", line=3)
mtext3d("MPG",          edge="z--", line=3)
```

See Also

For more on changing the appearance of the surface, see `?rgl.material`.

13.9. Saving a Three-Dimensional Plot

Problem

You want to save a three-dimensional plot created with the rgl package.

Solution

To save a bitmap image of a plot created with rgl, use `rgl.snapshot()`. This will capture the exact image that is on the screen:

```
library(rgl)
plot3d(mtcars$wt, mtcars$disp, mtcars$mpg, type="s", size=0.75, lit=FALSE)

rgl.snapshot('3dplot.png', fmt='png')
```

You can also use `rgl.postscript()` to save a PostScript or PDF file:

```
rgl.postscript('figs/miscgraph/3dplot.pdf', fmt='pdf')

rgl.postscript('figs/miscgraph/3dplot.ps', fmt='ps')
```

PostScript and PDF output does not support many features of the OpenGL library on which rgl is based. For example, it does not support transparency, and the sizes of objects such as points and lines may not be the same as what appears on the screen.

Discussion

To make the output more repeatable, you can save your current viewpoint and restore it later:

```
# Save the current viewpoint
view <- par3d("userMatrix")

# Restore the saved viewpoint
par3d(userMatrix = view)
```

To save `view` in a script, you can use `dput()`, then copy and paste the output into your script:

```
dput(view)

structure(c(0.907931625843048, 0.267511069774628, -0.322642296552658,
0, -0.410978674888611, 0.417272746562958, -0.810543060302734,
0, -0.0821993798017502, 0.868516683578491, 0.488796472549438,
0, 0, 0, 0, 1), .Dim = c(4L, 4L))
```

Once you have the text representation of the `userMatrix`, add the following to your script:

```
view <- structure(c(0.907931625843048, 0.267511069774628, -0.322642296552658,
0, -0.410978674888611, 0.417272746562958, -0.810543060302734,
0, -0.0821993798017502, 0.868516683578491, 0.488796472549438,
0, 0, 0, 0, 1), .Dim = c(4L, 4L))

par3d(userMatrix = view)
```

13.10. Animating a Three-Dimensional Plot

Problem

You want to animate a three-dimensional plot by moving the viewpoint around the plot.

Solution

Rotating a 3D plot can provide a more complete view of the data. To animate a 3D plot, use `play3d()` with `spin3d()`:

```
library(rgl)
plot3d(mtcars$wt, mtcars$disp, mtcars$mpg, type="s", size=0.75, lit=FALSE)

play3d(spin3d())
```

Discussion

By default, the graph will be rotated on the z (vertical) axis, until you send a `break` command to R.

You can change the rotation axis, rotation speed, and duration:

```
# Spin on x-axis, at 4 rpm, for 20 seconds
play3d(spin3d(axis=c(1,0,0), rpm=4), duration=20)
```

To save the movie, use the `movie3d()` function in the same way as `play3d()`. It will generate a series of *.png* files, one for each frame, and then attempt to combine them into a single animated *.gif* file using the `convert` program from the ImageMagick image utility.

This will spin the plot once in 15 seconds, at 50 frames per second:

```
# Spin on z axis, at 4 rpm, for 15 seconds
movie3d(spin3d(axis=c(0,0,1), rpm=4), duration=15, fps=50)
```

The output file will be saved in a temporary directory, and the name will be printed on the R console.

If you don't want to use ImageMagick to convert the output to a *.gif*, you can specify `convert=FALSE` and then convert the series of *.png* files to a movie using some other utility.

13.11. Creating a Dendrogram

Problem

You want to make a dendrogram to show how items are clustered.

Solution

Use `hclust()` and plot the output from it. This can require a fair bit of data preprocessing. For this example, we'll first take a subset of the `countries` data set from the year 2009. For simplicity, we'll also drop all rows that contain an `NA`, and then select a random 25 of the remaining rows:

```
library(gcookbook) # For the data set

# Get data from year 2009
c2 <- subset(countries, Year==2009)

# Drop rows that have any NA values
c2 <- c2[complete.cases(c2), ]

# Pick out a random 25 countries
# (Set random seed to make this repeatable)
set.seed(201)
c2 <- c2[sample(1:nrow(c2), 25), ]

c2
```

```
                  Name Code Year        GDP laborrate  healthexp infmortality
6731          Mongolia  MNG 2009  1690.4170      72.9   74.19826         27.8
1733            Canada  CAN 2009 39599.0418      67.8 4379.76084          5.2
...
5966   Macedonia, FYR  MKD 2009  4510.2380      54.0  313.68971         10.6
10148     Turkmenistan  TKM 2009  3710.4536      68.0   77.06955         48.0
```

Notice that the row names (the first column) are essentially random numbers, since the rows were selected randomly. We need to do a few more things to the data before making a dendrogram from it. First, we need to set the *row names*—right now there's a column called `Name`, but the row names are those random numbers (we don't often use row names, but for the `hclust()` function they're essential). Next, we'll need to drop all the columns that aren't values used for clustering. These columns are `Name`, `Code`, and `Year`:

```
rownames(c2) <- c2$Name
c2 <- c2[,4:7]
c2
```

```
                       GDP laborrate  healthexp infmortality
Mongolia         1690.4170      72.9   74.19826         27.8
Canada          39599.0418      67.8 4379.76084          5.2
...
Macedonia, FYR   4510.2380      54.0  313.68971         10.6
Turkmenistan     3710.4536      68.0   77.06955         48.0
```

The values for `GDP` are several orders of magnitude larger than the values for, say, `infmortality`. Because of this, the effect of `infmortality` on the clustering will be negligible compared to the effect of GDP.This probably isn't what we want. To address this issue, we'll scale the data:

```
c3 <- scale(c2)
c3

                      GDP   laborrate     healthexp infmortality
Mongolia       -0.6783472  1.15028714 -0.6341393599  -0.08334689
Canada          1.7504703  0.59747293  1.9736219974  -0.88014885
 ...
Macedonia, FYR -0.4976803 -0.89837729 -0.4890859471  -0.68976254
Turkmenistan   -0.5489228  0.61915192 -0.6324002997   0.62883892
attr(,"scaled:center")
         GDP    laborrate    healthexp infmortality
   12277.960       62.288     1121.198       30.164
attr(,"scaled:scale")
         GDP    laborrate    healthexp infmortality
15607.852864     9.225523  1651.056974    28.363384
```

By default the `scale()` function scales each column relative to its standard deviation, but other methods may be used.

Finally, we're ready to make the dendrogram, as shown in Figure 13-19:

```
hc <- hclust(dist(c3))

# Make the dendrogram
plot(hc)

# With text aligned
plot(hc, hang = -1)
```

Discussion

A cluster analysis is simply a way of assigning points to groups in an *n*-dimensional space (four dimensions, in this example). A hierarchical cluster analysis divides each group into two smaller groups, and can be represented with the dendrograms in this recipe. There are many different parameters you can control in the hierarchical cluster analysis process, and there may not be a single "right" way to do it for your data.

First, we normalized the data using `scale()` with its default settings. You can scale your data differently, or not at all. (With this data set, *not* scaling the data will lead to GDP overwhelming the other variables, as shown in Figure 13-20.)

For the distance calculation, we used the default method, `"euclidean"`, which calculates the Euclidean distance between the points. The other possible methods are `"maximum"`, `"manhattan"`, `"canberra"`, `"binary"`, and `"minkowski"`.

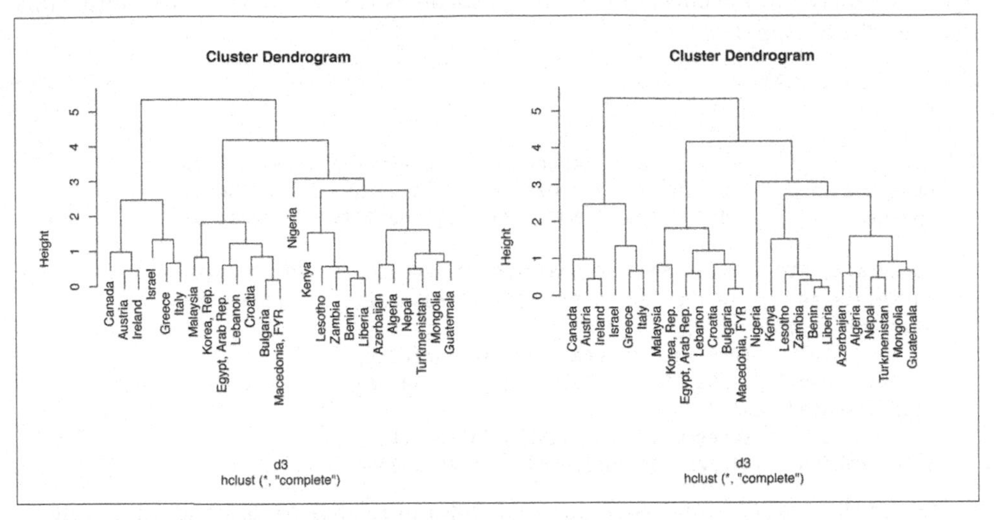

Figure 13-19. Left: a dendrogram; right: with text aligned

The `hclust()` function provides several methods for performing the cluster analysis. The default is `"complete"`; the other possible methods are `"ward"`, `"single"`, `"average"`, `"mcquitty"`, `"median"`, and `"centroid"`.

See Also

See `?hclust` for more information about the different clustering methods.

13.12. Creating a Vector Field

Problem

You want to make a vector field.

Solution

Use `geom_segment()`. For this example, we'll use the `isabel` data set:

```
library(gcookbook) # For the data set
isabel

         x        y      z        vx        vy           vz         t    speed
 -83.00000 41.70000  0.035        NA        NA           NA        NA       NA
 -83.00000 41.62786  0.035        NA        NA           NA        NA       NA
 -83.00000 41.55571  0.035        NA        NA           NA        NA       NA
 ...
 -62.04208 23.88036 18.035 -12.54371 -5.300128 -0.045253485 -66.96269 13.61749
```

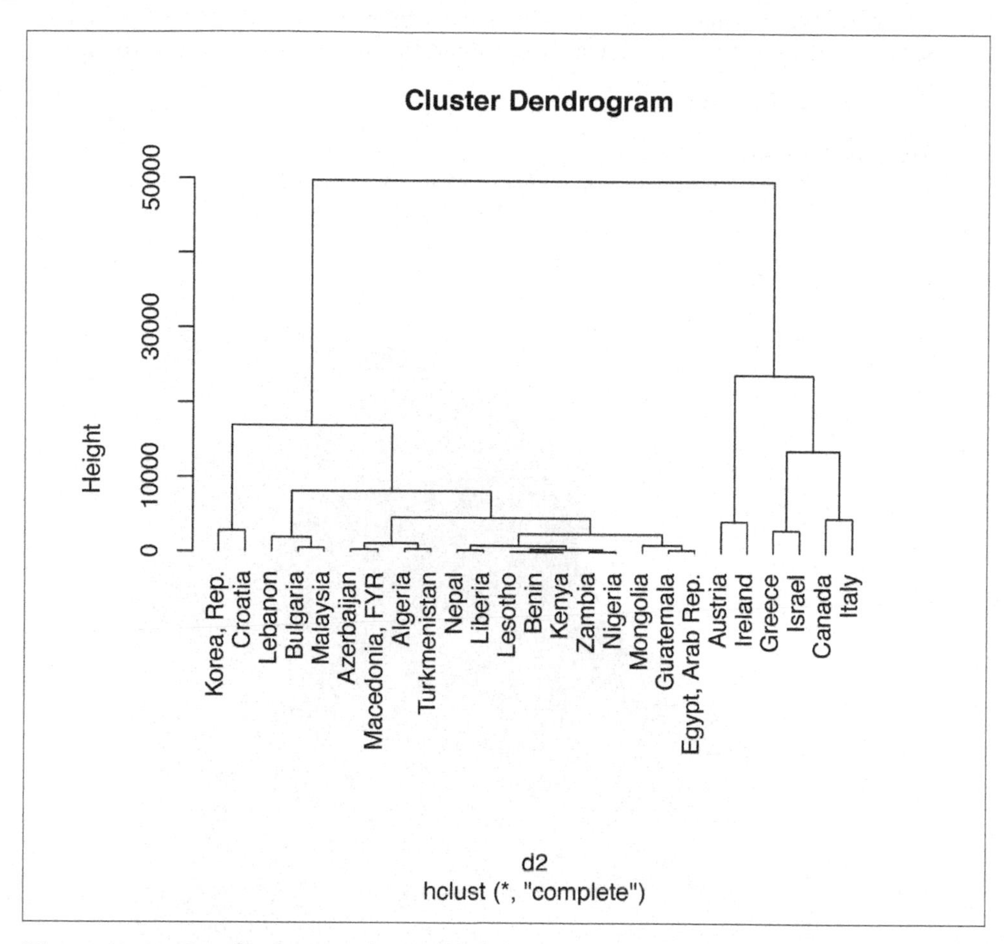

Figure 13-20. Dendrogram with unscaled data—notice the much larger Height values, which are largely due to the unscaled GDP values

```
-62.04208 23.80822 18.035 -12.56157 -5.254994 -0.020277001 -66.98840 13.61646
-62.04208 23.73607 18.035 -12.78071 -5.259613  0.005555035 -67.00575 13.82064
```

x and y are the longitude and latitude, respectively, and z is the height in kilometers. The vx, vy, and vz values are the wind speed components in each of these directions, in meters per second, and speed is the wind speed.

The height (z) ranges from 0.035 km to 18.035 km. For this example, we'll just use the lowest slice of data.

To draw the vectors (Figure 13-21), we'll use geom_segment(). Each segment has a starting point and an ending point. We'll use the x and y values as the starting points for each segment, then add a fraction of the vx and vy values to get the end points for each segment. If we didn't scale down these values, the lines would be much too long:

```
islice <- subset(isabel, z == min(z))

ggplot(islice, aes(x=x, y=y)) +
       geom_segment(aes(xend = x + vx/50, yend = y + vy/50),
                    size = 0.25)   # Make the line segments 0.25 mm thick
```

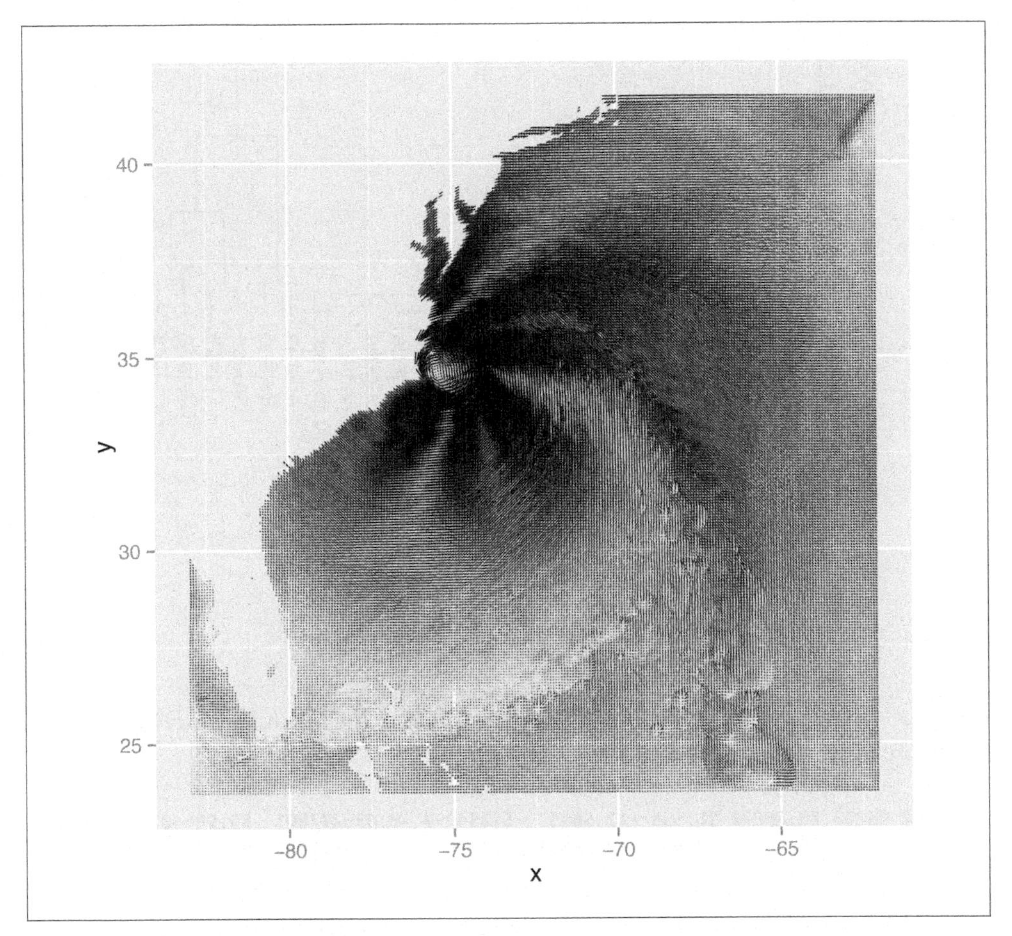

Figure 13-21. First attempt at a vector field—the resolution of the data is too high, but it does hint at some interesting patterns not visible in graphs with a lower data resolution

This vector field has two problems: the data is at too high a resolution to read, and the segments do not have arrows indicating the direction of the flow. To reduce the resolution of the data, we'll define a function every_n() that keeps one out of every n values in the data and drops the rest:

```
# Take a slice where z is equal to the minimum value of z
islice <- subset(isabel, z == min(z))

# Keep 1 out of every 'by' values in vector x
every_n <- function(x, by = 2) {
    x <- sort(x)
    x[seq(1, length(x), by = by)]
}

# Keep 1 of every 4 values in x and y
keepx <- every_n(unique(isabel$x), by=4)
keepy <- every_n(unique(isabel$y), by=4)

# Keep only those rows where x value is in keepx and y value is in keepy
islicesub <- subset(islice, x %in% keepx  &  y %in% keepy)
```

Now that we've taken a subset of the data, we can plot it, with arrowheads, as shown in Figure 13-22:

```
# Need to load grid for arrow() function
library(grid)

# Make the plot with the subset, and use an arrowhead 0.1 cm long
ggplot(islicesub, aes(x=x, y=y)) +
    geom_segment(aes(xend = x+vx/50, yend = y+vy/50),
                 arrow = arrow(length = unit(0.1, "cm")), size = 0.25)
```

Discussion

One effect of arrowheads is that short vectors appear with more ink than is proportional to their length. This could somewhat distort the interpretation of the data. To mitigate this effect, it may also be useful to map the speed to other properties, like size (line thickness), alpha, or colour. Here, we'll map speed to alpha (Figure 13-23, left):

```
# The existing 'speed' column includes the z component. We'll calculate
# speedxy, the horizontal speed.
islicesub$speedxy <- sqrt(islicesub$vx^2 + islicesub$vy^2)

# Map speed to alpha
ggplot(islicesub, aes(x=x, y=y)) +
    geom_segment(aes(xend = x+vx/50, yend = y+vy/50, alpha = speed),
                 arrow = arrow(length = unit(0.1,"cm")), size = 0.6)
```

Next, we'll map speed to colour. We'll also add a map of the United States and zoom in on the area of interest, as shown in the graph on the right in Figure 13-23, using coord_cartesian() (without this, the entire USA will be displayed):

```
# Get USA map data
usa <- map_data("usa")
```

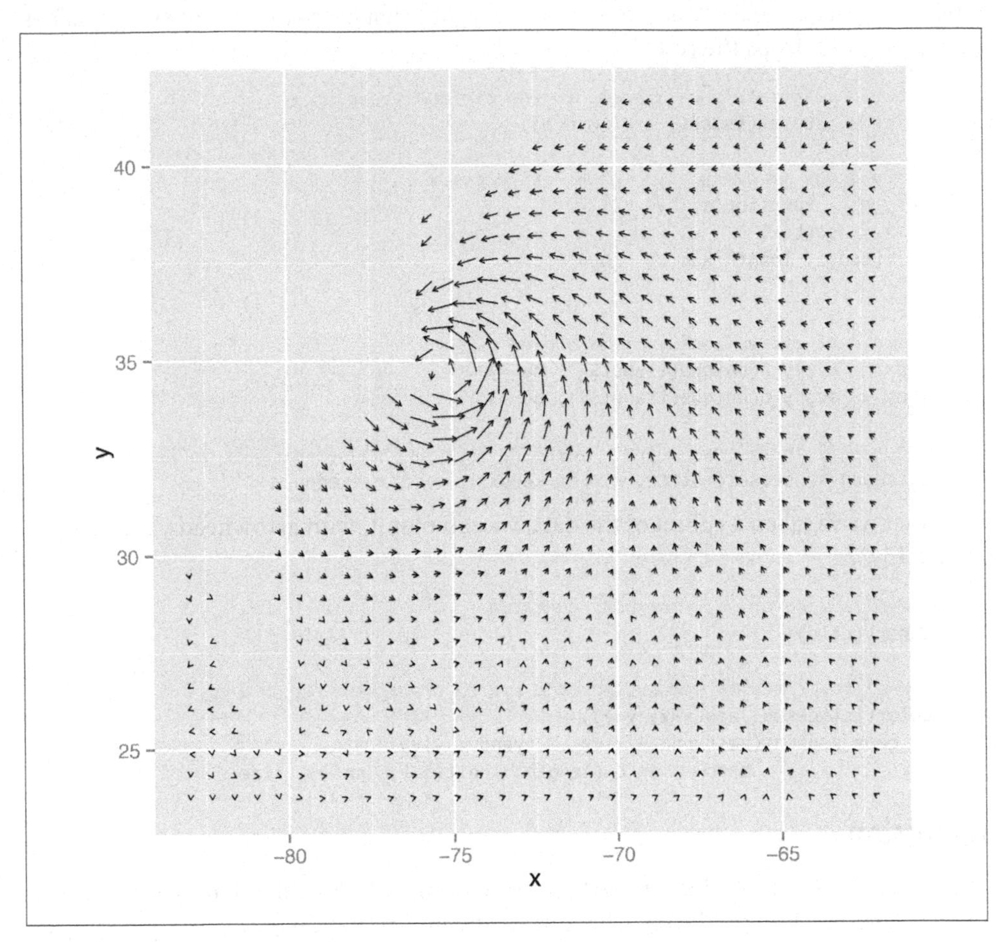

Figure 13-22. Vector field with arrowheads

```
# Map speed to colour, and set go from "grey80" to "darkred"
ggplot(islicesub, aes(x=x, y=y)) +
    geom_segment(aes(xend = x+vx/50, yend = y+vy/50, colour = speed),
                 arrow = arrow(length = unit(0.1,"cm")), size = 0.6) +
    scale_colour_continuous(low="grey80", high="darkred") +
    geom_path(aes(x=long, y=lat, group=group), data=usa) +
    coord_cartesian(xlim = range(islicesub$x), ylim = range(islicesub$y))
```

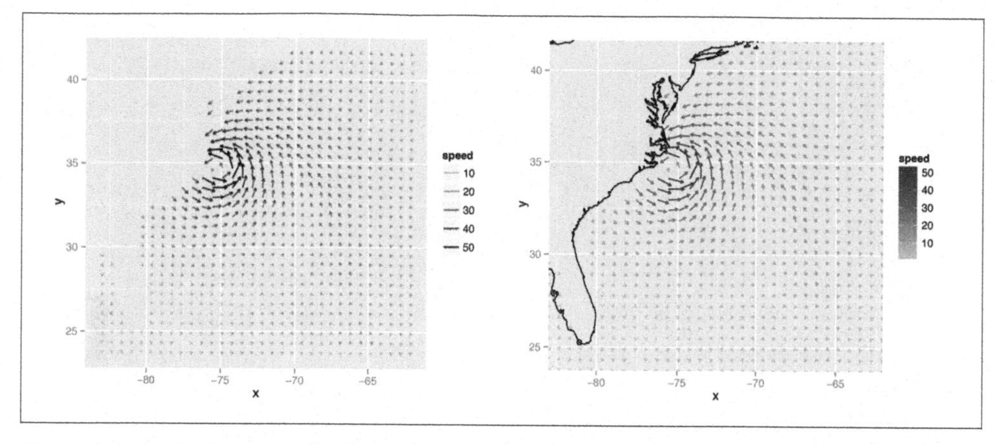

Figure 13-23. Left: vector field with speed mapped to alpha; right: with speed mapped to colour

The `isabel` data set has three-dimensional data, so we can also make a faceted graph of the data, as shown in Figure 13-24. Because each facet is small, we will use a sparser subset than before:

```
# Keep 1 out of every 5 values in x and y, and 1 in 2 values in z
keepx <- every_n(unique(isabel$x), by=5)
keepy <- every_n(unique(isabel$y), by=5)
keepz <- every_n(unique(isabel$z), by=2)

isub <- subset(isabel, x %in% keepx  &  y %in% keepy  &  z %in% keepz)

ggplot(isub, aes(x=x, y=y)) +
    geom_segment(aes(xend = x+vx/50, yend = y+vy/50, colour = speed),
                 arrow = arrow(length = unit(0.1,"cm")), size = 0.5) +
    scale_colour_continuous(low="grey80", high="darkred") +
    facet_wrap( ~ z)
```

See Also

If you want to use a different color palette, see Recipe 12.6.

See Recipe 8.2 for more information about zooming in on part of a graph.

13.13. Creating a QQ Plot

Problem

You want to make a quantile-quantile (QQ) plot to compare an empirical distribution to a theoretical distribution.

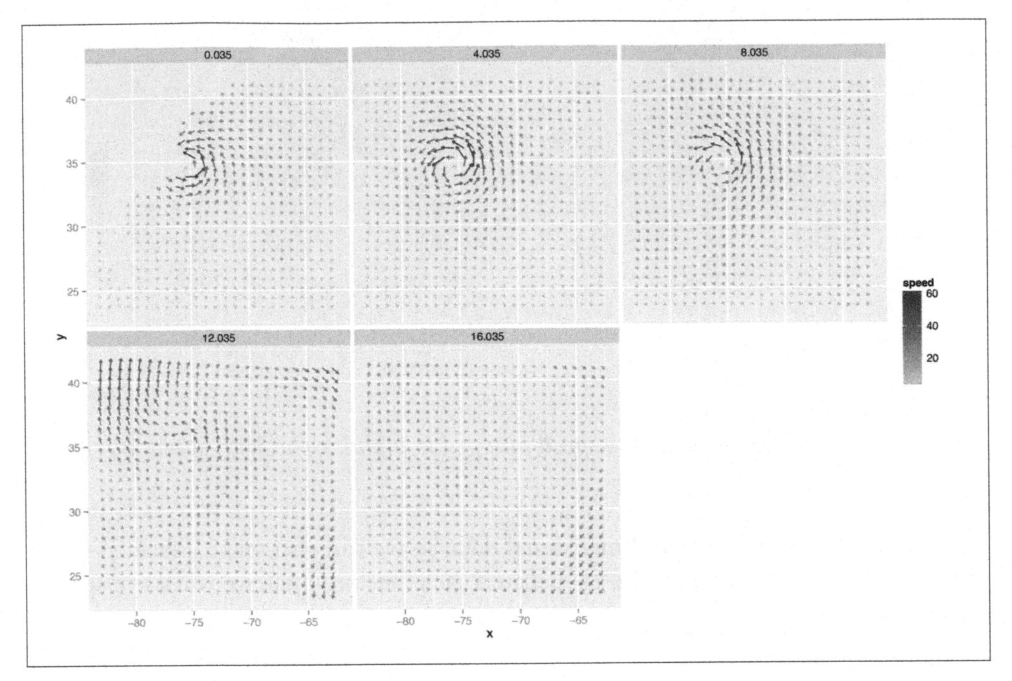

Figure 13-24. Vector field of wind speeds faceted on z

Solution

Use `qqnorm()` to compare to a normal distribution. Give `qqnorm()` a vector of numerical values, and add a theoretical distribution line with `qqline()` (Figure 13-25):

```
library(gcookbook) # For the data set

# QQ plot of height
qqnorm(heightweight$heightIn)
qqline(heightweight$heightIn)

# QQ plot of age
qqnorm(heightweight$ageYear)
qqline(heightweight$ageYear)
```

Discussion

The points for `heightIn` are close to the line, which means that the distribution is close to normal. In contrast, the points for `ageYear` veer far away from the line, especially on the left, indicating that the distribution is skewed. A histogram may also be useful for exploring how the data is distributed.

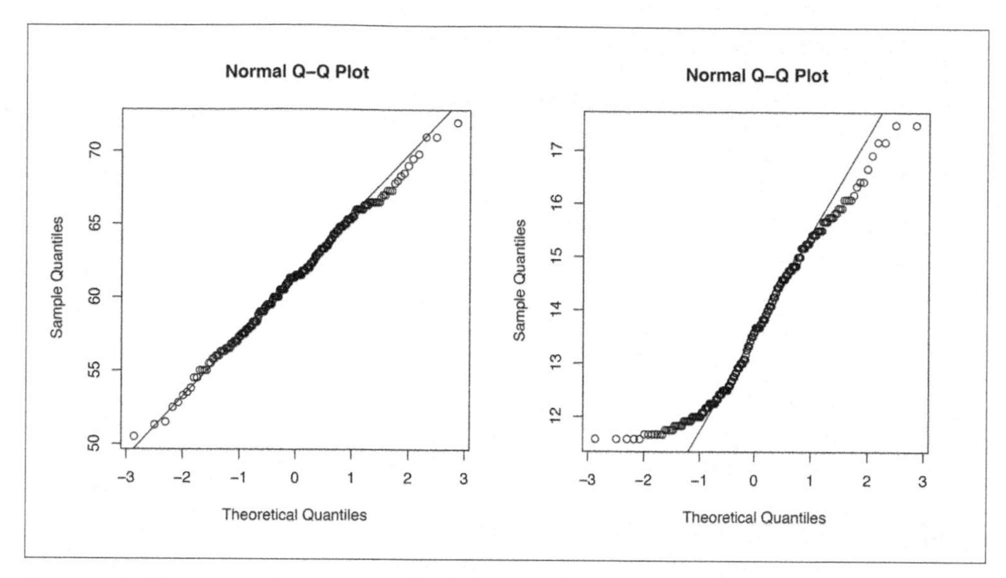

Figure 13-25. Left: QQ plot of height, which is close to normally distributed; right: QQ plot of age, which is not normally distributed

See Also

See `?qqplot` for information on comparing data to theoretical distributions other than the normal distribution.

ggplot2 has a `stat_qq()` function, but it doesn't provide an easy way to draw the QQ line.

13.14. Creating a Graph of an Empirical Cumulative Distribution Function

Problem

You want to graph the empirical cumulative distribution function (ECDF) of a data set.

Solution

Use `stat_ecdf()` (Figure 13-26):

```
library(gcookbook) # For the data set

# ecdf of heightIn
ggplot(heightweight, aes(x=heightIn)) + stat_ecdf()
```

```
# ecdf of ageYear
ggplot(heightweight, aes(x=ageYear)) + stat_ecdf()
```

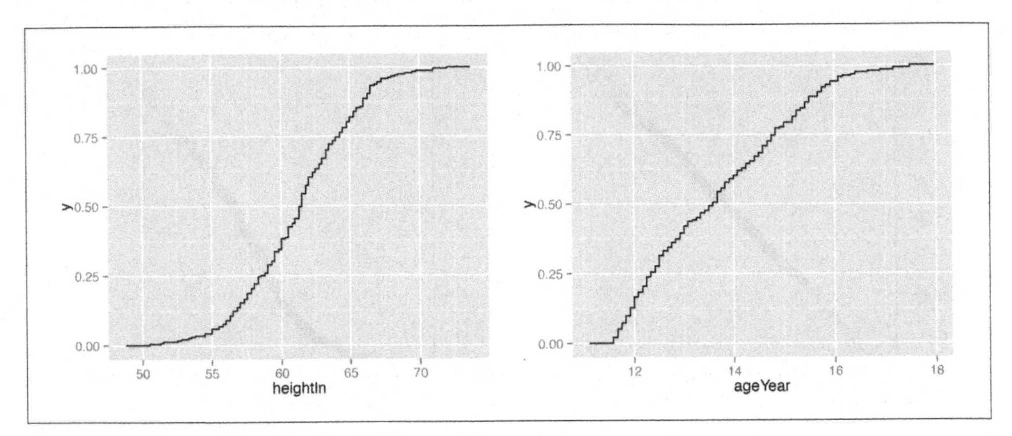

Figure 13-26. Left: ECDF of height; right: ECDF of age

Discussion

The ECDF shows what proportion of observations are at or below the given *x* value. Because it is *empirical*, the line takes a step up at each *x* value where there are one or more observations.

13.15. Creating a Mosaic Plot

Problem

You want to make a mosaic plot to visualize a contingency table.

Solution

Use the `mosaic()` function from the vcd package. For this example we'll use the `USBAdmissions` data set, which is a contingency table with three dimensions. We'll first take a look at the data in a few different ways:

```
UCBAdmissions

, , Dept = A

          Gender
Admit      Male Female
  Admitted  512     89
  Rejected  313     19

 ... [four other Depts]
```

```
, , Dept = F

          Gender
Admit      Male Female
  Admitted   22     24
  Rejected  351    317

# Print a "flat" contingency table
ftable(UCBAdmissions)

                Dept   A   B   C   D   E   F
Admit    Gender
Admitted Male        512 353 120 138  53  22
         Female       89  17 202 131  94  24
Rejected Male        313 207 205 279 138 351
         Female       19   8 391 244 299 317

dimnames(UCBAdmissions)

$Admit
[1] "Admitted" "Rejected"

$Gender
[1] "Male"   "Female"

$Dept
[1] "A" "B" "C" "D" "E" "F"
```

The three dimensions are `Admit`, `Gender`, and `Dept`. To visualize the relationships between the variables (Figure 13-27), use `mosaic()` and pass it a formula with the variables that will be used to split up the data:

```
# You may need to install first, with install.packages("vcd")
library(vcd)
# Split by Admit, then Gender, then Dept
mosaic( ~ Admit + Gender + Dept, data=UCBAdmissions)
```

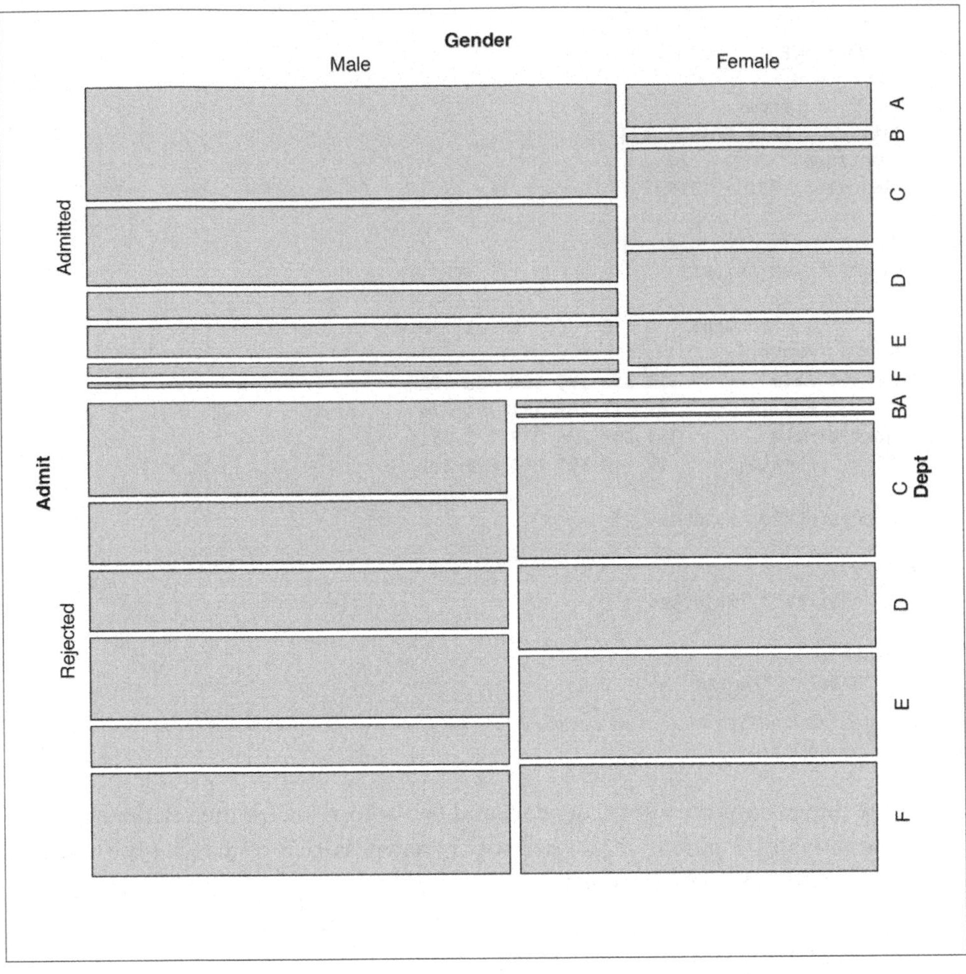

Figure 13-27. Mosaic plot of UC-Berkeley admissions data—the area of each rectangle is proportional to the number of cases in that cell

Notice that `mosaic()` splits the data in the order in which the variables are provided: first on admission status, then gender, then department. The resulting plot order makes it very clear that more applicants were rejected than admitted. It is also clear that within the admitted group there were many more men than women, while in the rejected group there were approximately the same number of men and women. It is difficult to make comparisons within each department, though. A different variable splitting order may reveal some other interesting information.

Another way of looking at the data is to split first by department, then gender, then admission status, as in Figure 13-28. This makes the admission status the last variable that is partitioned, so that *after* partitioning by department and gender, the admitted and rejected cells for each group are right next to each other:

```
mosaic( ~ Dept + Gender + Admit, data=UCBAdmissions,
    highlighting="Admit", highlighting_fill=c("lightblue", "pink"),
    direction=c("v","h","v"))
```

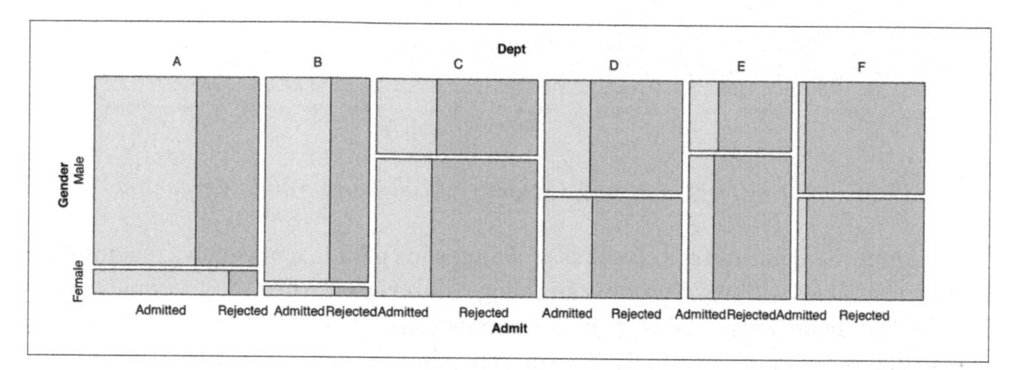

Figure 13-28. Mosaic plot with a different variable splitting order: first department, then gender, then admission status

We also specified a variable to highlight (`Admit`), and which colors to use in the highlighting.

Discussion

In the preceding example we also specified the *direction* in which each variable will be split. The first variable, `Dept`, is split vertically; the second variable, `Gender`, is split horizontally; and the third variable, `Admit`, is split vertically. The reason that we chose these directions is because, in this particular example, it makes it easy to compare the male and female groups within each department.

We can also use different splitting directions, as shown in Figures 13-29 and 13-30:

```
# Another possible set of splitting directions
mosaic( ~ Dept + Gender + Admit, data=UCBAdmissions,
    highlighting="Admit", highlighting_fill=c("lightblue", "pink"),
    direction=c("v", "v", "h"))

# This order makes it difficult to compare male and female
mosaic( ~ Dept + Gender + Admit, data=UCBAdmissions,
    highlighting="Admit", highlighting_fill=c("lightblue", "pink"),
    direction=c("v", "h", "h"))
```

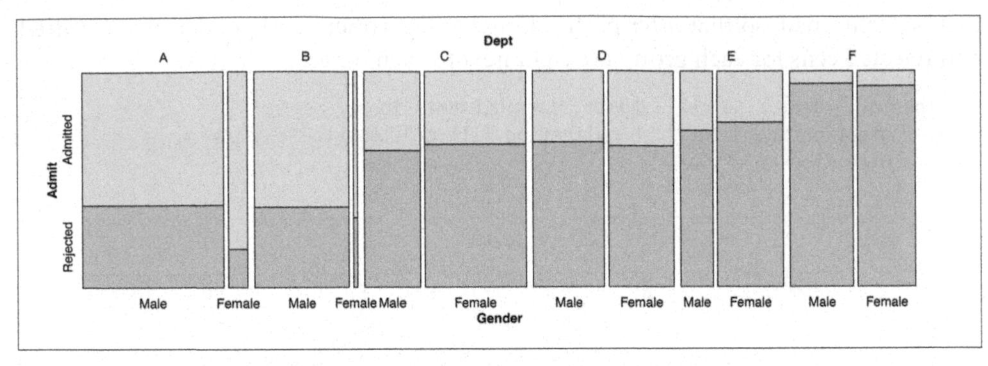

Figure 13-29. Splitting Dept vertically, Gender vertically, and Admit horizontally

The example here illustrates a classic case of Simpson's paradox, in which a relationship between variables within subgroups can change (or reverse!) when the groups are combined. The `UCBerkeley` table contains admissions data from the University of California-Berkeley in 1973. Overall, men were admitted at a higher rate than women, and because of this, the university was sued for gender bias. But when each department was examined separately, it was found that they each had approximately equal admission rates for men and women. The difference in overall admission rates was because women were more likely to apply to competitive departments with lower admission rates.

In Figures 13-28 and 13-29, you can see that within each department, admission rates were approximately equal between men and women. You can also see that departments with higher admission rates (A and B) were very imbalanced in the gender ratio of applicants: far more men applied to these departments than did women. As you can see, partitioning the data in different orders and directions can bring out different aspects of the data. In Figure 13-29, as in Figure 13-28, it's easy to compare male and female admission rates within each department and across departments. Splitting `Dept` vertically, `Gender` horizontally, and `Admit` horizontally, as in Figure 13-30, makes it difficult to compare male and female admission rates within each department, but it is easy to compare male and female application rates across departments.

See Also

See `?mosaicplot` for another function that can create mosaic plots.

P.J. Bickel, E.A. Hammel, and J.W. O'Connell, "Sex Bias in Graduate Admissions: Data from Berkeley," *Science* 187 (1975): 398–404.

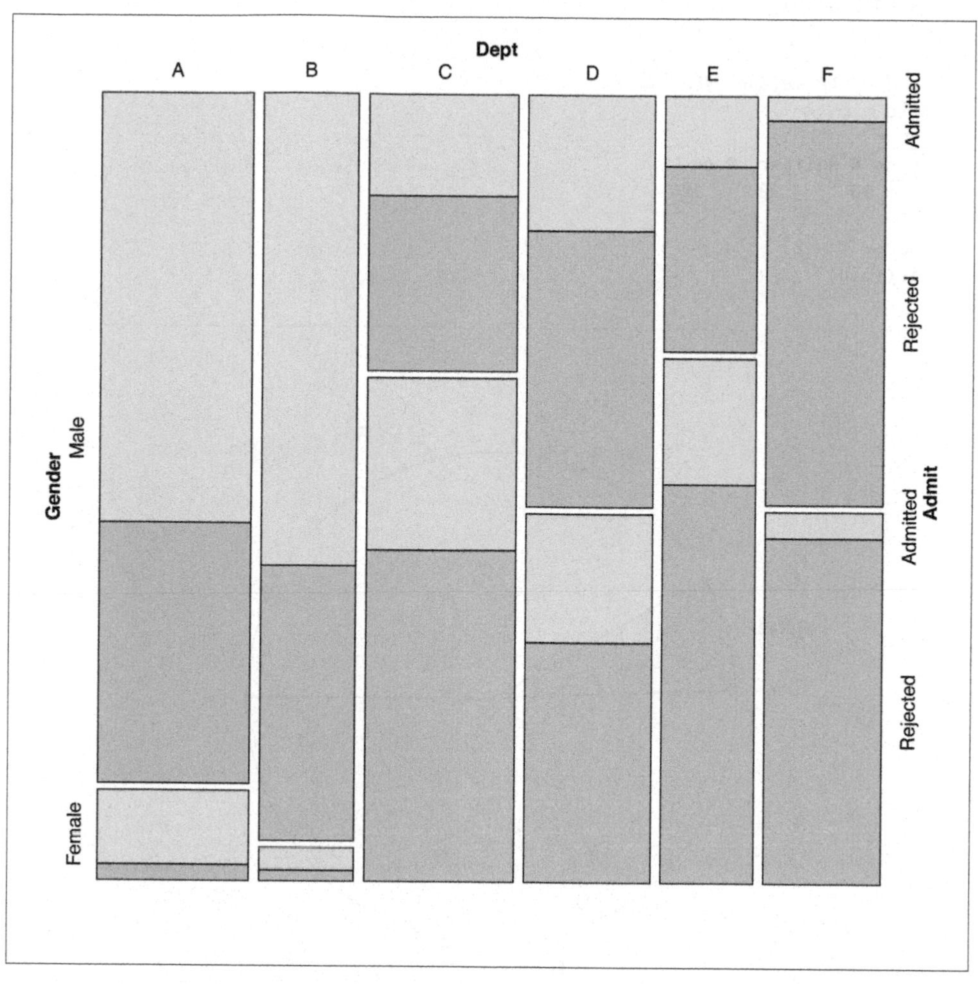

Figure 13-30. Splitting Dept vertically, Gender horizontally, and Admit horizontally

13.16. Creating a Pie Chart

Problem

You want to make a pie chart.

Solution

Use the `pie()` function. In this example (Figure 13-31), we'll use the `survey` data set from the MASS library:

```
library(MASS)  # For the data set

# Get a table of how many cases are in each level of fold
fold <- table(survey$Fold)
fold

 L on R Neither  R on L
     99      18     120

# Make the pie chart
pie(fold)
```

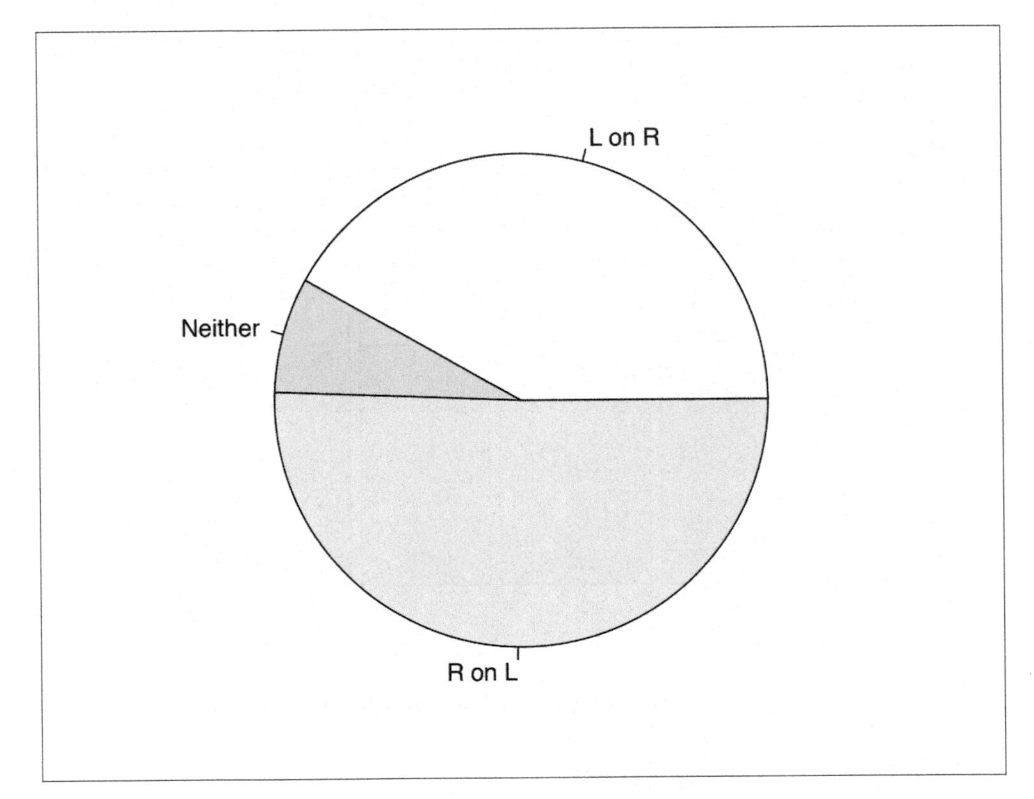

Figure 13-31. A pie chart

We passed `pie()` an object of class `table`. We could have instead given it a named vector, or a vector of values and a vector of labels, like this:

```
pie(c(99, 18, 120), labels=c("L on R", "Neither", "R on L"))
```

Discussion

The lowly pie chart is the subject of frequent abuse from data visualization experts. If you're thinking of using a pie chart, consider whether a bar graph (or stacked bar graph) would convey the information more effectively. Despite their faults, pie charts do have one important virtue: everyone knows how to read them.

13.17. Creating a Map

Problem

You want to create a geographical map.

Solution

Retrieve map data from the maps package and draw it with `geom_polygon()` (which can have a color fill) or `geom_path()` (which can't have a fill). By default, the latitude and longitude will be drawn on a Cartesian coordinate plane, but you can use `coord_map()` and specify a projection. The default projection is `"mercator"`, which, unlike the Cartesian plane, has a progressively changing spacing for latitude lines (Figure 13-32):

```
library(maps) # For map data
# Get map data for USA
states_map <- map_data("state") # ggplot2 must be loaded to use map_data()

ggplot(states_map, aes(x=long, y=lat, group=group)) +
    geom_polygon(fill="white", colour="black")

# geom_path (no fill) and Mercator projection
ggplot(states_map, aes(x=long, y=lat, group=group)) +
    geom_path() + coord_map("mercator")
```

Discussion

The `map_data()` function returns a data frame with the following columns:

`long`
: Longitude.

`lat`
: Latitude.

`group`
: This is a grouping variable for each polygon. A region or subregion might have multiple polygons, for example, if it includes islands.

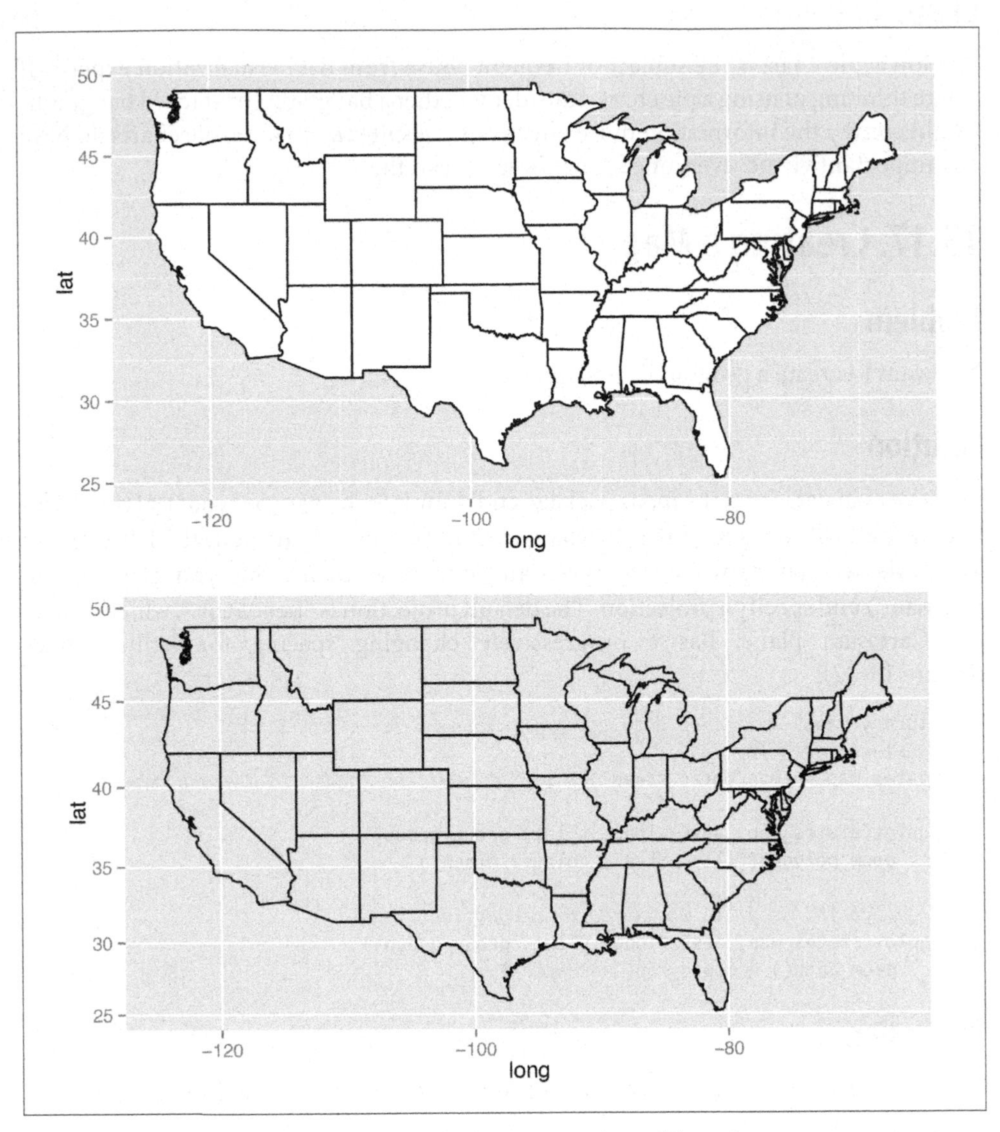

Figure 13-32. Top: a basic map with fill; bottom: with no fill, and Mercator projection

`order`
: The order to connect each point within a group.

`region`
: Roughly, the names of countries, although some other objects are present (such as some lakes).

`subregion`
: The names of subregions within a region, which can contain multiple groups. For example, the Alaska subregion includes many islands, each with its own group.

There are a number of different maps available, including `world`, `nz`, `france`, `italy`, `usa` (outline of the United States), `state` (each state in the USA), and `county` (each county in the USA). For example, to get map data for the world:

```
# Get map data for world
world_map <- map_data("world")
world_map

        long      lat group order      region subregion
 -133.3664 58.42416     1     1      Canada      <NA>
 -132.2681 57.16308     1     2      Canada      <NA>
 -132.0498 56.98610     1     3      Canada      <NA>
 ...
  124.7772 11.35419  2284 27634 Philippines     Leyte
  124.9697 11.30280  2284 27635 Philippines     Leyte
  125.0155 11.13887  2284 27636 Philippines     Leyte
```

If you want to draw a map of a region in the `world` map for which there isn't a separate map, you can first look for the region name, like so:

```
sort(unique(world_map$region))

 "Afghanistan"                 "Albania"                  "Algeria"
 "American Samoa"              "Andaman Islands"          "Andorra"
 "Angola"                      "Anguilla"                 "Antarctica"
  ...
 "USA"                         "USSR"                     "Vanuatu"
 "Venezuela"                   "Vietnam"                  "Virgin Islands"
 "Vislinskiy Zaliv"            "Wales"                    "West Bank"
 "Western Sahara"              "Yemen"                    "Yugoslavia"
 "Zaire"                       "Zambia"                   "Zimbabwe"

# You might have noticed that it's a little out of date!
```

It's possible to get data for specific regions from a particular map (Figure 13-33):

```
east_asia <- map_data("world", region=c("Japan", "China", "North Korea",
                                        "South Korea"))
# Map region to fill color
ggplot(east_asia, aes(x=long, y=lat, group=group, fill=region)) +
    geom_polygon(colour="black") +
    scale_fill_brewer(palette="Set2")
```

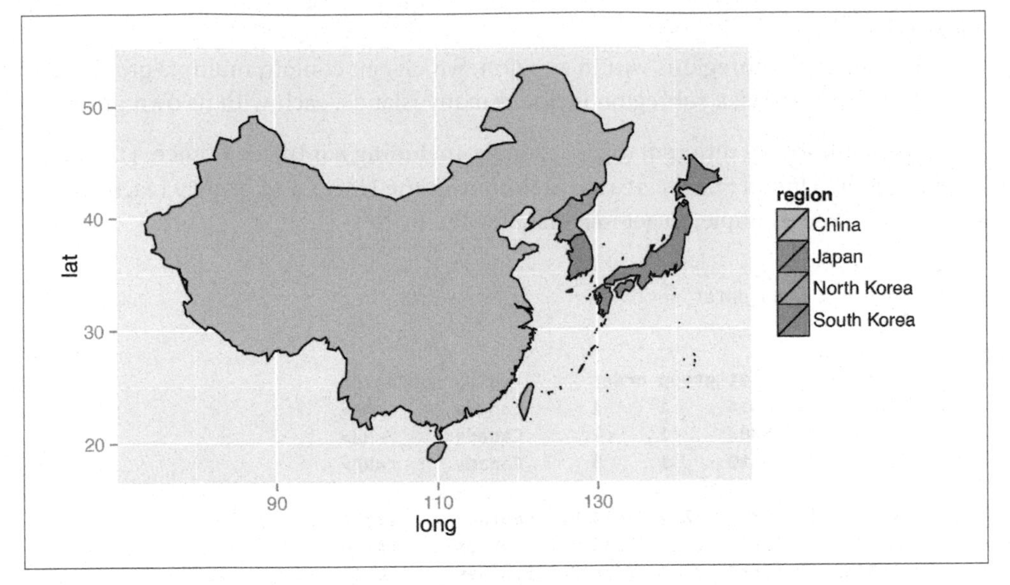

Figure 13-33. Specific regions from the world map

If there is a separate map available for a region, such as `nz` (New Zealand), that map data will be at a higher resolution than if you were to extract it from the `world` map, as shown in Figure 13-34:

```
# Get New Zealand data from world map
nz1 <- map_data("world", region="New Zealand")
nz1 <- subset(nz1, long > 0 & lat > -48)        # Trim off islands
ggplot(nz1, aes(x=long, y=lat, group=group)) + geom_path()

# Get New Zealand data from the nz map
nz2 <- map_data("nz")
ggplot(nz2, aes(x=long, y=lat, group=group)) + geom_path()
```

See Also

See the `mapdata` package for more map data sets. It includes maps of China and Japan, as well as a high-resolution world map, `worldHires`.

See the `map()` function, for quickly generating maps.

See `?mapproject` for a list of available map projections.

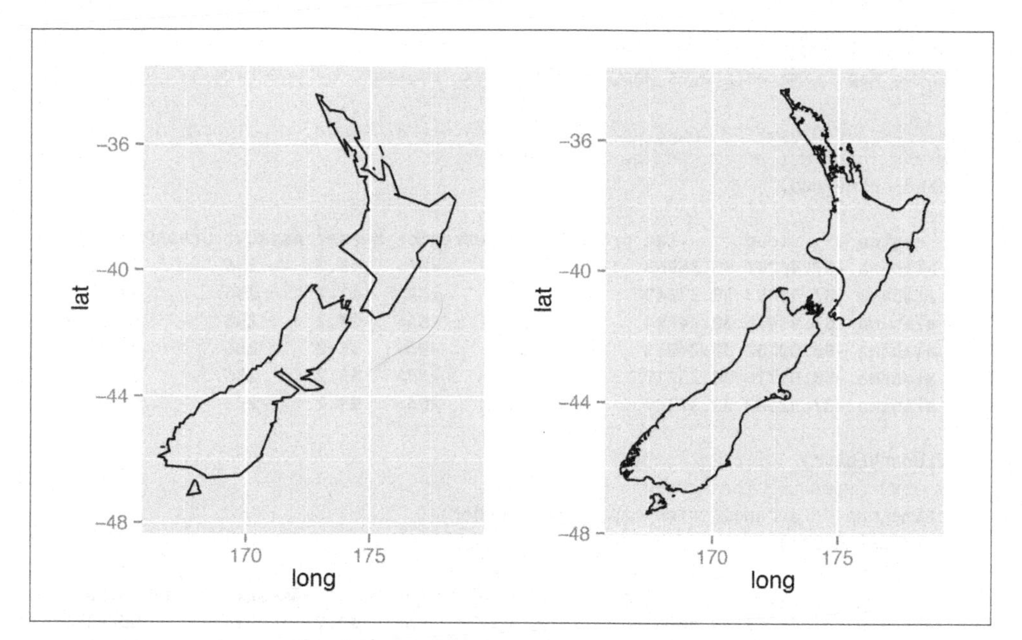

Figure 13-34. Left: New Zealand data taken from world map; right: data from nz map

13.18. Creating a Choropleth Map

Problem

You want to create a map with regions that are colored according to variable values.

Solution

Merge the value data with the map data, then map a variable to `fill`:

```
# Transform the USArrests data set to the correct format
crimes <- data.frame(state = tolower(rownames(USArrests)), USArrests)
crimes

                        state Murder Assault UrbanPop Rape
Alabama               alabama   13.2     236       58 21.2
Alaska                 alaska   10.0     263       48 44.5
Arizona               arizona    8.1     294       80 31.0
 ...
West Virginia   west virginia    5.7      81       39  9.3
Wisconsin           wisconsin    2.6      53       66 10.8
Wyoming               wyoming    6.8     161       60 15.6

library(maps) # For map data
states_map <- map_data("state")
```

```
# Merge the data sets together
crime_map <- merge(states_map, crimes, by.x="region", by.y="state")

# After merging, the order has changed, which would lead to polygons drawn in
# the incorrect order. So, we sort the data.
head(crime_map)

  region      long      lat group order subregion Murder Assault UrbanPop Rape
 alabama -87.46201 30.38968     1     1      <NA>   13.2     236       58 21.2
 alabama -87.48493 30.37249     1     2      <NA>   13.2     236       58 21.2
 alabama -87.95475 30.24644     1    13      <NA>   13.2     236       58 21.2
 alabama -88.00632 30.24071     1    14      <NA>   13.2     236       58 21.2
 alabama -88.01778 30.25217     1    15      <NA>   13.2     236       58 21.2
 alabama -87.52503 30.37249     1     3      <NA>   13.2     236       58 21.2

library(plyr)  # For arrange() function
# Sort by group, then order
crime_map <- arrange(crime_map, group, order)
head(crime_map)

  region      long      lat group order subregion Murder Assault UrbanPop Rape
 alabama -87.46201 30.38968     1     1      <NA>   13.2     236       58 21.2
 alabama -87.48493 30.37249     1     2      <NA>   13.2     236       58 21.2
 alabama -87.52503 30.37249     1     3      <NA>   13.2     236       58 21.2
 alabama -87.53076 30.33239     1     4      <NA>   13.2     236       58 21.2
 alabama -87.57087 30.32665     1     5      <NA>   13.2     236       58 21.2
 alabama -87.58806 30.32665     1     6      <NA>   13.2     236       58 21.2
```

Once the data is in the correct format, it can be plotted (Figure 13-35), mapping one of the columns with data values to `fill`:

```
ggplot(crime_map, aes(x=long, y=lat, group=group, fill=Assault)) +
    geom_polygon(colour="black") +
    coord_map("polyconic")
```

Discussion

The preceding example used the default color scale, which goes from dark to light blue. If you want to show how the values diverge from some middle value, you can use `scale_fill_gradient2()`, as shown in Figure 13-36:

```
ggplot(crimes, aes(map_id = state, fill=Assault)) +
    geom_map(map = states_map, colour="black") +
    scale_fill_gradient2(low="#559999", mid="grey90", high="#BB650B",
                         midpoint=median(crimes$Assault)) +
    expand_limits(x = states_map$long, y = states_map$lat) +
    coord_map("polyconic")
```

The previous example mapped continuous values to `fill`, but we could just as well use discrete values. It's sometimes easier to interpret the data if the values are discretized.

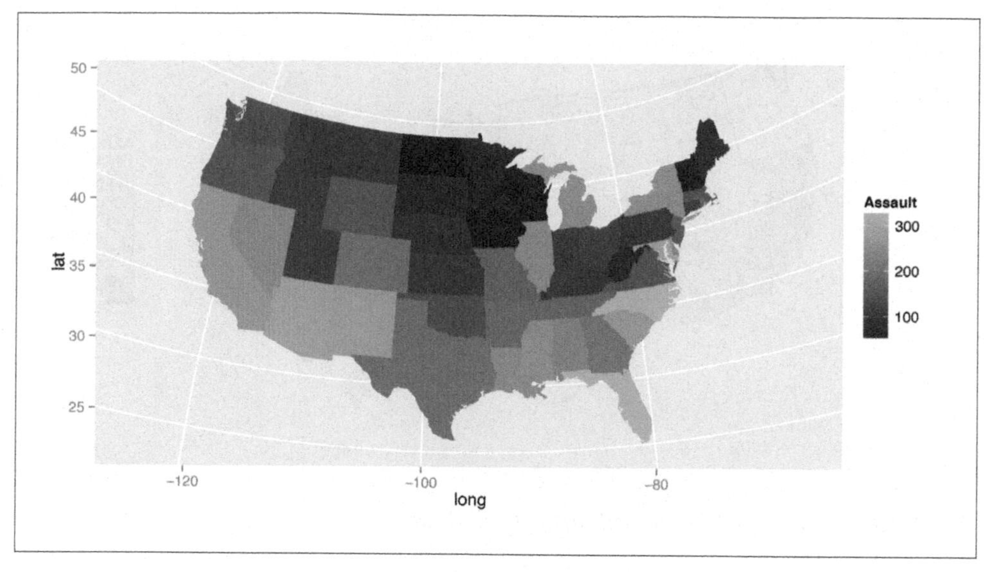

Figure 13-35. A map with a variable mapped to fill

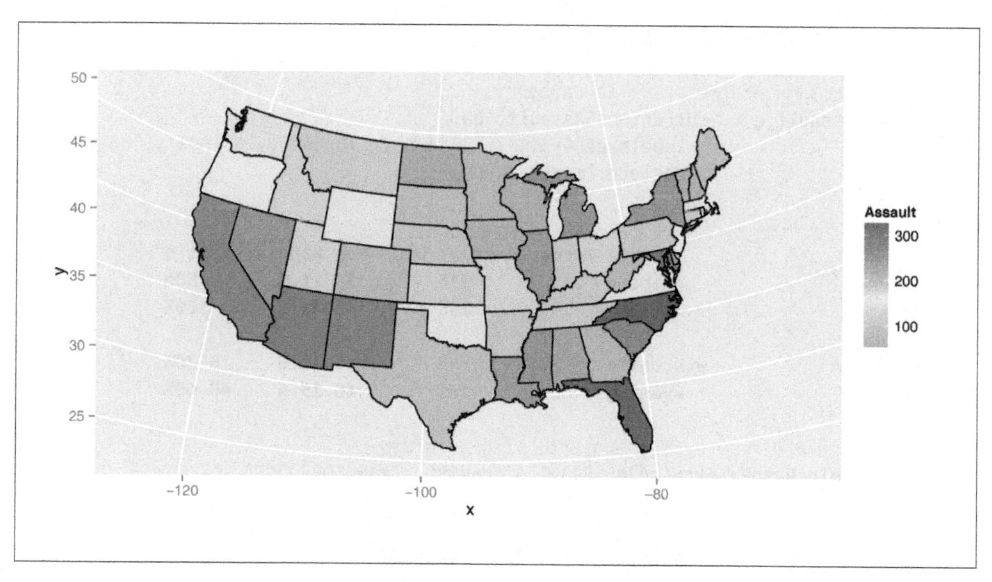

Figure 13-36. With a diverging color scale

For example, we can categorize the values into quantiles and show those quantiles, as in Figure 13-37:

```
# Find the quantile bounds
qa <- quantile(crimes$Assault, c(0, 0.2, 0.4, 0.6, 0.8, 1.0))
```

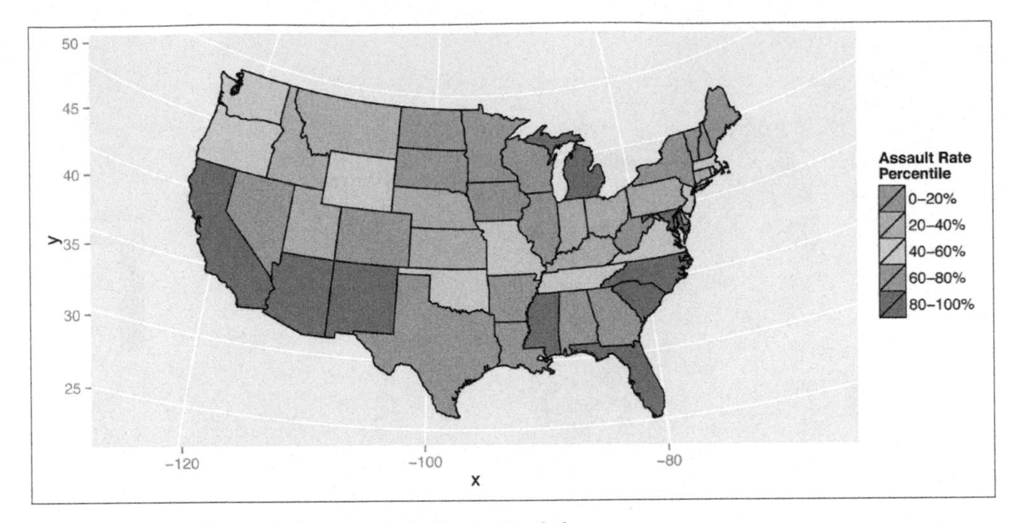

Figure 13-37. Choropleth map with discretized data

```
qa

   0%   20%   40%   60%   80%  100%
 45.0  98.8 135.0 188.8 254.2 337.0

# Add a column of the quantile category
crimes$Assault_q <- cut(crimes$Assault, qa,
                        labels=c("0-20%", "20-40%", "40-60%", "60-80%", "80-100%"),
                        include.lowest=TRUE)
crimes

                      state Murder Assault UrbanPop Rape Assault_q
Alabama             alabama   13.2     236       58 21.2    60-80%
Alaska               alaska   10.0     263       48 44.5   80-100%
 ...
Wisconsin         wisconsin    2.6      53       66 10.8     0-20%
Wyoming             wyoming    6.8     161       60 15.6    40-60%

# Generate a discrete color palette with 5 values
pal <- colorRampPalette(c("#559999", "grey80", "#BB650B"))(5)
pal

 "#559999" "#90B2B2" "#CCCCCC" "#C3986B" "#BB650B"

ggplot(crimes, aes(map_id = state, fill=Assault_q)) +
    geom_map(map = states_map, colour="black") +
    scale_fill_manual(values=pal) +
    expand_limits(x = states_map$long, y = states_map$lat) +
    coord_map("polyconic") +
    labs(fill="Assault Rate\nPercentile")
```

Another way to make a choropleth, but without needing to merge the map data with the value data, is to use `geom_map()`. As of this writing, this will render maps faster than the method just described.

For this method, the map data frame must have columns named `lat`, `long`, and `region`. In the value data frame, there must be a column that is matched to the `region` column in the map data frame, and this column is specified by mapping it to the `map_id` aesthetic. For example, this code will have the same output as the first example (Figure 13-35):

```
# The 'state' column in the crimes data is to be matched to the 'region' column
# in the states_map data
ggplot(crimes, aes(map_id = state, fill=Assault)) +
    geom_map(map = states_map) +
    expand_limits(x = states_map$long, y = states_map$lat) +
    coord_map("polyconic")
```

Notice that we also needed to use `expand_limits()`. This is because unlike most geoms, `geom_map()` doesn't automatically set the `x` and `y` limits; the use of `expand_limits()` makes it include those `x` and `y` values. (Another way to accomplish the same result is to use `ylim()` and `xlim()`.)

See Also

For an example of data overlaid on a map, see Recipe 13.12.

For more on using continuous colors, see Recipe 12.6.

13.19. Making a Map with a Clean Background

Problem

You want to remove background elements from a map.

Solution

First, save the following theme:

```
# Create a theme with many of the background elements removed
theme_clean <- function(base_size = 12) {
require(grid) # Needed for unit() function
  theme_grey(base_size) %+replace%
  theme(
    axis.title        = element_blank(),
    axis.text         = element_blank(),
    panel.background  = element_blank(),
    panel.grid        = element_blank(),
    axis.ticks.length = unit(0, "cm"),
```

```
        axis.ticks.margin = unit(0, "cm"),
        panel.margin      = unit(0, "lines"),
        plot.margin       = unit(c(0, 0, 0, 0), "lines"),
        complete = TRUE
      )
    }
```

Then add it to the map (Figure 13-38). In this example, we'll add it to one of the choropleths we created in Recipe 13.18:

```
ggplot(crimes, aes(map_id = state, fill=Assault_q)) +
    geom_map(map = states_map, colour="black") +
    scale_fill_manual(values=pal) +
    expand_limits(x = states_map$long, y = states_map$lat) +
    coord_map("polyconic") +
    labs(fill="Assault Rate\nPercentile") +
    theme_clean()
```

Figure 13-38. A map with a clean background

There's a bug in R versions 2.15.2 and earlier, which may throw an error that looks like this:

```
Error in grid.Call.graphics(L_setviewport, pvp, TRUE) :
  Non-finite location and/or size for viewport
```

This happens because some dimensions add up to having zero length, and the grid graphics engine has trouble handling this. This bug was fixed as of R 3.0. If you're using a version of R where this happens, you can work around it by changing the theme to use `axis.ticks.margin = unit(0.01, "cm")` instead of `axis.ticks.margin = unit(0, "cm")`.

Discussion

In some maps, it's important to include contextual information such as the latitude and longitude. In others, this information is unimportant and distracts from the information that's being conveyed. In Figure 13-38, it's unlikely that viewers will care about the latitude and longitude of the states. They can probably identify the states by shape and relative position, and even if they can't, having the latitude and longitude isn't really helpful.

13.20. Creating a Map from a Shapefile

Problem

You want to create a geographical map from an Esri shapefile.

Solution

Load the shapefile using `readShapePoly()` from the maptools package, convert it to a data frame with `fortify()`, then plot it (Figure 13-39):

```
library(maptools)

# Load the shapefile and convert to a data frame
taiwan_shp <- readShapePoly("TWN_adm/TWN_adm2.shp")
taiwan_map <- fortify(taiwan_shp)

ggplot(taiwan_map, aes(x = long, y = lat, group=group)) + geom_path()
```

Discussion

Esri shapefiles are a common format for map data. The `readShapePoly()` function reads a shape file and returns a `SpatialPolygonsDataFrame` object:

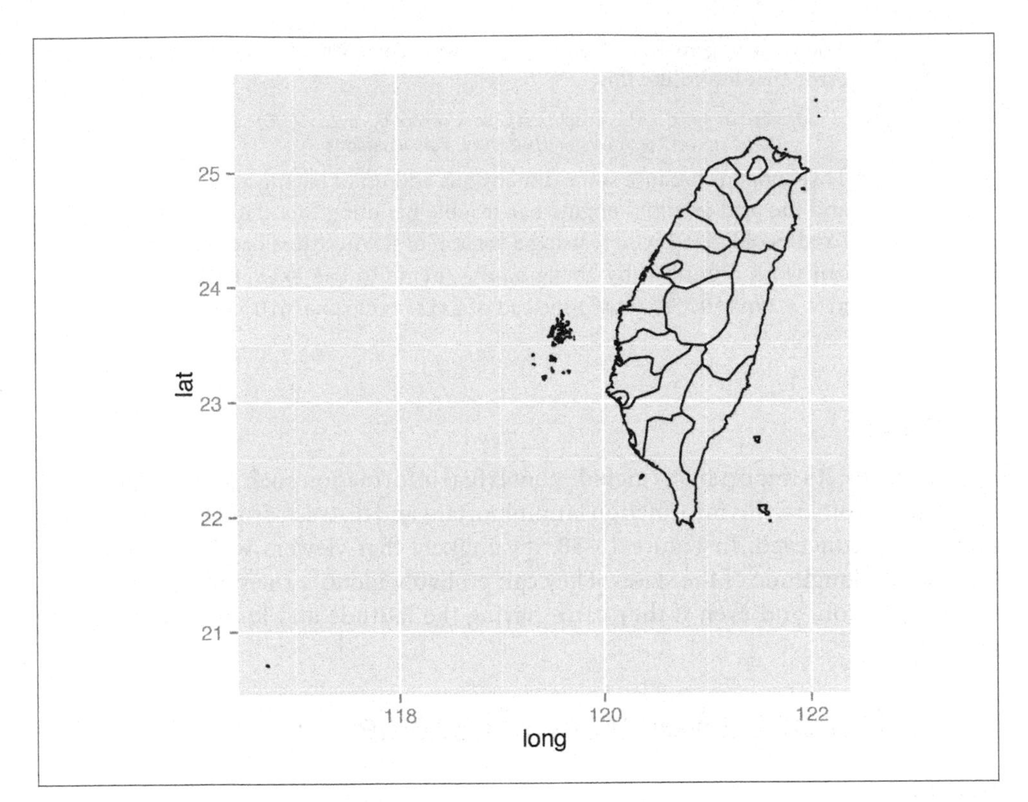

Figure 13-39. A map created from a shapefile

```
taiwan_shp <- readShapePoly("TWN_adm/TWN_adm2.shp")

# Look at the structure of the object
str(taiwan_shp)

Formal class 'SpatialPolygonsDataFrame' [package "sp"] with 5 slots
  ..@ data       :'data.frame': 22 obs. of  11 variables:
  .. ..$ ID_0     : int [1:22] 223 223 223 223 223 223 223 223 223 223 ...
  .. ..$ ISO      : Factor w/ 1 level "TWN": 1 1 1 1 1 1 1 1 1 1 ...
  .. ..$ NAME_0   : Factor w/ 1 level "Taiwan": 1 1 1 1 1 1 1 1 1 1 ...
  .. ..$ ID_1     : int [1:22] 1 2 3 4 4 4 4 4 4 4 ...
 ... [lots more stuff]
  ..@ proj4string:Formal class 'CRS' [package "sp"] with 1 slots
  .. .. ..@ projargs: chr NA
```

Converting it to a regular data frame gives the following:

```
taiwan_map <- fortify(taiwan_shp)
taiwan_map

     long      lat order  hole piece group id
```

```
120.2390 22.75155      1 FALSE     1   0.1  0
120.2701 22.74135      2 FALSE     1   0.1  0
120.2996 22.70920      3 FALSE     1   0.1  0
...
120.1340 23.61569   1236 FALSE     3  21.3 21
120.1340 23.61597   1237 FALSE     3  21.3 21
120.1365 23.61597   1238 FALSE     3  21.3 21
```

It's actually possible to pass the `SpatialPolygonsDataFrame` object directly to `ggplot()`, which will automatically `fortify()` it:

```
# Send the SpatialPolygonsDataFrame directly to ggplot()
ggplot(taiwan_shp, aes(x=long, y=lat, group=group)) + geom_path()
```

Even though this code is a bit simpler, you may still want to convert it yourself using `fortify()`. This will let you more easily inspect the data structure that is sent to `ggplot()`, or merge the data frame with another data set.

See Also

The shapefile used in this example is not included in the gcookbook package. It and many other shapefiles are available for download (*http://www.gadm.org*).

CHAPTER 14

Output for Presentation

Broadly speaking, visualizations of data serve two purposes: discovery and communication. In the discovery phase, you'll create exploratory graphics, and when you do this, it's important to be able try out different things quickly. In the communication phase, you'll present your graphics to others. When you do that, you'll need to tweak the appearance of the graphics (which I've written about in previous chapters), and you'll usually need to put them somewhere other than on your computer screen. This chapter is about that last part: *saving* your graphics so that they can be presented in documents.

14.1. Outputting to PDF Vector Files

Problem

You want to create a PDF of your plot.

Solution

There are two ways to output to PDF files. One method is to open the PDF graphics device with `pdf()`, make the plots, then close the device with `dev.off()`. This method works for most graphics in R, including base graphics and grid-based graphics like those created by ggplot2 and lattice:

```
# width and height are in inches
pdf("myplot.pdf", width=4, height=4)

# Make plots
plot(mtcars$wt, mtcars$mpg)
print(ggplot(mtcars, aes(x=wt, y=mpg)) + geom_point())

dev.off()
```

If you make more than one plot, each one will go on a separate page in the PDF output. Notice that we called `print()` on the `ggplot` object to make sure that it will be output even when this code is in a script.

The `width` and `height` are in inches, so to specify the dimensions in centimeters, you must do the conversion manually:

```
# 8x8 cm
pdf("myplot.pdf", width=8/2.54, height=8/2.54)
```

If you are creating plots from a script and it throws an error while creating one, R might not reach the call to `dev.off()`, and could be left in a state where the PDF device is still open. When this happens, the PDF file won't open properly until you manually call `dev.off()`.

If you are creating a graph with ggplot2, using `ggsave()` can be a little simpler. It simply saves the last plot created with `ggplot()`:

```
ggplot(mtcars, aes(x=wt, y=mpg)) + geom_point()

# Default is inches, but you can specify unit
ggsave("myplot.pdf", width=8, height=8, units="cm")
```

With `ggsave()`, you don't need to print the `ggplot` object, and if there is an error while creating or saving the plot, there's no need to manually close the graphic device. `ggsave()` can't be used to make multipage plots, though.

Discussion

PDF files are usually the best option when your goal is to output to printed documents. They work easily with LaTeX and can be used in presentations with Apple's Keynote, but Microsoft programs may have trouble importing them. (See Recipe 14.3 for details on creating vector images that can be imported into Microsoft programs.)

PDF files are also generally smaller than bitmap files such as portable network graphics (PNG) files, because they contain a set of instructions, such as "Draw a line from here to there," instead of information about the color of each pixel. However, there are cases where bitmap files are smaller. For example, if you have a scatter plot that is heavily overplotted, a PDF file can end up much larger than a PNG—even though most of the points are obscured, the PDF file will still contain instructions for drawing each and every point, whereas a bitmap file will not contain the redundant information. See Recipe 5.5 for an example.

See Also

If you want to manually edit the PDF or SVG file, see Recipe 14.4.

14.2. Outputting to SVG Vector Files

Problem

You want to create a scalable vector graphics (SVG) image of your plot.

Solution

SVG files can be created and used in much the same way as PDF files:

```
svg("myplot.svg", width=4, height=4)
plot(...)
dev.off()

# With ggsave()
ggsave("myplot.svg", width=8, height=8, units="cm")
```

Discussion

When it comes to importing images, some programs may handle SVG files better than PDFs, and vice versa. For example, web browsers tend to have better SVG support, while document-creation programs like LaTeX tend to have better PDF support.

14.3. Outputting to WMF Vector Files

Problem

You want to create a Windows metafile (WMF) image of your plot.

Solution

WMF files can be created and used in much the same way as PDF files—but they can only be created on Windows:

```
win.metafile("myplot.wmf", width=4, height=4)
plot(...)
dev.off()

# With ggsave()
ggsave("myplot.wmf", width=8, height=8, units="cm")
```

Discussion

Windows programs such as Microsoft Word and PowerPoint have poor support for importing PDF files, but they natively support WMF. One drawback is that WMF files do not support transparency (alpha).

14.4. Editing a Vector Output File

Problem

You want to open a vector output file for final editing.

Solution

Sometimes you need to make final tweaks to the appearance of a graph for presentation. You can open PDF and SVG files with the excellent free program Inkscape, or with the commercial program Adobe Illustrator.

Discussion

Font support can be a problem when you open a PDF file with Inkscape. Normally, point objects drawn with the PDF device will be written as symbols from the Zapf Dingbats font. This can be problematic if you want to open the file in an editor like Illustrator or Inkscape; for example, points may appear as the letter *q*, as in Figure 14-1, because that is the corresponding letter for a solid bullet in Zapf Dingbats.

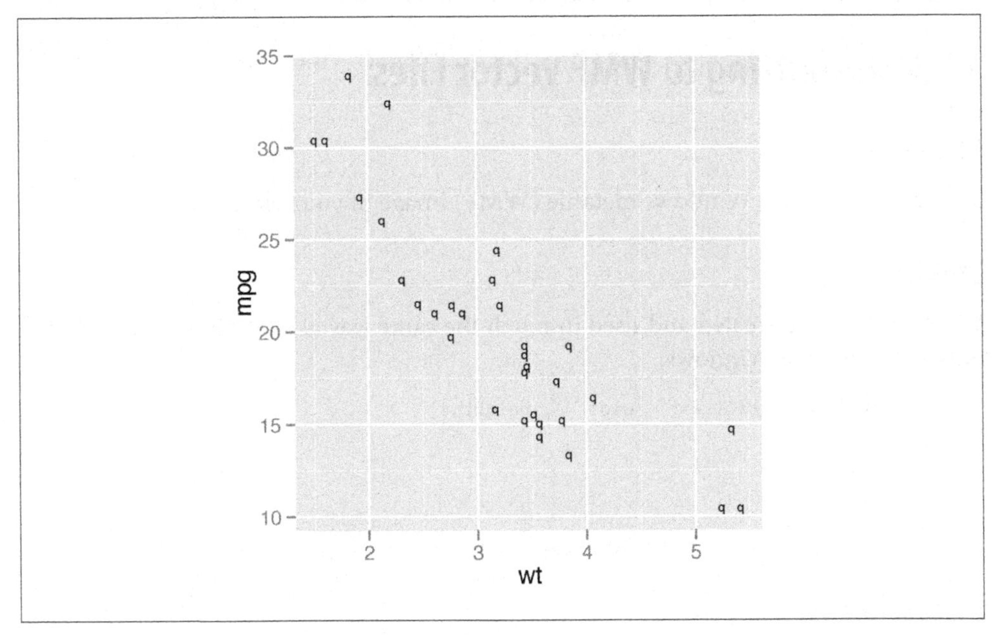

Figure 14-1. Bad conversion of point symbols after opening in Inkscape—also notice that the spacing of the fonts is slightly off

To avoid this problem, set `useDingbats=FALSE`. This will make the circles be drawn as circles instead of as font characters:

```
pdf("myplot.pdf", width=4, height=4, useDingbats=FALSE)

# or
ggsave("myplot.pdf", width=4, height=4, useDingbats=FALSE)
```

Inkscape might have some issues with fonts as well. You may have noticed that the fonts in Figure 14-1 don't look quite right. This is because Inkscape (version 0.48) couldn't find Helvetica, and substituted the font Bitstream Vera Sans instead. A workaround is to copy the Helvetica font file to your personal font library. For example, on Mac OS X, run `cp /System/Library/Fonts/Helvetica.dfont ~/Library/Fonts/` from a Terminal window to do this, then, when it says there is a font conflict, click "Ignore Conflict." After this, Inkscape should properly display the Helvetica font.

14.5. Outputting to Bitmap (PNG/TIFF) Files

Problem

You want to create a bitmap of your plot, writing to a PNG file.

Solution

There are two ways to output to PNG bitmap files. One method is to open the PNG graphics device with `png()`, make the plots, then close the device with `dev.off()`. This method works for most graphics in R, including base graphics and grid-based graphics like those created by ggplot2 and lattice:

```
# width and height are in pixels
png("myplot.png", width=400, height=400)

# Make plot
plot(mtcars$wt, mtcars$mpg)

dev.off()
```

For outputting multiple plots, put `%d` in the filename. This will be replaced with 1, 2, 3, and so on, for each subsequent plot:

```
# width and height are in pixels
png("myplot-%d.png", width=400, height=400)

plot(mtcars$wt, mtcars$mpg)
print(ggplot(mtcars, aes(x=wt, y=mpg)) + geom_point())
```

```
dev.off()
```

Notice that we called `print()` on the `ggplot` object to make sure that it will be output even when this code is in a script.

The `width` and `height` are in pixels, and the default is to output at 72 pixels per inch (ppi). This resolution is suitable for displaying on a screen, but will look pixelated and jagged in print.

For high-quality print output, use at least 300 ppi. Figure 14-2 shows portions of the same plot at different resolutions. In this example, we'll use 300 ppi and create a 4×4-inch PNG file:

```
ppi <- 300
# Calculate the height and width (in pixels) for a 4x4-inch image at 300 ppi
png("myplot.png", width=4*ppi, height=4*ppi, res=ppi)
plot(mtcars$wt, mtcars$mpg)
dev.off()
```

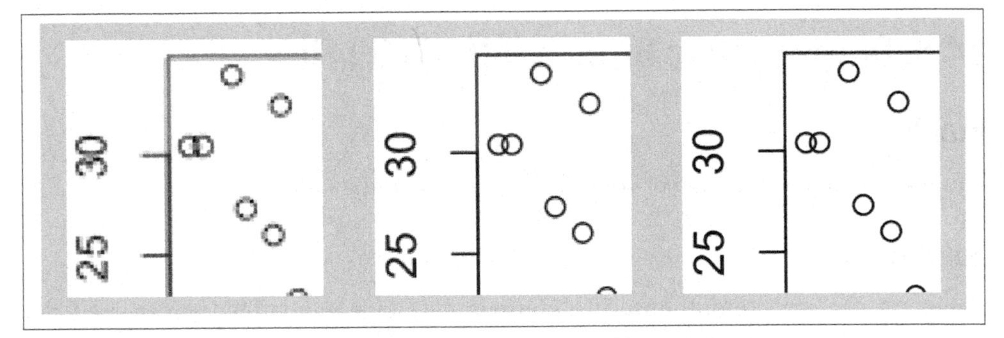

Figure 14-2. From left to right: PNG output at 72, 150, and 300 ppi (actual size)

If you are creating plots from a script and it throws an error while creating one, R might not reach the call to `dev.off()`, and could be left in a state where the PNG device is still open. When this happens, the PNG file won't open properly in a viewing program until you manually call `dev.off()`.

If you are creating a graph with ggplot2, using `ggsave()` can be a little simpler. It simply saves the last plot created with `ggplot()`. You specify the width and height in inches, not pixels, and tell it how many pixels per inch to use:

```
ggplot(mtcars, aes(x=wt, y=mpg)) + geom_point()

# Default dimensions are in inches, but you can specify the unit
ggsave("myplot.png", width=8, height=8, unit="cm", dpi=300)
```

With `ggsave()`, you don't need to print the `ggplot` object, and if there is an error while creating or saving the plot there's no need to manually close the graphic device.

Although the argument name is `dpi`, it really controls the *pixels* per inch (ppi), not the *dots* per inch. When a grey pixel is rendered in print, it is output with many smaller dots of black ink—and so print output has more dots per inch than pixels per inch.

Discussion

R supports other bitmap formats, like BMP, TIFF, and JPEG, but there's really not much reason to use them instead of PNG.

The exact appearance of the resulting bitmaps varies from platform to platform. Unlike R's PDF output device, which renders consistently across platforms, the bitmap output devices may render the same plot differently on Windows, Linux, and Mac OS X. There can even be variation within each of these operating systems.

Different platforms will render fonts differently, some platforms will antialias (smooth) lines while others will not, and some platforms support alpha (transparency) while others do not. If your platform lacks support for features like antialiasing and alpha, you can use the `CairoPNG()` device, from the Cairo package:

```
install.packages("Cairo")  # One-time installation
CairoPNG("myplot.png")
plot(...)
dev.off()
```

While `CairoPNG()` does not guarantee identical rendering across platforms (fonts may not be exactly the same), it does support features like antialiasing and alpha.

Changing the resolution affects the size (in pixels) of graphical objects like text, lines, and points. For example, a 6-by-6-inch image at 75 ppi has the same pixel dimensions as a 3-by-3-inch image at 150 ppi, but the appearance will be different, as shown in Figure 14-3. Both of these images are 450×450 pixels. When displayed on a computer screen, they may display at approximately the same size, as they do here.

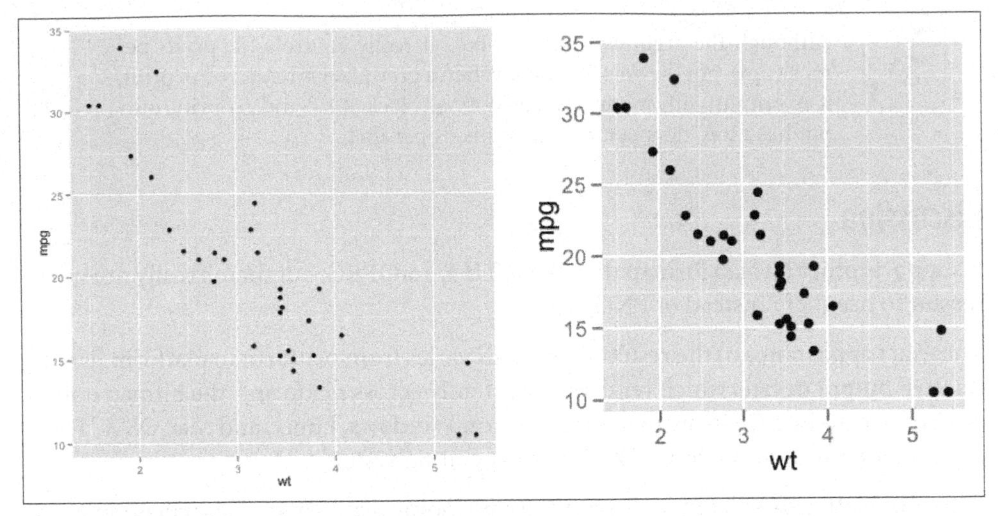

Figure 14-3. Left: 6×6 inch image at 75 ppi; right: 3×3 inch image at 150 ppi

14.6. Using Fonts in PDF Files

Problem

You want to use fonts other than the basic ones provided by R in a PDF file.

Solution

The extrafont package can be used to create PDF files with different fonts.

There are a number of steps involved, beginning with some one-time setup. Download and install Ghostscript (*http://www.ghostscript.com/download*), then run the following in R:

```
install.packages("extrafont")
library(extrafont)

# Find and save information about fonts installed on your system
font_import()

# List the fonts
fonts()
```

After the one-time setup is done, there are tasks you need to do in each R session:

```
library(extrafont)
# Register the fonts with R
loadfonts()

# On Windows, you may need to tell it where Ghostscript is installed
```

```
# (adjust the path to match your installation of Ghostscript)
Sys.setenv(R_GSCMD = "C:/Program Files/gs/gs9.05/bin/gswin32c.exe")
```

Finally, you can create a PDF file and embed fonts into it, as in Figure 14-4:

```
library(ggplot2)
ggplot(mtcars, aes(x=wt, y=mpg)) + geom_point() +
    ggtitle("Title text goes here") +
    theme(text = element_text(size = 16, family="Impact"))

ggsave("myplot.pdf", width=4, height=4)

embed_fonts("myplot.pdf")
```

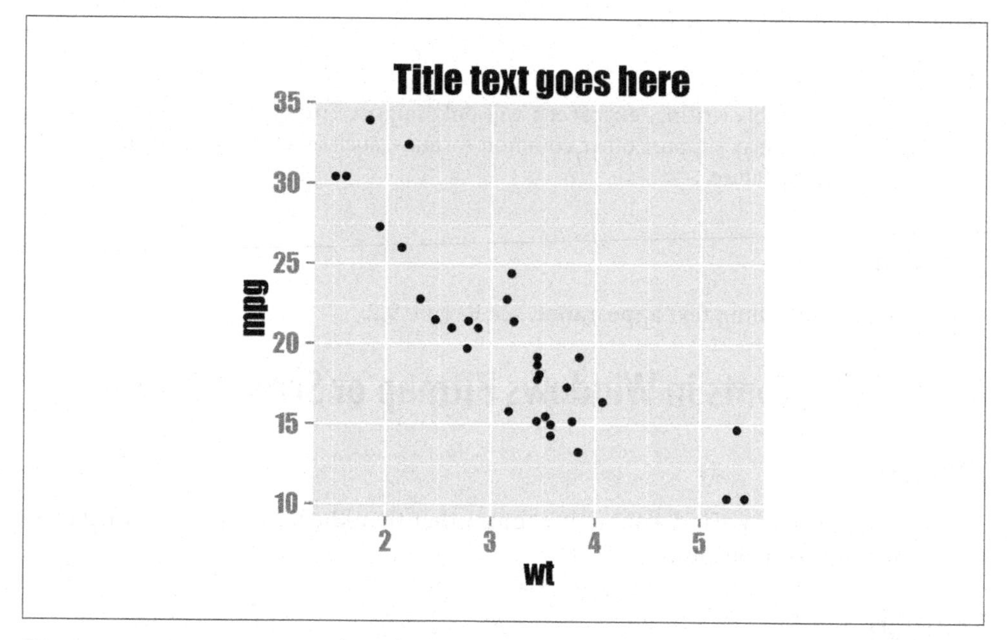

Figure 14-4. PDF output with embedded font Impact

Discussion

Fonts can be difficult to work with in R. Some output devices, such as the on-screen `quartz` device on Mac OS X, can display any font installed on the computer. Other output devices, such as the default `png` device on Windows, aren't able to display system fonts.

On top of this, PDF files have their own quirks when it comes to fonts. The PDF specification has 14 "core" fonts. These are fonts that every PDF renderer has, and they include standards such as Times, Helvetica, and Courier. If you create a PDF with these fonts, any PDF renderer should display it properly.

If you want to use a font that is *not* one of these core fonts, though, there's no guarantee that the PDF renderer on a given device will have that font, so you can't be sure that the font will display properly on another computer or printer. To solve this problem, non-core fonts can be *embedded* into the PDF; in other words, the PDF file can itself contain a copy of the font you want to use.

If you are putting multiple PDF figures in a PDF document, you may want to embed the fonts in the finished document instead of in each figure. This will make the final document smaller, since it will only have the font embedded once, instead of once for each figure.

Embedding fonts with R can be a tricky process, but the `extrafont` package handles many of the ugly details for you.

As of this writing, `extrafont` will only import TrueType (*.ttf*) fonts, but it may support other common formats, such as OpenType (*.otf*), in the future.

See Also

For more on controlling text appearance, see Recipe 9.2.

14.7. Using Fonts in Windows Bitmap or Screen Output

Problem

You are using Windows and want to use fonts other than the basic ones provided by R for bitmap or screen output.

Solution

The `extrafont` package can be used to create bitmap or screen output. The procedure is similar to using `extrafont` with PDF files (Recipe 14.6). The one-time setup is almost the same, except that Ghostscript is not required:

```
install.packages("extrafont")
library(extrafont)

# Find and save information about fonts installed on your system
font_import()

# List the fonts
fonts()
```

After the one-time setup is done, there are tasks you need to do in each R session:

```
library(extrafont)
# Register the fonts for Windows
loadfonts("win")
```

Finally, you can create each output file or display graphs on screen, as in Figure 14-5:

```
library(ggplot2)
ggplot(mtcars, aes(x=wt, y=mpg)) + geom_point() +
    ggtitle("Title text goes here") +
    theme(text = element_text(size = 16, family="Georgia", face="italic"))

ggsave("myplot.png", width=4, height=4, dpi=300)
```

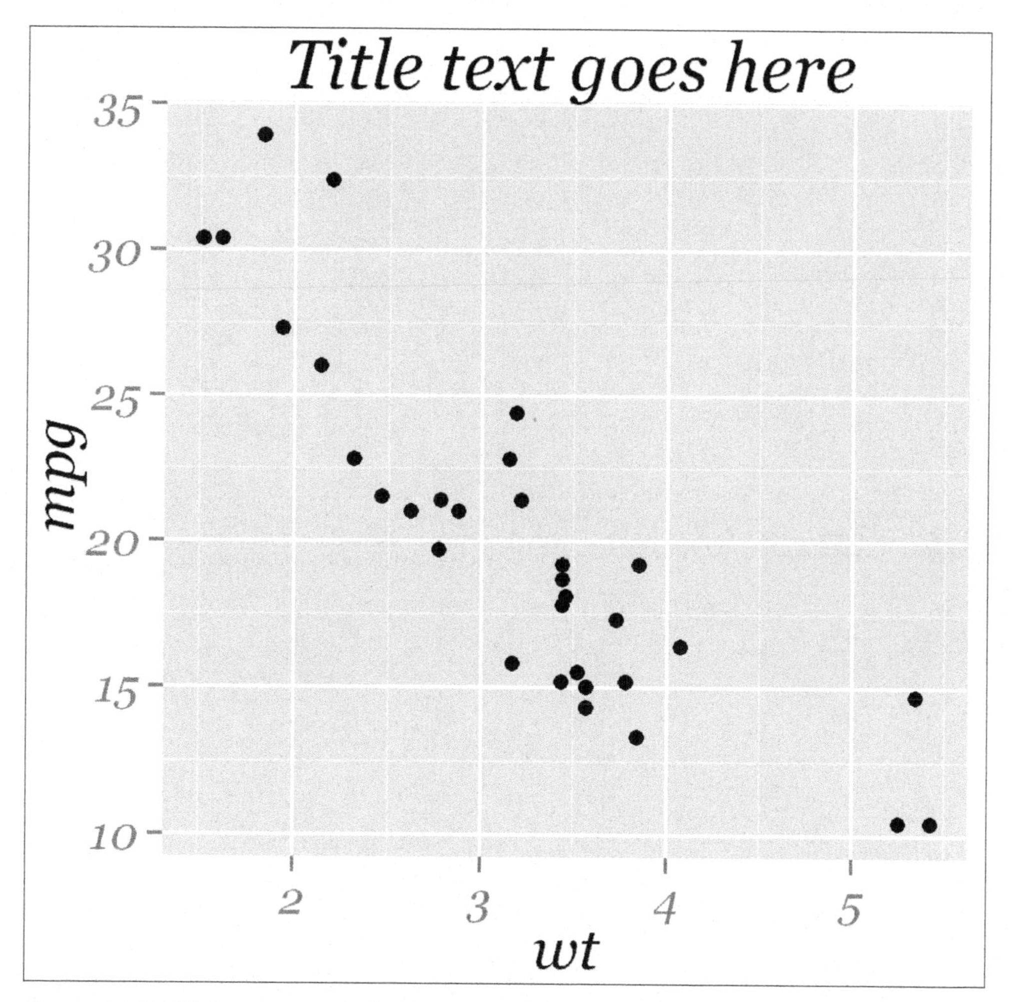

Figure 14-5. PNG output with font Georgia Italic

Discussion

Fonts are handled in a completely different way for bitmaps than they are for PDF files.

On Windows, for bitmap output it is necessary to register each font manually with R (`extrafont` makes this much easier). On Mac OS X and Linux, the fonts should already be available for bitmap output; it isn't necessary to register them manually.

CHAPTER 15

Getting Your Data into Shape

When it comes to making graphs, half the battle occurs before you call any graphing commands. Before you pass your data to the graphing functions, it must first be read in and given the correct structure. The data sets provided with R are ready to use, but when dealing with real-world data, this usually isn't the case: you'll have to clean up and restructure the data before you can visualize it.

Data sets in R are most often stored in data frames. They're typically used as two-dimensional data structures, with each row representing one case and each column representing one variable. Data frames are essentially lists of vectors and factors, all of the same length, where each vector or factor represents one column.

Here's the `heightweight` data set:

```
library(gcookbook) # For the data set
heightweight

sex ageYear ageMonth heightIn weightLb
  f   11.92      143     56.3     85.0
  f   12.92      155     62.3    105.0
...
  m   13.92      167     62.0    107.5
  m   12.58      151     59.3     87.0
```

It consists of five columns, with each row representing one case: a set of information about a single person. We can get a clearer idea of how it's structured by using the `str()` function:

```
str(heightweight)

'data.frame':   236 obs. of  5 variables:
 $ sex     : Factor w/ 2 levels "f","m": 1 1 1 1 1 1 1 1 1 1 ...
 $ ageYear : num  11.9 12.9 12.8 13.4 15.9 ...
 $ ageMonth: int  143 155 153 161 191 171 185 142 160 140 ...
```

```
 $ heightIn: num  56.3 62.3 63.3 59 62.5 62.5 59 56.5 62 53.8 ...
 $ weightLb: num  85 105 108 92 112 ...
```

The first column, sex, is a factor with two levels, "f" and "m", and the other four columns are vectors of numbers (one of them, ageMonth, is specifically a vector of integers, but for the purposes here, it behaves the same as any other numeric vector).

Factors and character vectors behave similarly in ggplot2—the main difference is that with character vectors, items will be displayed in lexicographical order, but with factors, items will be displayed in the same order as the factor levels, which you can control.

15.1. Creating a Data Frame

Problem

You want to create a data frame from vectors.

Solution

You can put vectors together in a data frame with `data.frame()`:

```
# Two starting vectors
g <- c("A", "B", "C")
x <- 1:3

dat <- data.frame(g, x)
dat

 g x
 A 1
 B 2
 C 3
```

Discussion

A data frame is essentially a list of vectors and factors. Each vector or factor can be thought of as a column in the data frame.

If your vectors are in a list, you can convert the list to a data frame with the `as.data.frame()` function:

```
lst <- list(group = g, value = x)    # A list of vectors

dat <- as.data.frame(lst)
```

15.2. Getting Information About a Data Structure

Problem

You want to find out information about an object or data structure.

Solution

Use the `str()` function:

```
str(ToothGrowth)

'data.frame':    60 obs. of  3 variables:
 $ len : num  4.2 11.5 7.3 5.8 6.4 10 11.2 11.2 5.2 7 ...
 $ supp: Factor w/ 2 levels "OJ","VC": 2 2 2 2 2 2 2 2 2 2 ...
 $ dose: num  0.5 0.5 0.5 0.5 0.5 0.5 0.5 0.5 0.5 0.5 ...
```

This tells us that `ToothGrowth` is a data frame with three columns, `len`, `supp`, and `dose`. `len` and `dose` contain numeric values, while `supp` is a factor with two levels.

Discussion

The `str()` function is very useful for finding out more about data structures. One common source of problems is a data frame where one of the columns is a character vector instead of a factor, or vice versa. This can cause puzzling issues with analyses or graphs.

When you print out a data frame the normal way, by just typing the name at the prompt and pressing Enter, factor and character columns appear exactly the same. The difference will be revealed only when you run `str()` on the data frame, or print out the column by itself:

```
tg <- ToothGrowth
tg$supp <- as.character(tg$supp)

str(tg)

'data.frame':    60 obs. of  3 variables:
 $ len : num  4.2 11.5 7.3 5.8 6.4 10 11.2 11.2 5.2 7 ...
 $ supp: chr  "VC" "VC" "VC" "VC" ...
 $ dose: num  0.5 0.5 0.5 0.5 0.5 0.5 0.5 0.5 0.5 0.5 ...

# Print out the columns by themselves

# From old data frame (factor)
ToothGrowth$supp

 [1] VC VC VC VC VC VC VC VC VC VC VC VC VC VC VC VC VC VC VC VC VC VC VC VC VC
[26] VC VC VC VC VC OJ OJ OJ OJ OJ OJ OJ OJ OJ OJ OJ OJ OJ OJ OJ OJ OJ OJ OJ OJ
[51] OJ OJ OJ OJ OJ OJ OJ OJ OJ OJ
```

```
Levels: OJ VC

# From new data frame (character)
tg$supp

 [1] "VC" "VC" "VC" "VC" "VC" "VC" "VC" "VC" "VC" "VC" "VC" "VC" "VC" "VC" "VC"
[16] "VC" "VC" "VC" "VC" "VC" "VC" "VC" "VC" "VC" "VC" "VC" "VC" "VC" "VC" "VC"
[31] "OJ" "OJ" "OJ" "OJ" "OJ" "OJ" "OJ" "OJ" "OJ" "OJ" "OJ" "OJ" "OJ" "OJ" "OJ"
[46] "OJ" "OJ" "OJ" "OJ" "OJ" "OJ" "OJ" "OJ" "OJ" "OJ" "OJ" "OJ" "OJ" "OJ" "OJ"
```

15.3. Adding a Column to a Data Frame

Problem

You want to add a column to a data frame.

Solution

Just assign some value to the new column.

If you assign a single value to the new column, the entire column will be filled with that value. This adds a column named `newcol`, filled with `NA`:

```
data$newcol <- NA
```

You can also assign a vector to the new column:

```
data$newcol <- vec
```

If the length of the vector is less than the number of rows in the data frame, then the vector is repeated to fill all the rows.

Discussion

Each "column" of a data frame is a vector or factor. R handles them slightly differently from standalone vectors, because all the columns in a data frame have the same length.

15.4. Deleting a Column from a Data Frame

Problem

You want to delete a column from a data frame.

Solution

Assign `NULL` to that column:

```
data$badcol <- NULL
```

Discussion

You can also use the subset() function and put a - (minus sign) in front of the column(s) to drop:

```
# Return data without badcol
data <- subset(data, select = -badcol)

# Exclude badcol and othercol
data <- subset(data, select = c(-badcol, -othercol))
```

See Also

Recipe 15.7 for more on getting a subset of a data frame.

15.5. Renaming Columns in a Data Frame

Problem

You want to rename the columns in a data frame.

Solution

Use the names(dat) <- function:

```
names(dat) <- c("name1", "name2", "name3")
```

Discussion

If you want to rename the columns by name:

```
library(gcookbook) # For the data set
names(anthoming)   # Print the names of the columns

 "angle" "expt"  "ctrl"

names(anthoming)[names(anthoming) == "ctrl"] <- c("Control")
names(anthoming)[names(anthoming) == "expt"] <- c("Experimental")
names(anthoming)

 "angle"        "Experimental" "Control"
```

They can also be renamed by numeric position:

```
names(anthoming)[1] <- "Angle"
names(anthoming)

 "Angle"        "Experimental" "Control"
```

15.6. Reordering Columns in a Data Frame

Problem

You want to change the order of columns in a data frame.

Solution

To reorder columns by their numeric position:

```
dat <- dat[c(1,3,2)]
```

To reorder by column name:

```
dat <- dat[c("col1", "col3", "col2")]
```

Discussion

The previous examples use list-style indexing. A data frame is essentially a list of vectors, and indexing into it as a list will return another data frame. You can get the same effect with matrix-style indexing:

```
library(gcookbook) # For the data set
anthoming

 angle expt ctrl
   -20    1    0
   -10    7    3
     0    2    3
    10    0    3
    20    0    1

anthoming[c(1,3,2)]     # List-style indexing

 angle ctrl expt
   -20    0    1
   -10    3    7
     0    3    2
    10    3    0
    20    1    0

# Putting nothing before the comma means to select all rows
anthoming[, c(1,3,2)]   # Matrix-style indexing

 angle ctrl expt
   -20    0    1
   -10    3    7
     0    3    2
    10    3    0
    20    1    0
```

In this case, both methods return the same result, a data frame. However, when retrieving a single column, list-style indexing will return a data frame, while matrix-style indexing will return a vector, unless you use `drop=FALSE`:

```
anthoming[3]      # List-style indexing

 ctrl
    0
    3
    3
    3
    1

anthoming[, 3]   # Matrix-style indexing

 0 3 3 3 1

anthoming[, 3, drop=FALSE]   # Matrix-style indexing with drop=FALSE

 ctrl
    0
    3
    3
    3
    1
```

15.7. Getting a Subset of a Data Frame

Problem

You want to get a subset of a data frame.

Solution

Use the `subset()` function. It can be used to pull out rows that satisfy a set of conditions and to select particular columns.

We'll use the `climate` data set for the examples here:

```
library(gcookbook) # For the data set
climate

   Source Year Anomaly1y Anomaly5y Anomaly10y Unc10y
 Berkeley 1800        NA        NA     -0.435  0.505
 Berkeley 1801        NA        NA     -0.453  0.493
 Berkeley 1802        NA        NA     -0.460  0.486
  ...
  CRUTEM3 2009    0.7343        NA         NA     NA
  CRUTEM3 2010    0.8023        NA         NA     NA
  CRUTEM3 2011    0.6193        NA         NA     NA
```

The following will pull out only rows where Source is "Berkeley" and only the columns named Year and Anomaly10y:

```
subset(climate, Source == "Berkeley", select = c(Year, Anomaly10y))

Year Anomaly10y
1800     -0.435
1801     -0.453
1802     -0.460
...
2002      0.856
2003      0.869
2004      0.884
```

Discussion

It is possible to use multiple selection criteria, by using the | (OR) and & (AND) operators. For example, this will pull out only those rows where source is "Berkeley", between the years 1900 and 2000:

```
subset(climate, Source == "Berkeley"  &  Year >= 1900  &  Year <= 2000,
       select = c(Year, Anomaly10y))

Year Anomaly10y
1900     -0.171
1901     -0.162
1902     -0.177
...
1998      0.680
1999      0.734
2000      0.748
```

You can also get a subset of data by indexing into the data frame with square brackets, although this approach is somewhat less elegant. The following code has the same effect as the code we just saw. The part before the comma picks out the rows, and the part after the comma picks out the columns:

```
climate[climate$Source=="Berkeley" & climate$Year >= 1900 & climate$Year <= 2000,
        c("Year", "Anomaly10y")]
```

If you grab just a single column this way, it will be returned as a vector instead of a data frame. To prevent this, use drop=FALSE, as in:

```
climate[climate$Source=="Berkeley" & climate$Year >= 1900 & climate$Year <= 2000,
        c("Year", "Anomaly10y"), drop=FALSE]
```

Finally, it's also possible to pick out rows and columns by their numeric position. This gets the second and fifth columns of the first 100 rows:

```
climate[1:100, c(2, 5)]
```

I generally recommend indexing using names rather than numbers when possible. It makes the code easier to understand when you're collaborating with others or when you come back to it months or years after writing it, and it makes the code less likely to break when there are changes to the data, such as when columns are added or removed.

15.8. Changing the Order of Factor Levels

Problem

You want to change the order of levels in a factor.

Solution

The level order can be specified explicitly by passing the factor to `factor()` and specifying `levels`. In this example, we'll create a factor that initially has the wrong ordering:

```
# By default, levels are ordered alphabetically
sizes <- factor(c("small", "large", "large", "small", "medium"))
sizes

small  large  large  small  medium
Levels: large medium small

# Change the order of levels
sizes <- factor(sizes, levels = c("small", "medium", "large"))
sizes

small  large  large  small  medium
Levels: small medium large
```

The order can also be specified with `levels` when the factor is first created.

Discussion

There are two kinds of factors in R: ordered factors and regular factors. In both types, the levels are arranged in *some* order; the difference is that the order is meaningful for an ordered factor, but it is arbitrary for a regular factor—it simply reflects how the data is stored. For graphing data, the distinction between ordered and regular factors is generally unimportant, and they can be treated the same.

The order of factor levels affects graphical output. When a factor variable is mapped to an aesthetic property in ggplot2, the aesthetic adopts the ordering of the factor levels. If a factor is mapped to the x-axis, the ticks on the axis will be in the order of the factor levels, and if a factor is mapped to color, the items in the legend will be in the order of the factor levels.

To reverse the level order, you can use `rev(levels())`:

```
factor(sizes, levels = rev(levels(sizes)))

small  large  large  small  medium
Levels: small medium large
```

See Also

To reorder a factor based on the value of another variable, see Recipe 15.9.

Reordering factor levels is useful for controlling the order of axes and legends. See Recipes 8.4 and 10.3 for more information.

15.9. Changing the Order of Factor Levels Based on Data Values

Problem

You want to change the order of levels in a factor based on values in the data.

Solution

Use `reorder()` with the factor that has levels to reorder, the values to base the reordering on, and a function that aggregates the values:

```
# Make a copy since we'll modify it
iss <- InsectSprays
iss$spray

 [1] A A A A A A A A A A A A B B B B B B B B B B B B C C C C C C C C C C C C D D
[39] D D D D D D D D D D E E E E E E E E E E E E F F F F F F F F F F F F
Levels: A B C D E F

iss$spray <- reorder(iss$spray, iss$count, FUN=mean)
iss$spray

 [1] A A A A A A A A A A A A B B B B B B B B B B B B C C C C C C C C C C C C D D
[39] D D D D D D D D D D E E E E E E E E E E E E F F F F F F F F F F F F
attr(,"scores")
        A         B         C         D         E         F
14.500000 15.333333  2.083333  4.916667  3.500000 16.666667
Levels: C E D A B F
```

Notice that the original levels were ABCDEF, while the reordered levels are CEDABF. The new order is determined by splitting `iss$count` into pieces according to the values in `iss$spray`, and then taking the mean of each group.

Discussion

The usefulness of `reorder()` might not be obvious from just looking at the raw output. Figure 15-1 shows three graphs made with `reorder()`. In these graphs, the order in which the items appear is determined by their values.

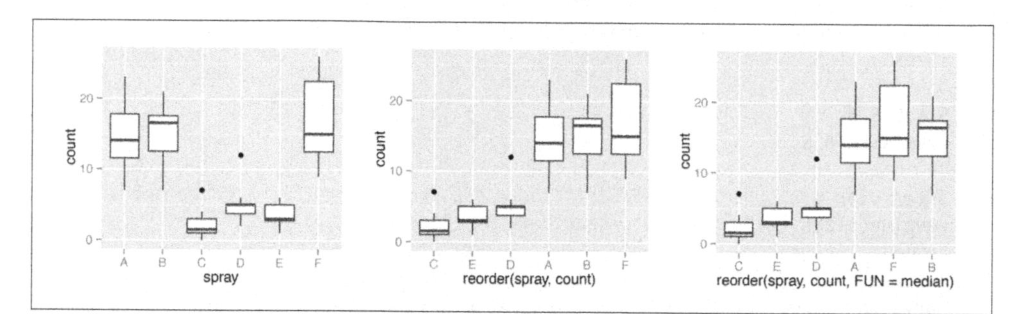

Figure 15-1. Left: original data; middle: reordered by the mean of each group; right: reordered by the median of each group

In the middle graph in Figure 15-1, the boxes are sorted by the mean. The horizontal line that runs across each box represents the *median* of the data. Notice that these values do not increase strictly from left to right. That's because with this particular data set, sorting by the mean gives a different order than sorting by the median. To make the median lines increase from left to right, as in the graph on the right in Figure 15-1, we used the `median()` function in `reorder()`.

See Also

Reordering factor levels is also useful for controlling the order of axes and legends. See Recipes 8.4 and 10.3 for more information.

15.10. Changing the Names of Factor Levels

Problem

You want to change the names of levels in a factor.

Solution

Use `revalue()` or `mapvalues()` from the plyr package:

```
sizes <- factor(c( "small", "large", "large", "small", "medium"))
sizes

small  large  large  small  medium
```

```
Levels: large medium small

levels(sizes)

"large"  "medium" "small"

# With revalue(), pass it a named vector with the mappings
sizes1 <- revalue(sizes, c(small="S", medium="M", large="L"))
sizes1

S L L S M
Levels: L M S

# Can also use quotes -- useful if there are spaces or other strange characters
revalue(sizes, c("small"="S", "medium"="M", "large"="L"))

# mapvalues() lets you use two separate vectors instead of a named vector
mapvalues(sizes, c("small", "medium", "large"), c("S", "M", "L"))
```

Discussion

The `revalue()` and `mapvalues()` functions are convenient, but for a more traditional (and clunky) R method for renaming factor levels, use the `levels()<-` function:

```
sizes <- factor(c( "small", "large", "large", "small", "medium"))

# Index into the levels and rename each one
levels(sizes)[levels(sizes)=="large"]  <- "L"
levels(sizes)[levels(sizes)=="medium"] <- "M"
levels(sizes)[levels(sizes)=="small"]  <- "S"
sizes

S L L S M
Levels: L M S
```

If you are renaming *all* your factor levels, there is a simpler method. You can pass a list to `levels()<-`:

```
sizes <- factor(c("small", "large", "large", "small", "medium"))
levels(sizes) <- list(S="small", M="medium", L="large")
sizes

S L L S M
Levels: L M S
```

With this method, all factor levels must be specified in the list; if any are missing, they will be replaced with `NA`.

It's also possible to rename factor levels by position, but this is somewhat inelegant:

```
# By default, levels are ordered alphabetically
sizes <- factor(c("small", "large", "large", "small", "medium"))
```

```
small  large  large  small  medium
Levels: large medium small

levels(sizes)[1] <- "L"
sizes

small  L      L      small  medium
Levels: L medium small

# Rename all levels at once
levels(sizes) <- c("L", "M", "S")
sizes

[1] S L L S M
Levels: L M S
```

It's safer to rename factor levels by name rather than by position, since you will be less likely to make a mistake (and mistakes here may be hard to detect). Also, if your input data set changes to have more (or fewer) levels, the numeric positions of the existing levels could change, which could cause serious but nonobvious problems for your analysis.

See Also

If, instead of a factor, you have a character vector with items to rename, see Recipe 15.12.

15.11. Removing Unused Levels from a Factor

Problem

You want to remove unused levels from a factor.

Solution

Sometimes, after processing your data you will have a factor that contains levels that are no longer used. Here's an example:

```
sizes <- factor(c("small", "large", "large", "small", "medium"))
sizes <- sizes[1:3]
sizes

small  large  large
Levels: large medium small
```

To remove them, use `droplevels()`:

```
sizes <- droplevels(sizes)
sizes
```

```
small  large  large
Levels: large small
```

Discussion

The `droplevels()` function preserves the order of factor levels.

You can use the `except` argument to keep particular levels.

15.12. Changing the Names of Items in a Character Vector

Problem

You want to change the names of items in a character vector.

Solution

Use `revalue()` or `mapvalues()` from the plyr package:

```
sizes <- c("small", "large", "large", "small", "medium")
sizes

 "small"  "large"  "large"  "small"  "medium"

# With revalue(), pass it a named vector with the mappings
sizes1 <- revalue(sizes, c(small="S", medium="M", large="L"))
sizes1

 "S" "L" "L" "S" "M"

# Can also use quotes -- useful if there are spaces or other strange characters
revalue(sizes, c("small"="S", "medium"="M", "large"="L"))

# mapvalues() lets you use two separate vectors instead of a named vector
mapvalues(sizes, c("small", "medium", "large"), c("S", "M", "L"))
```

Discussion

A more traditional R method is to use square-bracket indexing to select the items and rename them:

```
sizes <- c("small", "large", "large", "small", "medium")
sizes

 "small"  "large"  "large"  "small"  "medium"

sizes[sizes=="small"]  <- "S"
sizes[sizes=="medium"] <- "M"
sizes[sizes=="large"]  <- "L"
```

```
sizes

  "S" "L" "L" "S" "M"
```

See Also

If, instead of a character vector, you have a factor with levels to rename, see Recipe 15.10.

15.13. Recoding a Categorical Variable to Another Categorical Variable

Problem

You want to recode a categorical variable to another variable.

Solution

For the examples here, we'll use a subset of the `PlantGrowth` data set:

```
# Work on a subset of the PlantGrowth data set
pg <- PlantGrowth[c(1,2,11,21,22), ]
pg

weight group
  4.17  ctrl
  5.58  ctrl
  4.81  trt1
  6.31  trt2
  5.12  trt2
```

In this example, we'll recode the categorical variable `group` into another categorical variable, `treatment`. If the old value was `"ctrl"`, the new value will be `"No"`, and if the old value was `"trt1"` or `"trt2"`, the new value will be `"Yes"`.

This can be done with the `match()` function:

```
pg <- PlantGrowth

oldvals <- c("ctrl", "trt1", "trt2")
newvals <- factor(c("No",   "Yes",  "Yes"))

pg$treatment <- newvals[ match(pg$group, oldvals) ]
```

It can also be done (more awkwardly) by indexing in the vectors:

```
pg$treatment[pg$group == "ctrl"] <- "no"
pg$treatment[pg$group == "trt1"] <- "yes"
pg$treatment[pg$group == "trt2"] <- "yes"

# Convert to a factor
```

```
pg$treatment <- factor(pg$treatment)
pg

weight group treatment
  4.17  ctrl        no
  5.58  ctrl        no
  4.81  trt1       yes
  6.31  trt2       yes
  5.12  trt2       yes
```

Here, we combined two of the factor levels and put the result into a new column. If you simply want to rename the levels of a factor, see Recipe 15.10.

Discussion

The coding criteria can also be based on values in multiple columns, by using the & and | operators:

```
pg$newcol[pg$group == "ctrl"  &  pg$weight <  5] <- "no_small"
pg$newcol[pg$group == "ctrl"  &  pg$weight >= 5] <- "no_large"
pg$newcol[pg$group == "trt1"] <- "yes"
pg$newcol[pg$group == "trt2"] <- "yes"

pg$newcol <- factor(pg$newcol)
pg

weight group weightcat treatment   newcol
  4.17  ctrl     small        no no_small
  5.58  ctrl     large        no no_large
  4.81  trt1     small       yes      yes
  4.17  trt1     small       yes      yes
  6.31  trt2     large       yes      yes
  5.12  trt2     large       yes      yes
```

It's also possible to combine two columns into one using the `interaction()` function, which appends the values with a `"."` in between. This combines the `weightcat` and `treatment` columns into a new column, `weighttrt`:

```
pg$weighttrt <- interaction(pg$weightcat, pg$treatment)
pg

weight group weightcat treatment   newcol weighttrt
  4.17  ctrl     small        no no_small  small.no
  5.58  ctrl     large        no no_large  large.no
  4.81  trt1     small       yes      yes small.yes
  4.17  trt1     small       yes      yes small.yes
  6.31  trt2     large       yes      yes large.yes
  5.12  trt2     large       yes      yes large.yes
```

See Also

For more on renaming factor levels, see Recipe 15.10.

See Recipe 15.14 for recoding continuous values to categorical values.

15.14. Recoding a Continuous Variable to a Categorical Variable

Problem

You want to recode a continuous variable to another variable.

Solution

For the examples here, we'll use a subset of the `PlantGrowth` data set.

```
# Work on a subset of the PlantGrowth data set
pg <- PlantGrowth[c(1,2,11,21,22), ]
pg

weight group
  4.17  ctrl
  5.58  ctrl
  4.81  trt1
  6.31  trt2
  5.12  trt2
```

In this example, we'll recode the continuous variable `weight` into a categorical variable, `wtclass`, using the `cut()` function:

```
pg$wtclass <- cut(pg$weight, breaks = c(0, 5, 6, Inf))
pg

 weight group wtclass
   4.17  ctrl   (0,5]
   5.58  ctrl   (5,6]
   4.81  trt1   (0,5]
   4.17  trt1   (0,5]
   6.31  trt2 (6,Inf]
   5.12  trt2   (5,6]
```

Discussion

For three categories we specify four bounds, which can include `Inf` and `-Inf`. If a data value falls outside of the specified bounds, it's categorized as `NA`. The result of `cut()` is a factor, and you can see from the example that the factor levels are named after the bounds.

To change the names of the levels, set the `labels`:

```
pg$wtclass <- cut(pg$weight, breaks = c(0, 5, 6, Inf),
                  labels = c("small", "medium", "large"))
pg
```

```
weight group wtclass
  4.17  ctrl   small
  5.58  ctrl  medium
  4.81  trt1   small
  4.17  trt1   small
  6.31  trt2   large
  5.12  trt2  medium
```

As indicated by the factor levels, the bounds are by default *open* on the left and *closed* on the right. In other words, they don't include the lowest value, but they do include the highest value. For the smallest category, you can have it include both the lower and upper values by setting `include.lowest=TRUE`. In this example, this would result in 0 values going into the `small` category; otherwise, 0 would be coded as `NA`.

If you want the categories to be closed on the left and open on the right, set `right = FALSE`:

```
cut(pg$weight, breaks = c(0, 5, 6, Inf), right = FALSE)
```

See Also

To recode a categorical variable to another categorical variable, see Recipe 15.13.

15.15. Transforming Variables

Problem

You want to transform a variable in a data frame.

Solution

Reference the new column with the `$` operator, and assign some values to it. For this example, we'll use a copy of the `heightweight` data set:

```
library(gcookbook) # For the data set
# Make a copy of the data
hw <- heightweight
hw
```

```
sex ageYear ageMonth heightIn weightLb
  f   11.92      143     56.3     85.0
  f   12.92      155     62.3    105.0
 ...
```

```
  m   13.92        167      62.0    107.5
  m   12.58        151      59.3     87.0
```

This will convert `heightIn` to centimeters and store it in a new column, `heightCm`:

```
hw$heightCm <- hw$heightIn * 2.54
hw

sex ageYear ageMonth heightIn weightLb heightCm
  f   11.92      143     56.3     85.0  143.002
  f   12.92      155     62.3    105.0  158.242
 ...
  m   13.92      167     62.0    107.5  157.480
  m   12.58      151     59.3     87.0  150.622
```

Discussion

For slightly easier-to-read code, you can use `transform()` or `mutate()` from the plyr package. You only need to specify the data frame once, as the first argument to the function, meaning these provide a cleaner syntax, especially if you are transforming multiple variables:

```
hw <- transform(hw, heightCm = heightIn * 2.54, weightKg = weightLb / 2.204)
library(plyr)
hw <- mutate(hw, heightCm = heightIn * 2.54, weightKg = weightLb / 2.204)
hw

sex ageYear ageMonth heightIn weightLb heightCm weightKg
  f   11.92      143     56.3     85.0  143.002 38.56624
  f   12.92      155     62.3    105.0  158.242 47.64065
 ...
  m   13.92      167     62.0    107.5  157.480 48.77495
  m   12.58      151     59.3     87.0  150.622 39.47368
```

It is also possible to calculate a new variable based on multiple variables:

```
# These all have the same effect:
hw <- transform(hw, bmi = weightKg / (heightCm / 100)^2)
hw <- mutate(hw, bmi = weightKg / (heightCm / 100)^2)
hw$bmi <- hw$weightKg / (hw$heightCm/100)^2
hw

sex ageYear ageMonth heightIn weightLb heightCm weightKg      bmi
  f   11.92      143     56.3     85.0  143.002 38.56624 18.85919
  f   12.92      155     62.3    105.0  158.242 47.64065 19.02542
 ...
  m   13.92      167     62.0    107.5  157.480 48.77495 19.66736
  m   12.58      151     59.3     87.0  150.622 39.47368 17.39926
```

The main functional difference between `transform()` and `mutate()` is that `transform()` calculates the new columns simultaneously, while `mutate()` calculates the new columns sequentially, allowing you to base one new column on another new column.

Since `bmi` is calculated from `heightCm` and `weightKg`, it is not possible to calculate all of them in a single call to `transform()`; `heightCm` and `weightKg` must be calculated first, and then `bmi`, as shown here.

With `mutate()`, however, we can calculate them all in one go. The following code has the same effect as the previous separate blocks:

```
hw <- heightweight
hw <- mutate(hw,
    heightCm = heightIn * 2.54,
    weightKg = weightLb / 2.204,
    bmi = weightKg / (heightCm / 100)^2)
```

See Also

See Recipe 15.16 for how to perform group-wise transformations on data.

15.16. Transforming Variables by Group

Problem

You want to transform variables by performing operations on groups of data, as specified by a grouping variable.

Solution

Use `ddply()` from the plyr package with the `transform()` function, and specify the operations:

```
library(MASS) # For the data set
library(plyr)
cb <- ddply(cabbages, "Cult", transform, DevWt = HeadWt - mean(HeadWt))
```

```
 Cult Date HeadWt VitC       DevWt
  c39  d16    2.5   51 -0.40666667
  c39  d16    2.2   55 -0.70666667
 ...
  c52  d21    1.5   66 -0.78000000
  c52  d21    1.6   72 -0.68000000
```

Discussion

Let's take a closer look at the `cabbages` data set. It has two grouping variables (factors): `Cult`, which has levels `c39` and `c52`, and `Date`, which has levels `d16`, `d20`, and `d21`. It also has two measured numeric variables, `HeadWt` and `VitC`:

```
cabbages
```

```
 Cult Date HeadWt VitC
  c39  d16    2.5   51
  c39  d16    2.2   55
...
  c52  d21    1.5   66
  c52  d21    1.6   72
```

Suppose we want to find, for each case, the deviation of `HeadWt` from the overall mean. All we have to do is take the overall mean and subtract it from the observed value for each case:

```
transform(cabbages, DevWt = HeadWt - mean(HeadWt))

 Cult Date HeadWt VitC        DevWt
  c39  d16    2.5   51 -0.093333333
  c39  d16    2.2   55 -0.393333333
...
  c52  d21    1.5   66 -1.093333333
  c52  d21    1.6   72 -0.993333333
```

You'll often want to do separate operations like this for each group, where the groups are specified by one or more grouping variables. Suppose, for example, we want to normalize the data within each group by finding the deviation of each case from the mean *within the group*, where the groups are specified by `Cult`. In these cases, we can use `ddply()` from the plyr package with the `transform()` function:

```
library(plyr)
cb <- ddply(cabbages, "Cult", transform, DevWt = HeadWt - mean(HeadWt))
cb

 Cult Date HeadWt VitC       DevWt
  c39  d16    2.5   51 -0.40666667
  c39  d16    2.2   55 -0.70666667
...
  c52  d21    1.5   66 -0.78000000
  c52  d21    1.6   72 -0.68000000
```

First it splits `cabbages` into separate data frames based on the value of `Cult`. There are two levels of `Cult`, `c39` and `c52`, so there are two data frames. It then applies the `transform()` function, with the remaining arguments, to each data frame.

Notice that the call to `ddply()` has all the same parts as the previous call to `transform()`. The only differences are that the parts are slightly rearranged and it adds the splitting variable, in this case, `Cult`.

The before and after results are shown in Figure 15-2:

```
# The data before normalizing
ggplot(cb, aes(x=Cult, y=HeadWt)) + geom_boxplot()

# After normalizing
ggplot(cb, aes(x=Cult, y=DevWt)) + geom_boxplot()
```

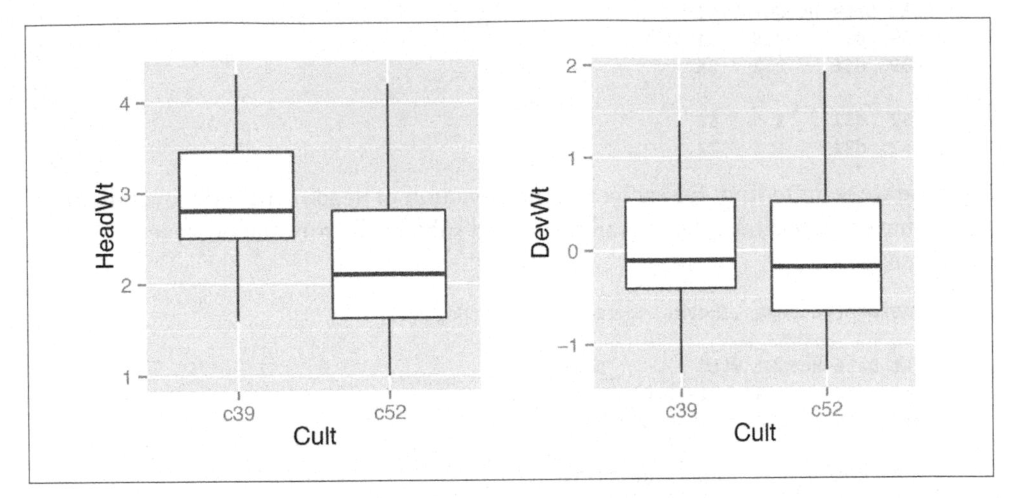

Figure 15-2. Left: before normalizing; right: after normalizing

You can also split the data frame on multiple variables and perform operations on multiple variables. This will split by `Cult` and `Date`, forming a group for each distinct combination of the two variables, and then it will calculate the deviation from the mean of `HeadWt` and `VitC` within each group:

```
ddply(cabbages, c("Cult", "Date"), transform,
      DevWt = HeadWt - mean(HeadWt), DevVitC = VitC - mean(VitC))
```

```
 Cult Date HeadWt VitC DevWt DevVitC
  c39  d16    2.5   51 -0.68     0.7
  c39  d16    2.2   55 -0.98     4.7
 ...
  c52  d21    1.5   66  0.03    -5.8
  c52  d21    1.6   72  0.13     0.2
```

See Also

To summarize data by groups, see Recipe 15.17.

15.17. Summarizing Data by Groups

Problem

You want to summarize your data, based on one or more grouping variables.

Solution

Use `ddply()` from the plyr package with the `summarise()` function, and specify the operations to do:

```
library(MASS) # For the data set
library(plyr)

ddply(cabbages, c("Cult", "Date"), summarise, Weight = mean(HeadWt),
          VitC = mean(VitC))
```

```
 Cult Date Weight VitC
  c39  d16   3.18 50.3
  c39  d20   2.80 49.4
  c39  d21   2.74 54.8
  c52  d16   2.26 62.5
  c52  d20   3.11 58.9
  c52  d21   1.47 71.8
```

Discussion

Let's take a closer look at the `cabbages` data set. It has two factors that can be used as grouping variables: `Cult`, which has levels `c39` and `c52`, and `Date`, which has levels `d16`, `d20`, and `d21`. It also has two numeric variables, `HeadWt` and `VitC`:

```
cabbages
```

```
 Cult Date HeadWt VitC
  c39  d16    2.5   51
  c39  d16    2.2   55
...
  c52  d21    1.5   66
  c52  d21    1.6   72
```

Finding the overall mean of `HeadWt` is simple. We could just use the `mean()` function on that column, but for reasons that will soon become clear, we'll use the `summarise()` function instead:

```
library(plyr)
summarise(cabbages, Weight = mean(HeadWt))
```

```
   Weight
 2.593333
```

The result is a data frame with one row and one column, named `Weight`.

Often we want to find information about each subset of the data, as specified by a grouping variable. For example, suppose we want to find the mean of each `Cult` group. To do this, we can use `ddply()` with `summarise()`. Notice how the arguments get shifted around when we use them together:

```
library(plyr)
ddply(cabbages, "Cult", summarise, Weight = mean(HeadWt))

 Cult   Weight
 c39 2.906667
 c52 2.280000
```

The command first splits the data frame `cabbages` into separate data frames based on the value of `Cult`. There are two levels of `Cult`, `c39` and `c52`, so there are two data frames. It then applies the `summarise()` function to each of these data frames; it calculates `Weight` by taking the `mean()` of the `HeadWt` column in each of the data frames. The resulting summarized data frames each have one row, and `ddply()` puts them back together into one data frame, which is then returned.

Summarizing the data frame by splitting it up with more variables (or columns) is simple: just use a vector that names the additional variables. It's also possible to get more than one summary value by specifying more calculated columns. Here we'll summarize each `Cult` and `Date` group, getting the average of `HeadWt` and `VitC`:

```
ddply(cabbages, c("Cult", "Date"), summarise, Weight = mean(HeadWt),
      VitC = mean(VitC))

 Cult Date Weight VitC
  c39  d16   3.18 50.3
  c39  d20   2.80 49.4
  c39  d21   2.74 54.8
  c52  d16   2.26 62.5
  c52  d20   3.11 58.9
  c52  d21   1.47 71.8
```

It's possible to do more than take the mean. You may, for example, want to compute the standard deviation and count of each group. To get the standard deviation, use the `sd()` function, and to get a count, use the `length()` function:

```
ddply(cabbages, c("Cult", "Date"), summarise,
      Weight = mean(HeadWt),
      sd = sd(HeadWt),
      n = length(HeadWt))

Cult Date Weight        sd  n
  c39  d16   3.18 0.9566144 10
  c39  d20   2.80 0.2788867 10
  c39  d21   2.74 0.9834181 10
  c52  d16   2.26 0.4452215 10
  c52  d20   3.11 0.7908505 10
  c52  d21   1.47 0.2110819 10
```

Other useful functions for generating summary statistics include `min()`, `max()`, and `median()`.

Dealing with NAs

One potential pitfall is that NAs in the data will lead to NAs in the output. Let's see what happens if we sprinkle a few NAs into HeadWt:

```
c1 <- cabbages                    # Make a copy
c1$HeadWt[c(1,20,45)] <- NA       # Set some values to NA

ddply(c1, c("Cult", "Date"), summarise,
      Weight = mean(HeadWt),
      sd = sd(HeadWt),
      n = length(HeadWt))

Cult Date Weight        sd  n
  c39  d16     NA        NA 10
  c39  d20     NA        NA 10
  c39  d21   2.74 0.9834181 10
  c52  d16   2.26 0.4452215 10
  c52  d20     NA        NA 10
  c52  d21   1.47 0.2110819 10
```

There are two problems here. The first problem is that `mean()` and `sd()` simply return NA if any of the input values are NA. Fortunately, these functions have an option to deal with this very issue: setting `na.rm=TRUE` will tell them to ignore the NAs.

The second problem is that `length()` counts NAs just like any other value, but since these values represent missing data, they should be excluded from the count. The `length()` function doesn't have an `na.rm` flag, but we can get the same effect by using `sum(!is.na(...))`. The `is.na()` function returns a logical vector: it has a TRUE for each NA item, and a FALSE for all other items. It is inverted by the `!`, and then `sum()` adds up the number of TRUEs. The end result is a count of non-NAs:

```
ddply(c1, c("Cult", "Date"), summarise,
      Weight = mean(HeadWt, na.rm=TRUE),
      sd = sd(HeadWt, na.rm=TRUE),
      n = sum(!is.na(HeadWt)))

 Cult Date   Weight        sd  n
  c39  d16 3.255556 0.9824855  9
  c39  d20 2.722222 0.1394433  9
  c39  d21 2.740000 0.9834181 10
  c52  d16 2.260000 0.4452215 10
  c52  d20 3.044444 0.8094923  9
  c52  d21 1.470000 0.2110819 10
```

Missing combinations

If there are any empty combinations of the grouping variables, they will not appear in the summarized data frame. These missing combinations can cause problems when making graphs. To illustrate, we'll remove all entries that have levels c52 and d21. The graph on the left in Figure 15-3 shows what happens when there's a missing combination in a bar graph:

```
# Copy cabbages and remove all rows with both c52 and d21
c2 <- subset(c1, !( Cult=="c52" & Date=="d21" ) )

c2a <- ddply(c2, c("Cult", "Date"), summarise,
        Weight = mean(HeadWt, na.rm=TRUE),
        sd = sd(HeadWt, na.rm=TRUE),
        n = sum(!is.na(HeadWt)))
c2a

 Cult Date   Weight        sd  n
  c39  d16 3.255556 0.9824855  9
  c39  d20 2.722222 0.1394433  9
  c39  d21 2.740000 0.9834181 10
  c52  d16 2.260000 0.4452215 10
  c52  d20 3.044444 0.8094923  9

# Make the graph
ggplot(c2a, aes(x=Date, fill=Cult, y=Weight)) + geom_bar(position="dodge")
```

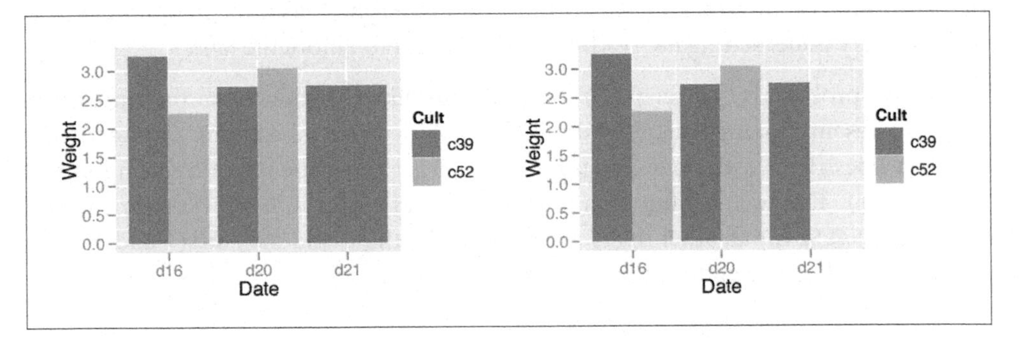

Figure 15-3. Left: bar graph with a missing combination; right: with missing combination filled

To fill in the missing combination (Figure 15-3, right), give `ddply()` the `.drop=FALSE` flag:

```
c2b <- ddply(c2, c("Cult", "Date"), .drop=FALSE, summarise,
        Weight = mean(HeadWt, na.rm=TRUE),
        sd = sd(HeadWt, na.rm=TRUE),
        n = sum(!is.na(HeadWt)))
c2b

 Cult Date   Weight        sd  n
  c39  d16 3.255556 0.9824855  9
  c39  d20 2.722222 0.1394433  9
  c39  d21 2.740000 0.9834181 10
  c52  d16 2.260000 0.4452215 10
```

```
 c52  d20 3.044444 0.8094923  9
 c52  d21      NaN        NA  0
```

```
# Make the graph
ggplot(c2b, aes(x=Date, fill=Cult, y=Weight)) + geom_bar(position="dodge")
```

See Also

If you want to calculate standard errors and confidence intervals, see Recipe 15.18.

See Recipe 6.8 for an example of using `stat_summary()` to calculate means and overlay them on a graph.

To perform transformations on data by groups, see Recipe 15.16.

15.18. Summarizing Data with Standard Errors and Confidence Intervals

Problem

You want to summarize your data with the standard error of the mean and/or confidence intervals.

Solution

Getting the standard error of the mean involves two steps: first get the standard deviation and count for each group, then use those values to calculate the standard error. The standard error for each group is just the standard deviation divided by the square root of the sample size:

```
library(MASS) # For the data set
library(plyr)

ca <- ddply(cabbages, c("Cult", "Date"), summarise,
            Weight = mean(HeadWt, na.rm=TRUE),
            sd = sd(HeadWt, na.rm=TRUE),
            n = sum(!is.na(HeadWt)),
            se = sd/sqrt(n))

ca
```

```
 Cult Date Weight        sd  n         se
  c39  d16   3.18 0.9566144 10 0.30250803
  c39  d20   2.80 0.2788867 10 0.08819171
  c39  d21   2.74 0.9834181 10 0.31098410
  c52  d16   2.26 0.4452215 10 0.14079141
  c52  d20   3.11 0.7908505 10 0.25008887
  c52  d21   1.47 0.2110819 10 0.06674995
```

In versions of plyr before 1.8, summarise() created all the new columns simultaneously, so you would have to create the se column separately, after creating the sd and n columns.

Discussion

Another method is to calculate the standard error in the call ddply. It's not possible to refer to the sd and n columns inside of the ddply call, so we'll have to recalculate them to get se. This will do the same thing as the two-step version shown previously:

```
ddply(cabbages, c("Cult", "Date"), summarise,
      Weight = mean(HeadWt, na.rm=TRUE),
      sd = sd(HeadWt, na.rm=TRUE),
      n = sum(!is.na(HeadWt)),
      se = sd / sqrt(n) )
```

Confidence Intervals

Confidence intervals are calculated using the standard error of the mean and the degrees of freedom. To calculate a confidence interval, use the qt() function to get the quantile, then multiply that by the standard error. The qt() function will give quantiles of the *t*-distribution when given a probability level and degrees of freedom. For a 95% confidence interval, use a probability level of .975; for the bell-shaped *t*-distribution, this will in essence cut off 2.5% of the area under the curve at either end. The degrees of freedom equal the sample size minus one.

This will calculate the multiplier for each group. There are six groups and each has the same number of observations (10), so they will all have the same multiplier:

```
ciMult <- qt(.975, ca$n-1)
ciMult

# 2.262157 2.262157 2.262157 2.262157 2.262157 2.262157
```

Now we can multiply that vector by the standard error to get the 95% confidence interval:

```
ca$ci <- ca$se * ciMult

 Cult Date Weight        sd  n         se        ci
  c39  d16   3.18 0.9566144 10 0.30250803 0.6843207
  c39  d20   2.80 0.2788867 10 0.08819171 0.1995035
  c39  d21   2.74 0.9834181 10 0.31098410 0.7034949
  c52  d16   2.26 0.4452215 10 0.14079141 0.3184923
  c52  d20   3.11 0.7908505 10 0.25008887 0.5657403
  c52  d21   1.47 0.2110819 10 0.06674995 0.1509989
```

We could have done this all in one line, like this:

```
ca$ci95 <- ca$se * qt(.975, ca$n)
```

For a 99% confidence interval, use `.995`.

Error bars that represent the standard error of the mean and confidence intervals serve the same general purpose: to give the viewer an idea of how good the estimate of the population mean is. The standard error is the standard deviation of the sampling distribution. Confidence intervals are easier to interpret. Very roughly, a 95% confidence interval means that there's a 95% chance that the true population mean is within the interval (actually, it doesn't mean this at all, but this seemingly simple topic is way too complicated to cover here; if you want to know more, read up on Bayesian statistics).

This function will perform all the steps of calculating the standard deviation, count, standard error, and confidence intervals. It can also handle `NA`s and missing combinations, with the `na.rm` and `.drop` options. By default, it provides a 95% confidence interval, but this can be set with the `conf.interval` argument:

```
summarySE <- function(data=NULL, measurevar, groupvars=NULL,
                      conf.interval=.95, na.rm=FALSE, .drop=TRUE) {
    require(plyr)

    # New version of length that can handle NAs: if na.rm==T, don't count them
    length2 <- function (x, na.rm=FALSE) {
        if (na.rm) sum(!is.na(x))
        else       length(x)
    }

    # This does the summary
    datac <- ddply(data, groupvars, .drop=.drop,
                   .fun = function(xx, col, na.rm) {
                           c( n    = length2(xx[,col], na.rm=na.rm),
                              mean = mean   (xx[,col], na.rm=na.rm),
                              sd   = sd     (xx[,col], na.rm=na.rm)
                              )
                          },
                   measurevar,
                   na.rm
            )

    # Rename the "mean" column
    datac <- rename(datac, c("mean" = measurevar))

    datac$se <- datac$sd / sqrt(datac$n)  # Calculate standard error of the mean

    # Confidence interval multiplier for standard error
    # Calculate t-statistic for confidence interval:
    # e.g., if conf.interval is .95, use .975 (above/below), and use
    #  df=n-1, or if n==0, use df=0
    ciMult <- qt(conf.interval/2 + .5, datac$n-1)
    datac$ci <- datac$se * ciMult
```

```
    return(datac)
}
```

The following usage example has a 99% confidence interval and handles NAs and missing combinations:

```
# Remove all rows with both c52 and d21
c2 <- subset(cabbages, !( Cult=="c52" & Date=="d21" ) )

# Set some values to NA
c2$HeadWt[c(1,20,45)] <- NA


summarySE(c2, "HeadWt", c("Cult", "Date"), conf.interval=.99,
          na.rm=TRUE, .drop=FALSE)

 Cult Date  n   HeadWt        sd         se        ci
  c39  d16  9 3.255556 0.9824855 0.32749517 1.0988731
  c39  d20  9 2.722222 0.1394433 0.04648111 0.1559621
  c39  d21 10 2.740000 0.9834181 0.31098410 1.0106472
  c52  d16 10 2.260000 0.4452215 0.14079141 0.4575489
  c52  d20  9 3.044444 0.8094923 0.26983077 0.9053867
  c52  d21  0      NaN        NA         NA        NA
Warning message:
In qt(p, df, lower.tail, log.p) : NaNs produced
```

It will give this warning message when there are missing combinations. This isn't a problem; it just indicates that it couldn't calculate a quantile for a group with no observations.

See Also

See Recipe 7.7 to use the values calculated here to add error bars to a graph.

15.19. Converting Data from Wide to Long

Problem

You want to convert a data frame from "wide" format to "long" format.

Solution

Use `melt()` from the `reshape2` package. In the `anthoming` data set, for each angle, there are two measurements: one column contains measurements in the experimental condition and the other contains measurements in the control condition:

```
library(gcookbook) # For the data set
anthoming

 angle expt ctrl
```

```
  -20    1    0
  -10    7    3
    0    2    3
   10    0    3
   20    0    1
```

We can reshape the data so that all the measurements are in one column. This will put the values from `expt` and `ctrl` into one column, and put the names into a different column:

```
library(reshape2)
melt(anthoming, id.vars="angle", variable.name="condition", value.name="count")

 angle condition count
   -20      expt     1
   -10      expt     7
     0      expt     2
    10      expt     0
    20      expt     0
   -20      ctrl     0
   -10      ctrl     3
     0      ctrl     3
    10      ctrl     3
    20      ctrl     1
```

This data frame represents the same information as the original one, but it is structured in a way that is more conducive to some analyses.

Discussion

In the source data, there are *ID* variables and *measure* variables. The ID variables are those that specify which values go together. In the source data, the first row holds measurements for when `angle` is –20. In the output data frame, the two measurements, for `expt` and `ctrl`, are no longer in the same row, but we can still tell that they belong together because they have the same value of `angle`.

The measure variables are by default all the non-ID variables. The names of these variables are put into a new column specified by `variable.name`, and the values are put into a new column specified by `value.name`.

If you don't want to use all the non-ID columns as measure variables, you can specify `measure.vars`. For example, in the `drunk` data set, we can use just the `0-29` and `30-39` groups:

```
drunk

    sex 0-29 30-39 40-49 50-59 60+
   male  185   207   260   180  71
 female    4    13    10     7  10
```

```
melt(drunk, id.vars="sex", measure.vars=c("0-29", "30-39"),
     variable.name="age", value.name="count")

    sex   age count
   male  0-29   185
 female  0-29     4
   male 30-39   207
 female 30-39    13
```

It's also possible to use more than one column as the ID variables:

```
plum_wide

 length       time dead alive
   long    at_once   84   156
   long in_spring  156    84
  short    at_once  133   107
  short in_spring  209    31

melt(plum_wide, id.vars=c("length","time"), variable.name="survival",
                              value.name="count")

 length       time survival count
   long    at_once     dead    84
   long in_spring     dead   156
  short    at_once     dead   133
  short in_spring     dead   209
   long    at_once    alive   156
   long in_spring    alive    84
  short    at_once    alive   107
  short in_spring    alive    31
```

Some data sets don't come with a column with an ID variable. For example, in the `corneas` data set, each row represents one pair of measurements, but there is no ID variable. Without an ID variable, you won't be able to tell how the values are meant to be paired together. In these cases, you can add an ID variable before using `melt()`:

```
# Make a copy of the data
co <- corneas
co

 affected notaffected
      488         484
      478         478
      480         492
      426         444
      440         436
      410         398
      458         464
      460         476

# Add an ID column
co$id <- 1:nrow(co)
```

```
melt(co, id.vars="id", variable.name="eye", value.name="thickness")

 id         eye thickness
  1    affected       488
  2    affected       478
  3    affected       480
  4    affected       426
  5    affected       440
  6    affected       410
  7    affected       458
  8    affected       460
  1 notaffected       484
  2 notaffected       478
  3 notaffected       492
  4 notaffected       444
  5 notaffected       436
  6 notaffected       398
  7 notaffected       464
  8 notaffected       476
```

Having numeric values for the ID variable may be problematic for subsequent analyses, so you may want to convert `id` to a character vector with `as.character()`, or a factor with `factor()`.

See Also

See Recipe 15.20 to do conversions in the other direction, from long to wide.

See the `stack()` function for another way of converting from wide to long.

15.20. Converting Data from Long to Wide

Problem

You want to convert a data frame from "long" format to "wide" format.

Solution

Use the `dcast()` function from the `reshape2` package. In this example, we'll use the `plum` data set, which is in a long format:

```
library(gcookbook) # For the data set
plum

 length      time survival count
   long   at_once     dead    84
   long in_spring     dead   156
  short   at_once     dead   133
```

```
short in_spring     dead   209
 long   at_once    alive   156
 long in_spring    alive    84
short   at_once    alive   107
short in_spring    alive    31
```

The conversion to wide format takes each unique value in one column and uses those values as headers for new columns, then uses another column for source values. For example, we can "move" values in the `survival` column to the top and fill them with values from `count`:

```
library(reshape2)
dcast(plum, length + time ~ survival, value.var="count")
```

```
length      time dead alive
  long   at_once   84   156
  long in_spring  156    84
 short   at_once  133   107
 short in_spring  209    31
```

Discussion

The `dcast()` function requires you to specify the *ID* variables (those that remain in columns) and the *variable* variables (those that get "moved to the top"). This is done with a formula where the ID variables are before the tilde (`~`) and the variable variables are after it.

In the preceding example, there are two ID variables and one variable variable. In the next one, there is one ID variable and two variable variables. When there is more than one variable variable, the values are combined with an underscore:

```
dcast(plum, time ~ length + survival, value.var="count")
```

```
      time long_dead long_alive short_dead short_alive
   at_once        84        156        133         107
 in_spring       156         84        209          31
```

See Also

See Recipe 15.19 to do conversions in the other direction, from wide to long.

See the `unstack()` function for another way of converting from long to wide.

15.21. Converting a Time Series Object to Times and Values

Problem

You have a time series object that you wish to convert to numeric vectors representing the time and values at each time.

Solution

Use the `time()` function to get the time for each observation, then convert the times and values to numeric vectors with `as.numeric()`:

```
# Look at nhtemp Time Series object
nhtemp

Time Series:
Start = 1912
End = 1971
Frequency = 1
 [1] 49.9 52.3 49.4 51.1 49.4 47.9 49.8 50.9 49.3 51.9 50.8 49.6 49.3 50.6 48.4
[16] 50.7 50.9 50.6 51.5 52.8 51.8 51.1 49.8 50.2 50.4 51.6 51.8 50.9 48.8 51.7
[31] 51.0 50.6 51.7 51.5 52.1 51.3 51.0 54.0 51.4 52.7 53.1 54.6 52.0 52.0 50.9
[46] 52.6 50.2 52.6 51.6 51.9 50.5 50.9 51.7 51.4 51.7 50.8 51.9 51.8 51.9 53.0

# Get times for each observation
as.numeric(time(nhtemp))

 [1] 1912 1913 1914 1915 1916 1917 1918 1919 1920 1921 1922 1923 1924 1925 1926
[16] 1927 1928 1929 1930 1931 1932 1933 1934 1935 1936 1937 1938 1939 1940 1941
[31] 1942 1943 1944 1945 1946 1947 1948 1949 1950 1951 1952 1953 1954 1955 1956
[46] 1957 1958 1959 1960 1961 1962 1963 1964 1965 1966 1967 1968 1969 1970 1971

# Get value of each observation
as.numeric(nhtemp)

 [1] 49.9 52.3 49.4 51.1 49.4 47.9 49.8 50.9 49.3 51.9 50.8 49.6 49.3 50.6 48.4
[16] 50.7 50.9 50.6 51.5 52.8 51.8 51.1 49.8 50.2 50.4 51.6 51.8 50.9 48.8 51.7
[31] 51.0 50.6 51.7 51.5 52.1 51.3 51.0 54.0 51.4 52.7 53.1 54.6 52.0 52.0 50.9
[46] 52.6 50.2 52.6 51.6 51.9 50.5 50.9 51.7 51.4 51.7 50.8 51.9 51.8 51.9 53.0

# Put them in a data frame
nht <- data.frame(year=as.numeric(time(nhtemp)), temp=as.numeric(nhtemp))
nht

 year temp
 1912 49.9
 1913 52.3
 ...
```

```
1970 51.9
1971 53.0
```

Discussion

Time series objects efficiently store information when there are observations at regular time intervals, but for use with ggplot2, they need to be converted to a format that separately represents times and values for each observation.

Some time series objects are cyclical. The `presidents` data set, for example, contains four observations per year, one for each quarter:

```
presidents

     Qtr1 Qtr2 Qtr3 Qtr4
1945   NA   87   82   75
1946   63   50   43   32
1947   35   60   54   55
...
1972   49   61   NA   NA
1973   68   44   40   27
1974   28   25   24   24
```

To convert it to a two-column data frame with one column representing the year with fractional values, we can do the same as before:

```
pres_rating <- data.frame(
    year   = as.numeric(time(presidents)),
    rating = as.numeric(presidents)
)
pres_rating

    year rating
 1945.00     NA
 1945.25     87
 1945.50     82
 ...
 1974.25     25
 1974.50     24
 1974.75     24
```

It is also possible to store the year and quarter in separate columns, which may be useful in some visualizations:

```
pres_rating2 <- data.frame(
    year    = as.numeric(floor(time(presidents))),
    quarter = as.numeric(cycle(presidents)),
    rating  = as.numeric(presidents)
)
pres_rating2

  year quarter rating
```

```
1945        1       NA
1945        2       87
1945        3       82
...
1974        2       25
1974        3       24
1974        4       24
```

See Also

The zoo package is also useful for working with time series objects.

APPENDIX A
Introduction to ggplot2

Most of the recipes in this book involve the ggplot2 package, written by Hadley Wickham. ggplot2 has only been around for a few years, but in that short time it has attracted many users in the R community because of its versatility, clear and consistent interface, and beautiful output.

ggplot2 takes a different approach to graphics than other graphing packages in R. It gets its name from Leland Wilkinson's *grammar of graphics*, which provides a formal, structured perspective on how to describe data graphics.

Even though this book deals largely with ggplot2, I don't mean to say that it's the be-all and end-all of graphics in R. For example, I sometimes find it faster and easier to inspect and explore data with R's base graphics, especially when the data isn't already structured properly for use with ggplot2. There are some things that ggplot2 can't do, or can't do as well as other graphing packages. There are other things that ggplot2 can do, but that specialized packages are better suited to handling. For most purposes, though, I believe that ggplot2 gives the best return on time invested, and it provides beautiful, publication-ready results.

Another excellent package for general-purpose graphs is lattice, by Deepyan Sarkar, which is an implementation of *trellis* graphics. It is included as part of the base installation of R.

If you want a deeper understanding of ggplot2, read on!

Background

In a data graphic, there is a mapping (or correspondence) from properties of the data to visual properties in the graphic. The data properties are typically numerical or categorical values, while the visual properties include the *x* and *y* positions of points, colors of lines, heights of bars, and so on. A data visualization that didn't map the data to visual properties wouldn't be a data visualization. On the surface, representing a number with an *x* coordinate may seem very different from representing a number with a color of a point, but at an abstract level, they are the same. Everyone who has made data graphics has at least an implicit understanding of this. For most of us, that's where our understanding remains.

In the grammar of graphics, this deep similarity is not just recognized, but made central. In R's base graphics functions, each mapping of data properties to visual properties is its own special case, and changing the mappings may require restructuring your data, issuing completely different graphing commands, or both.

To illustrate, I'll show a graph made from the `simpledat` data set from the gcookbook package:

```
library(gcookbook) # For the data set
simpledat
```

```
   A1 A2 A3
B1 10  7 12
B2  9 11  6
```

This will make a simple grouped bar graph, with the As going along the x-axis and the bars grouped by the Bs (Figure A-1):

```
barplot(simpledat, beside=TRUE)
```

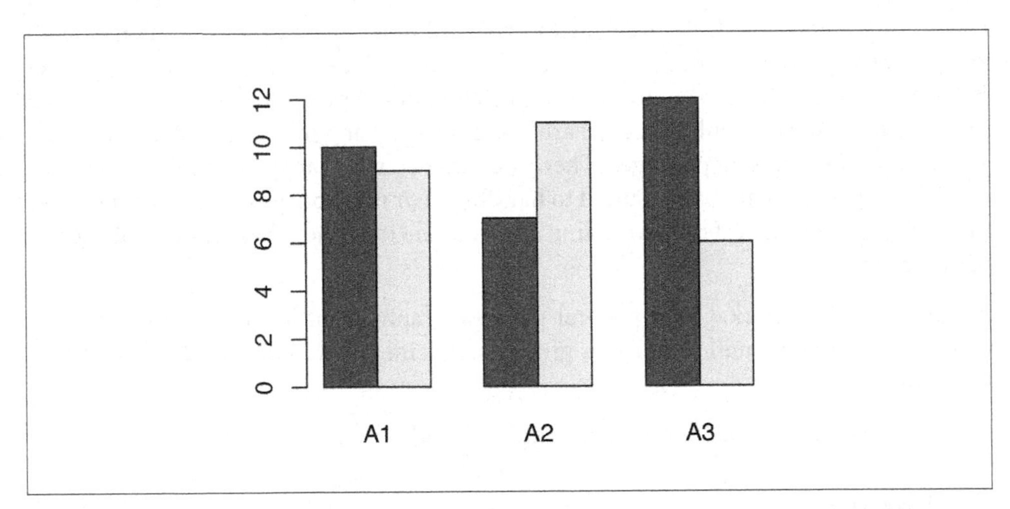

Figure A-1. A bar graph made with barplot()

One thing we might want to do is switch things up so the Bs go along the x-axis and the As are used for grouping. To do this, we need to restructure the data by transposing the matrix:

```
t(simpledat)

   B1 B2
A1 10  9
A2  7 11
A3 12  6
```

With the restructured data, we can create the graph the same way as before (Figure A-2):

```
barplot(t(simpledat), beside=TRUE)
```

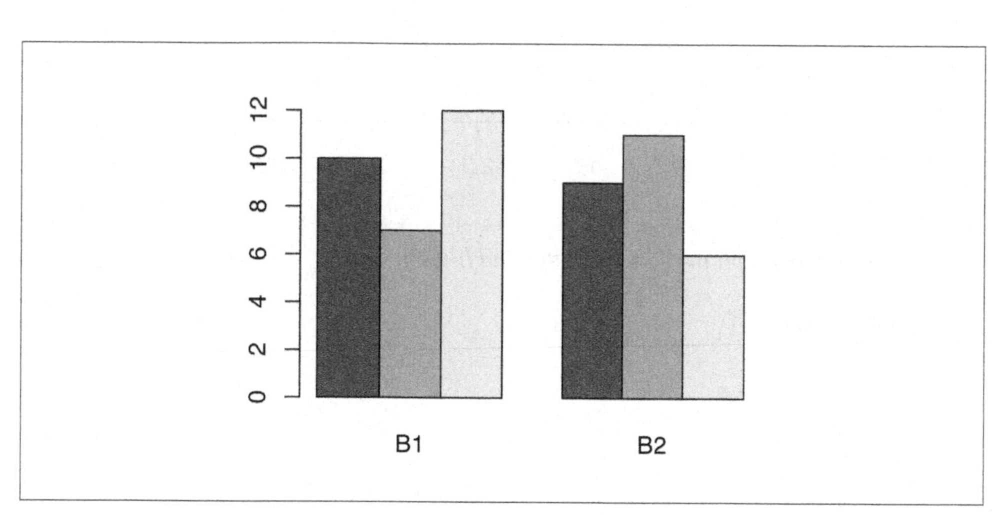

Figure A-2. A bar graph with transposed data

Another thing we might want to do is to represent the data with lines instead of bars, as shown in Figure A-3. To do this with base graphics, we need to use a completely different set of commands. First we call `plot()`, which tells R to create a new graph and draw a line for one row of data. Then we tell it to draw a second row with `lines()`:

```
plot(simpledat[1,], type="l")
lines(simpledat[2,], type="l", col="blue")
```

The resulting graph has a few quirks. The second (blue) line runs below the visible range, because the *y* range was set only for the first line, when the `plot()` function was called. Additionally, the x-axis is numbered instead of categorical.

Now let's take a look at the corresponding code and graphs with ggplot2. With ggplot2, the structure of the data is always the same: it requires a data frame in "long" format, as opposed to the "wide" format used previously. When the data is in long format, each row represents one item. Instead of having their groups determined by their *positions* in the matrix, the items have their groups specified in a separate column. Here is `simpledat`, converted to long format:

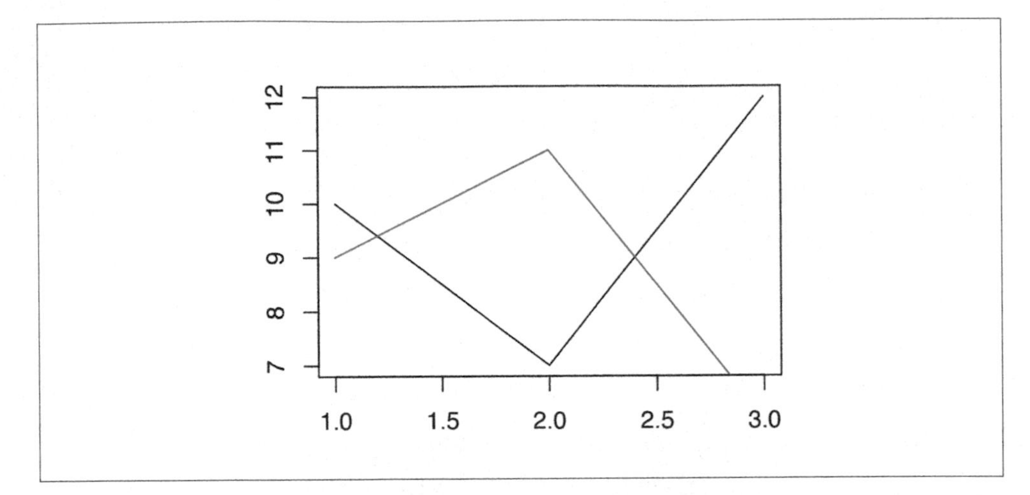

Figure A-3. A line graph made with plot() and lines()

```
simpledat_long

 Aval Bval value
   A1   B1    10
   A1   B2     9
   A2   B1     7
   A2   B2    11
   A3   B1    12
   A3   B2     6
```

This represents the same information, but with a different structure. There are advantages and disadvantages to the long format, but on the whole, I find that it makes things simpler when dealing with complicated data sets. See Recipes 15.19 and 15.20 for information about converting between wide and long data formats.

To make the first grouped bar graph (Figure A-4), we first have to load the ggplot2 library. Then we tell it to map `Aval` to the *x* position with `x=Aval`, and `Bval` to the fill color with `fill=Bval`. This will make the As run along the x-axis and the Bs determine the grouping. We also tell it to map `value` to the *y* position, or height, of the bars, with `y=value`. Finally, we tell it to draw bars with `geom_bar()` (don't worry about the other details yet; we'll get to those later):

```
library(ggplot2)
ggplot(simpledat_long, aes(x=Aval, y=value, fill=Bval)) +
    geom_bar(stat="identity", position="dodge")
```

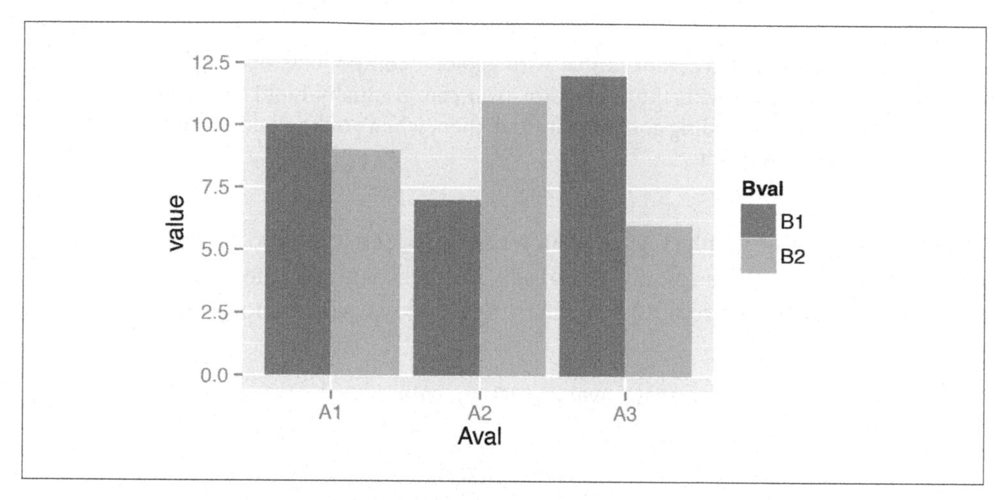

Figure A-4. A bar graph made with ggplot() and geom_bar()

To switch things so that the Bs go along the x-axis and the As determine the grouping (Figure A-5), we simply swap the mapping specification, with `x=Bval` and `fill=Aval`. Unlike with base graphics, we don't have to change the data; we just change the commands for making the graph:

```
ggplot(simpledat_long, aes(x=Bval, y=value, fill=Aval)) +
    geom_bar(stat="identity", position="dodge")
```

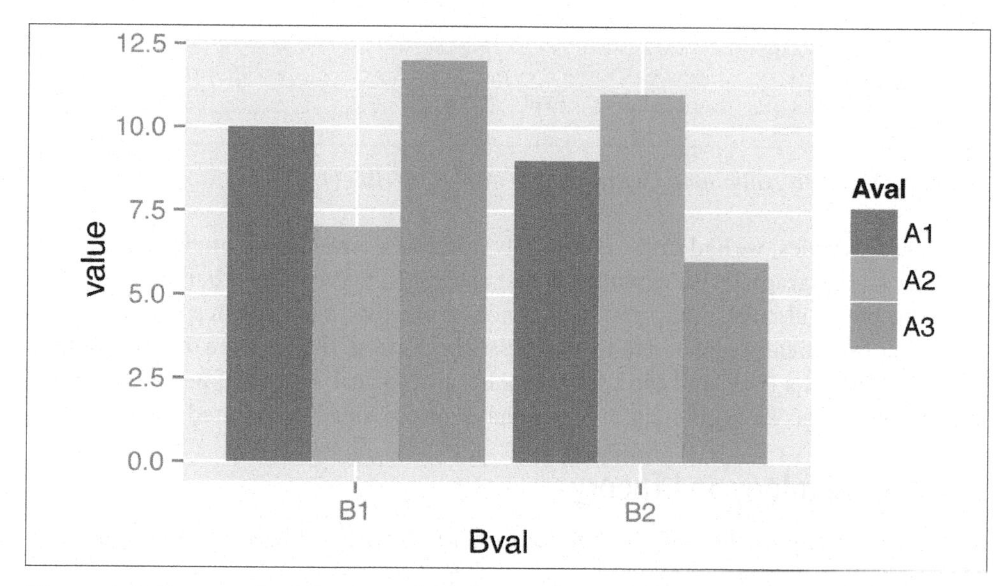

Figure A-5. Bar graph of the same data, but with x and fill mappings switched

You may have noticed that with ggplot2, components of the plot are combined with the + operator. You can gradually build up a ggplot object by adding components to it, then, when you're all done, you can tell it to print.

To change it to a line graph (Figure A-6), we change `geom_bar()` to `geom_line()`. We'll also map `Bval` to the *line* color, with `colour`, instead of the *fill* colour (note the British spelling—the author of ggplot2 is a Kiwi). Again, don't worry about the other details yet:

```
ggplot(simpledat_long, aes(x=Aval, y=value, colour=Bval, group=Bval)) +
    geom_line()
```

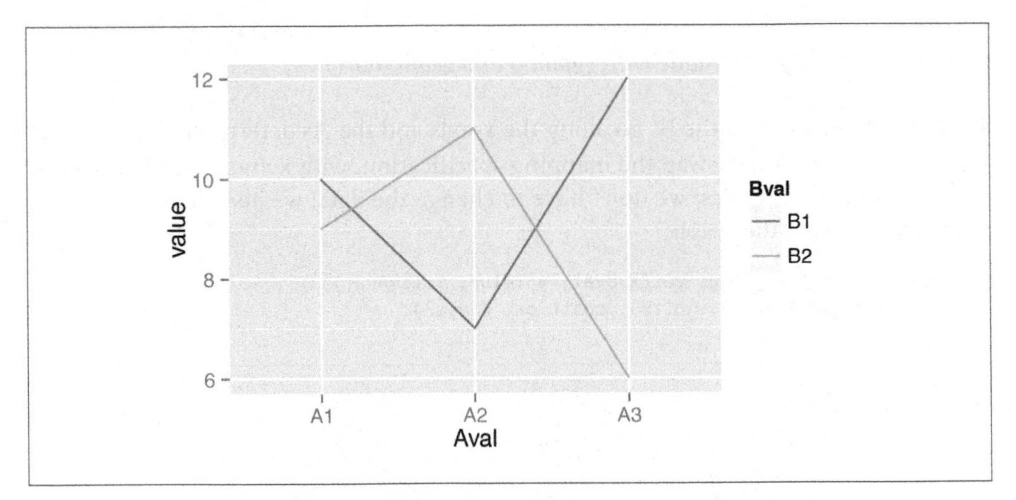

Figure A-6. A line graph made with ggplot() and geom_line()

With base graphics, we had to use completely different commands to make a line graph instead of a bar graph. With ggplot2, we just changed the *geom* from bars to lines. The resulting graph also has important differences from the base graphics version: the *y* range is automatically adjusted to fit all the data because all the lines are drawn together instead of one at a time, and the x-axis remains categorical instead of being converted to a numeric axis. The ggplot2 graphs also have automatically generated legends.

Some Terminology and Theory

Before we go any further, it'll be helpful to define some of the terminology used in ggplot2:

- The *data* is what we want to visualize. It consists of *variables*, which are stored as columns in a data frame.
- *Geoms* are the geometric objects that are drawn to represent the data, such as bars, lines, and points.
- Aesthetic attributes, or *aesthetics*, are visual properties of geoms, such as *x* and *y* position, line color, point shapes, etc.
- There are *mappings* from data values to aesthetics.
- *Scales* control the mapping from the values in the data space to values in the aesthetic space. A continuous *y* scale maps larger numerical values to vertically higher positions in space.
- *Guides* show the viewer how to map the visual properties back to the data space. The most commonly used guides are the tick marks and labels on an axis.

Here's an example of how a typical mapping works. You have *data*, which is a set of numerical or categorical values. You have *geoms* to represent each observation. You have an *aesthetic*, such as *y* (vertical) position. And you have a *scale*, which defines the mapping from the data space (numeric values) to the aesthetic space (vertical position). A typical linear *y*-scale might map the value 0 to the baseline of the graph, 5 to the middle, and 10 to the top. A logarithmic *y* scale would place them differently.

These aren't the only kinds of data and aesthetic spaces possible. In the abstract grammar of graphics, the data and aesthetics could be anything; in the ggplot2 implementation, there are some predetermined types of data and aesthetics. Commonly used data types include numeric values, categorical values, and text strings. Some commonly used aesthetics include horizontal and vertical position, color, size, and shape.

To interpret the graph, viewers refer to the *guides*. An example of a guide is the y-axis, including the tick marks and labels. The viewer refers to this guide to interpret what it means when a point is in the middle of the scale. A *legend* is another type of scale. A legend might show people what it means for a point to be a circle or a triangle, or what it means for a line to be blue or red.

Some aesthetics can only work with categorical variables, such as the shape of a point: triangles, circles, squares, etc. Some aesthetics work with categorical or continuous variables, such as *x* (horizontal) position. For a bar graph, the variable must be categorical—it would make no sense for there to be a continuous variable on the x-axis. For a scatter plot, the variable must be numeric. Both of these types of data (categorical and numeric) can be mapped to the aesthetic space of *x* position, but they require different types of scales.

In ggplot2 terminology, categorical variables are called *discrete*, and numeric variables are called *continuous*. These terms may not always correspond to how they're used elsewhere. Sometimes a variable that is continuous in the ggplot2 sense is discrete in the ordinary sense. For example, the number of visible sunspots must be an integer, so it's numeric (*continuous* to ggplot2) and discrete (in ordinary language).

Building a Simple Graph

Ggplot2 has a simple requirement for data structures: they must be stored in data frames, and each type of variable that is mapped to an aesthetic must be stored in its own column. In the `simpledat` examples we looked at earlier, we first mapped one variable to the `x` aesthetic and another to the `fill` aesthetic; then we changed the mapping specification to change which variable was mapped to which aesthetic.

We'll walk through a simple example here. First, we'll make a data frame of some sample data:

```
dat <- data.frame(xval=1:4, yval=c(3,5,6,9), group=c("A","B","A","B"))
dat

 xval yval group
    1    3     A
    2    5     B
    3    6     A
    4    9     B
```

A basic `ggplot()` specification looks like this:

```
ggplot(dat, aes(x=xval, y=yval))
```

This creates a `ggplot` object using the data frame `dat`. It also specifies default *aesthetic mappings* within `aes()`:

- `x=xval` maps the column `xval` to the *x* position.
- `y=yval` maps the column `yval` to the *y* position.

After we've given `ggplot()` the data frame and the aesthetic mappings, there's one more critical component: we need to tell it what *geometric objects* to put there. At this point, ggplot2 doesn't know if we want bars, lines, points, or something else to be drawn on the graph. We'll add `geom_point()` to draw points, resulting in a scatter plot:

```
ggplot(dat, aes(x=xval, y=yval)) + geom_point()
```

If you're going to reuse some of these components, you can store them in variables. We can save the `ggplot` object in `p`, and then add `geom_point()` to it. This has the same effect as the preceding code:

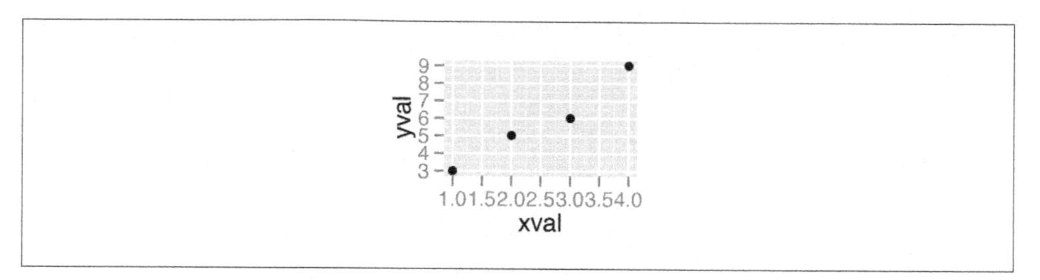

Figure A-7. A basic scatter plot

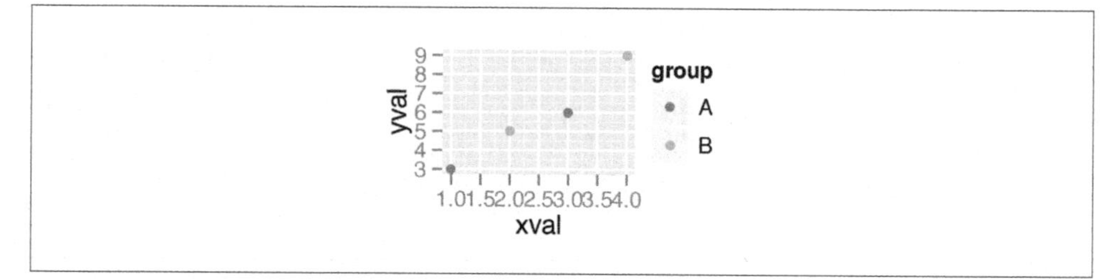

Figure A-8. A scatter plot with a variable mapped to colour

```
p <- ggplot(dat, aes(x=xval, y=yval))
p + geom_point()
```

We can also map the variable `group` to the color of the points, by putting `aes()` inside the call to `geom_point()`, and specifying `colour=group`:

```
p + geom_point(aes(colour=group))
```

This doesn't alter the *default* aesthetic mappings that we defined previously, inside of `ggplot(...)`. What it does is add an aesthetic mapping for this particular geom, `geom_point()`. If we added other geoms, this mapping would not apply to them.

Contrast this aesthetic *mapping* with aesthetic *setting*. This time, we won't use `aes()`; we'll just set the value of `colour` directly:

```
p + geom_point(colour="blue")
```

We can also modify the *scales*; that is, the mappings from data to visual attributes. Here, we'll change the *x* scale so that it has a larger range:

```
p + geom_point() + scale_x_continuous(limits=c(0,8))
```

If we go back to the example with the `colour=group` mapping, we can also modify the color scale:

```
p + geom_point() +
    scale_colour_manual(values=c("orange","forestgreen"))
```

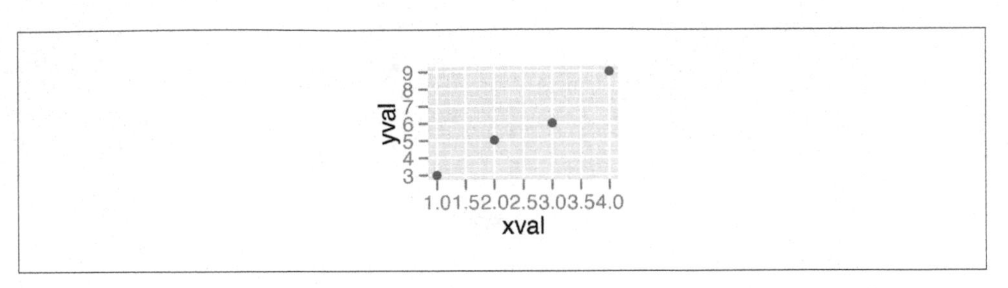

Figure A-9. A scatter plot with colors set

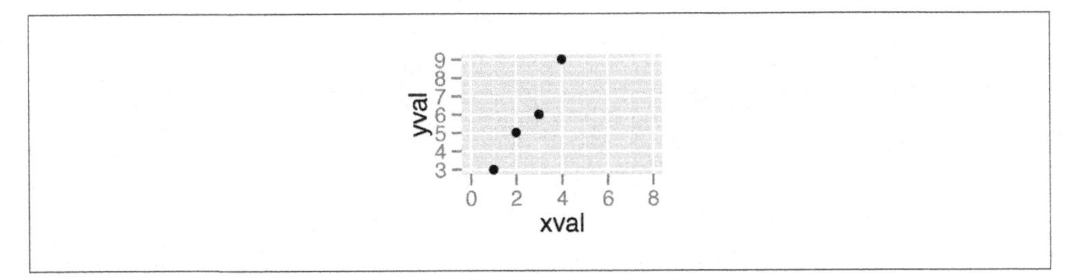

Figure A-10. A scatter plot with increased x-range

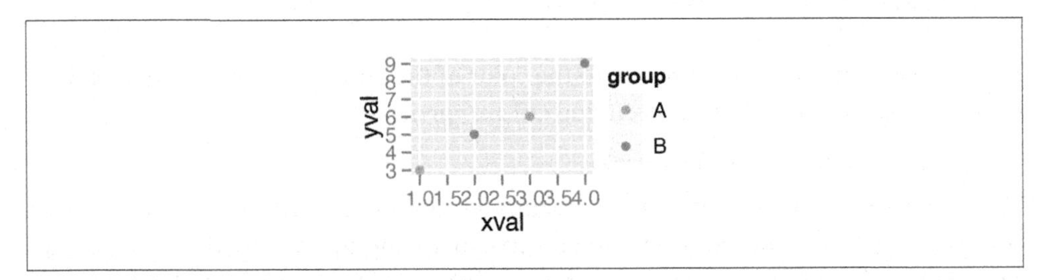

Figure A-11. A scatter plot with modified colors and a different palette

Both times when we modified the scale, the *guide* also changed. With the *x* scale, the guide was the markings along the x-axis. With the color scale, the guide was the legend.

Notice that we've used + to join together the pieces. In this last example, we ended a line with +, then added more on the next line. If you are going to have multiple lines, you have to put the + at the end of each line, instead of at the beginning of the next line. Otherwise, R's parser won't know that there's more stuff coming; it'll think you've finished the expression and evaluate it.

Printing

In R's base graphics, the graphing functions tell R to draw graphs to the output device (the screen or a file). Ggplot2 is a little different. The commands don't directly draw to the output device. Instead, the functions build plot *objects*, and the graphs aren't drawn until you use the `print()` function, as in `print(object)`. You might be thinking, "But wait, I haven't told R to print anything, yet it's made these graphs!" Well, that's not exactly true. In R, when you issue a command at the prompt, it really does two things: first it runs the command, then it runs `print()` with the returned result of that command.

The behavior at the interactive R prompt is different from when you run a script or function. In scripts, commands aren't automatically printed. The same is true for functions, but with a slight catch: the result of the last command in a function is returned, so if you call the function from the R prompt, the result of that last command will be printed because it's the result of the function.

Some introductions to ggplot2 make use of a function called `qplot()`, which is intended as a convenient interface for making graphs. It does require a little less typing than using `ggplot()` plus a geom, but I've found it a bit confusing to use because it has a slightly different way of specifying certain graphing parameters. I think it's simpler and easier to just use `ggplot()`.

Stats

Sometimes your data must be transformed or summarized before it is mapped to an aesthetic. This is true, for example, with a histogram, where the samples are grouped into bins and counted. The counts for each bin are then used to specify the height of a bar. Some geoms, like `geom_histogram()`, automatically do this for you, but sometimes you'll want to do this yourself, using various `stat_xx` functions.

Themes

Some aspects of a graph's appearance fall outside the scope of the grammar of graphics. These include the color of the background and grid lines in the graphing area, the fonts used in the axis labels, and the text in the graph title. These are controlled with the `theme()` function, explored in Chapter 9.

End

Hopefully you now have an understanding of the concepts behind ggplot2. The rest of this book shows you how to use it!

Index

Symbols

A

We'd like to hear your suggestions for improving our indexes. Send email to index@oreilly.com.

B

C

D

E

F

G

H

I

K

L

M

N

O

P

Q

R

S

T

U

V

W

X

Y

Z

About the Author

Winston Chang is a software engineer at RStudio, where he works on data visualization and software development tools for R. He holds a Ph.D. in Psychology from Northwestern University. During his time as a graduate student, he created a website called "Cookbook for R," which contains recipes for handling common tasks in R. In previous lives, he was a philosophy graduate student and a computer programmer.

Colophon

The animal on the cover of *R Graphics Cookbook* is a reindeer (*Rangifer tarandus*), also known as caribou in North America, which is a species of deer native to Arctic and Subarctic regions. Reindeer are ideally designed for life in hostile, cold environments, as their fur, antlers, noses, hooves, and vision have adapted to the low temperatures.

Their fur coat consists of an outer layer of straight, hollow, tubular hairs, which provide insulation from the cold and buoyancy in water, and a woolly undercoat. The coat is such an efficient insulator that when they lay on the snow, the snow does not melt. Reindeer are the only species of deer in which both male and female (and even calves) have antlers, and they have the largest antlers relative to body size among living deer species. Their antlers are shed annually and new antler growth occurs in the spring and summer.

Reindeer hooves adapt to the season: in the summer, when the tundra is soft and wet, the footpads become sponge-like and provide extra traction. In the winter, the pads shrink and tighten, exposing the rim of the hoof, which cuts into the ice and crusted snow to keep the deer from slipping. This also enables them to dig down (an activity known as cratering) through the snow to their favorite food, a lichen known as reindeer moss.

In 2012, researchers at University College London discovered reindeer are the only mammals that can see ultraviolet light. While human vision cuts off at wavelengths around 400 nm, reindeer can see up to 320 nm. This range only covers the part of the spectrum we can see with the help of a black light, but it is still enough to help reindeer see things in the glowing white of the Arctic that they would otherwise miss.

In the Santa Claus tale, Santa Claus's sleigh is pulled by flying reindeer. These were first named in the 1823 poem "A Visit from St. Nicholas," where they are called Dasher, Dancer, Prancer, Vixen, Comet, Cupid, Dunder, and Blixem.

The cover image is from Shaw's *Zoology*. The cover font is Adobe ITC Garamond. The text font is Adobe Minion Pro; the heading font is Adobe Myriad Condensed; and the code font is Dalton Maag's Ubuntu Mono.

CPSIA information can be obtained
at www.ICGtesting.com
Printed in the USA
LVOW02s1500091215
466129LV00005B/26/P